NAVAL INSTITUTE PRESS

SUSAN ARTIGIANI
Publicity Manager

291 Wood Road • Annapolis, MD 21402-5034
Voice: 410-295-1081 • Fax: 410-295-1084
www.usni.org • Email: sartigiani@usni.org

For review from

NAVAL INSTITUTE PRESS

THE WARTHOG
AND THE
CLOSE AIR SUPPORT DEBATE
BY DOUGLAS N. CAMPBELL

PUBLICATION DATE: SEPT. 2003

336 PAGES. 20 PHOTOS. NOTES. ABBREVIATIONS. BIBLIOG. INDEX.

ISBN: 1-55750-232-3. PRICE: $34.95

Available at bookstores, online,
or direct: www.NavalInstitute.org
800-233-8764 / 410-224-3378

*Please send two copies of
your published review to:*

Susan Artigiani, Publicity Manager
NAVAL INSTITUTE PRESS
291 Wood Road
Annapolis, MD 21402
410-295-1081 / Fax: 410-295-1084

The Warthhog and the Close Air Support Debate

The
WARTHOG
and the Close Air Support Debate

Douglas N. Campbell

Naval Institute Press
Annapolis, Maryland

Naval Institute Press
291 Wood Road
Annapolis, MD 21402

Library of Congress Cataloging-in-Publication Data
Campbell, Douglas, 1955—
The warthog and the close air support debate / /Douglas N. Campbell.
        p.   cm.
Includes bibliographical references and index.
    ISBN 1-55750-232-3 (hardcover : alk. paper)
1. A-10 (Jet attack plane) 2. Close air support. I. Title.
UG1242.A28C35 2003
358.4'3—dc21
                                        2003011180

Printed in the United States of America on acid-free paper ∞
10 09 08 07 06 05 04 03   9 8 7 6 5 4 3 2
First printing

To Teri, Patty, Mom, Ray Lancaster, and Muck Brown

CONTENTS

PREFACE

The inspiration for this book came from my experience as a military attack jet pilot and subsequent work writing a Ph.D. dissertation about the Air Force's close air support plane. In the late 1970s and early 1980s, I flew A-7s in the U.S. Navy, and heard many of my colleagues express concerns about the plane's survivability in both war and peacetime. A few even turned in their wings rather than fly an A-7. After four years flying the A-7 with combat-experienced "old salts," I saw that aircraft performance was important to air combat success, but it was not the only determinant. Other, sometimes less tangible, factors—such as desire, training quality, tactics, avionics, weapons, and specific combat conditions—played large roles. Even then, the impact of these factors was determined by each side's relative strengths in these areas. Thus, the replacement of the A-7 by the F/A-18 was not the unalloyed good that many pilots anticipated. The new plane's fuel consumption and weapons carriage weaknesses were glaring. Also, the A-7's multiple attack-related missions and complex avionics had affected the pilot's ability to excel at all of the associated responsibilities. Many fatal accidents involved some form of task saturation and loss of situational awareness. The F/A-18 added air superiority missions to the task list, which further challenged pilots' attention and training priorities.

In the mid-1980s I transferred to the U.S. Air Force and chose to fly the A-10. I enjoyed this type of flying, and believed in the A-10's ability to accomplish its specific mission of close air support. I flew A-10s in Europe as part of a wing that trained intensively to accomplish this mission in that region's demanding weather conditions and anticipated war-fighting environment. A-10 pilots understood their plane's speed and thrust deficiencies, but were confident because they made a deliberate effort to master the other air combat factors involved. Still, I occasionally heard the mantra "A-10s are not survivable in a high-threat environment," so reminiscent of the A-7 community on the eve of the F/A-18's arrival. If the dismal intelligence estimates of Warsaw Pact air defense were true, then it seemed to me that supersonic F-16s and F-111s would fare no better on their interdiction missions.

In the late 1980s I flew A-10 weapons and tactics tests as part of the Air Force's fighter test unit at Nellis Air Force Base, Nevada. In so doing, I went from relative isolation inside my own A-10 wing to active exposure to the rest of the tactical air force. The encounter was unpleasant, because the A-10 was not popular with many fighter pilots. To them, its slow speed and ugliness made it a tactical liability and a visual embarrassment. I also arrived as the Air Force tried to replace A-10s with modified F-16 fighters for the close air support role. The effort sparked a debate in the defense press and the political arena over the ideal close air support plane. This inspired at Nellis a public relations–oriented tactics test of the modified F-16, and pilot briefings that were veiled warnings to support the Air Force position.

These briefings and other official comments often featured the words *always, never,* and *all future wars* in explaining the A-10's alleged demise. In "all future wars," the F-16 would "always" be a better attack plane, and the type of air support environment in which the A-10 flew would "never" recur anyway. Given my background, I found such sweeping assertions ludicrous, and subsequent events reinforced my impressions. During this debate, Air Force leaders also proposed using modified A-7s for close air support. These were supposed to be better suited to a high-threat environment, but they were essentially the same planes that some Navy pilots had refused to fly ten years before. Further, A-10s did yeoman service in Desert Storm in a variety of missions. According to prewar Air Force assertions, some of these were "never" supposed to be flown again.

Public relations efforts also formed part of my exposure to the world of complex weapons technology development. Certainly tests required definable and provable goals instead of an absence of preconceived ideas that risked incoherent tests and garbled results. However, to paraphrase Shakespeare's Hamlet, the native hue of test integrity when creating "definable and provable" goals was sicklied o'er with the pale cast of political, financial, and doctrinal thought. The Nellis test leadership was as sharply attuned to the happenings at Tactical Air Command Headquarters, the Pentagon, the U.S. Congress, and even the White House as any defense magazine editor. Given that condition and the ongoing air support debate, the A-10 test office worked on a minimal budget, and any test that might buck the party line was not approved.

I recall informally proposing a test to disprove a debate-inspired claim that the A-10's anti-tank cannon would "never" destroy tanks in "all future wars." The test would have cost a few hundred thousand dollars. The colonel hearing this proposal told me something like this: When the Air Force needed a new plane—and its reasons were usually valid—it understood that opposition could still arise for various reasons, financial, ideological, or parochial. Therefore, the service had to make a strong case, and part of this involved demonstrating that the current plane was dangerously unsatisfactory. In this case, the A-10 was that current model, and thus, did I really think that the Air Force would give me money to prove it wrong?

From such a frustrating experience, I swore that I would tell this plane's story, and eventually it became the topic of my history Ph.D. dissertation at Texas Tech University. It matched two of my graduate study fields, military history and history of technology. The military connection was obvious, but the CAS plane story also well met technology history concerns such as technological constituencies and usage issues.

When I began my research, I was an A-10 partisan, and I still am. However, a funny thing happened on the way to the publication forum, as historical research revealed a story with nuances and unexpected developments. Commonly accepted A-10 opponents and supporters actually switched positions at times. Both sides cited historical accounts that did not stand up to scrutiny. Compelling logic supported the Air Force's ambivalence concerning close air support and associated planes. For one thing, the Air Force's top tactical priority, air superiority, is indeed important because any other mission may become lethally difficult to fly without it. Also, close air support is hard to accomplish for various reasons. Radio contact between the pilot and controller "middleman" who authorizes ordnance release can be difficult. When I flew the mission, the controller's radio often sounded like the scratchy, garbled speakers at a small-town hamburger drive-in. The target can be hard to identify, which creates more frustration. Pilots may prefer missions such as interdiction, where they can attack a clearly defined enemy without a lot of airborne administrative coordination.

These and other items revealed by my experience and research often forced me to examine why I supported the close air support mission and the dedicated-plane concept. Obviously, neither fit well into the Air Force's doctrinal and force structure scheme, which accounted for financial constraints while assuming the worst-case war-fighting scenario: fighting a modern industrial nation and a potent air force. However, this country encountered too many other situations where such a plane and its specially trained pilots ensured our ground troops' victory, and even their survival.

Thus, my story will reflect two other titles I considered for this book. The Air Force CAS plane was a "plane in the middle," whose precarious existence was based upon the changing circumstances and mutual interest of two armed services—indeed, two different war-fighting mediums and perspectives. Its creation and operational survival were not a given; instead, it relied upon many favorable factors. Even so, America's foreign policy and war-fighting approach created the trump card that made the CAS mission and its dedicated plane necessary. The CAS plane was the "winged covenant" whose existence promised air support to American ground troops.

ACKNOWLEDGMENTS

There are a great many people who are responsible for this work besides myself. First, my wife Teri and daughter Patty put up with a lot during this effort, and their patience and support cannot be repaid. My mother also supported this task with encouragement and occasional financial aid to this cash-strapped grad student.

My Texas Tech graduate committee chairman, Dr. James Reckner, was the finest chairman a student could have. His and his wife Middy's patience with my numerous calls about details great and small was a big help. My other committee members, Drs. James Harper, Robert Hayes, Otto Nelson, and Ron Rainger, helped inspire this work through discussions and suggestions. Tom Cutler and Lys Ann Shore provided both moral and valuable technical support in turning this manuscript into a book. Finally, the Texas Tech University Library interlibrary loan office supplied the first documentary materials that showed me that this work was academically feasible.

Other library and research staffs were also critical to this effort. The harried front-desk folks at the Air Force Historical Research Center at Maxwell Air Force Base, Alabama, worked through a Tuskegee Airmen reunion and a Society for Military History symposium to help me gather a wealth of information. Material not available there was made available by the Air University Library. Dan Keogh and Pamela Cheney at the Army Military History Institute at Carlisle Barracks, Pennsylvania, were the very models of professionalism. Jean August and others at the Air Force Material Command's History Office rendered critical support by declassifying and releasing documents that gave me some of the proverbial "original sources." Jeff Thurman at the Defense Visual Information Center responded with alacrity to last-minute photo requests. Lois Knowles and the staff at the library of New Mexico State University, Alamogordo, helped with source searches. Last, but not least, Marsha Buffin and the other librarians at the Holloman Air Force Base library obtained books that allowed me to update sources after I completed the dissertation.

John Moring, his wife Nan, and their wonderful family made their house my house during my research work in the Washington, D.C., area. Air Force officer and fellow Texas Tech graduate school pilgrim Matt Rodman provided moral support and was able to glean some materials out of the Office of Air Force History for my work. Retired Air Force colonel and workplace boss Klaus Klause accommodated my writing needs. The folks at Falcon Copiers in Alamogordo helped me reproduce a lot of papers.

I am also grateful to the interviewees—several dozen of them. This topic is current enough that one can choose between waiting decades for source documents that may not even exist, or hunting down key participants who are willing to talk. Although I do not want to offend others by praising a particular few, I do want to cite retired U.S. Air Force Col. Ray Lancaster for unsolicited words of encouragement throughout this effort. However, all the interviewees gave their time and knowledge, and often pointed out others who could help. Some of them devoted hours to recount and explain patiently what they knew, and if one hears on tape my verbose interview style, one will respect their patience!

1

INTRODUCTION

A joke appeared in the *Reader's Digest* in July 1967, as U.S. involvement in the Vietnam War deepened, and Americans became more aware of the military's tactics. A boy returned from Sunday school and his mother asked him what he had learned. He said that he heard about how Moses and his people escaped from Pharaoh's Egypt. He told her that as Pharaoh's tanks approached the Israelites, Moses got on the radio and called in an air strike, which knocked the tanks out of action. The puzzled mother asked, "Is that really the way the teacher told the story?" The boy replied, "If I told it her way, you'd never believe it!"[1]

Thus it is with close air support, known also by its acronym, CAS. Just as missiles, tanks, and machine guns offer an apparently obvious battlefield edge, logic dictates that ground troops use airplanes for extra firepower. Planes move over the battlefield in a way the ground forces cannot. Their high speed and lethal onboard weaponry enable them to surprise enemy troops with a hard knock. Airpower can destroy the enemy currently vexing the ground commander, and can also locate and hit threatening forces just beyond the front lines. It all seems so easy; and as the joke implied, the U.S.

Air Force came to deliver CAS so well in Vietnam that people took it for granted.

But CAS is not easy to accomplish. Indeed, the term itself defies a simple definition. How close is close? For much of its existence, the U.S. Air Force used a close air support definition that went something like this: support of troops engaged or nearly engaged in combat so that close coordination is required with the ground units. Details of the definition changed over the years. After the Vietnam War, where the Air Force accomplished CAS so well, a survey conducted among Army and Air Force officers revealed no agreement about its exact definition or other related terms.

Even if one accepts the definition given above, other problems arise. The difference in visual perspective between the airman and the soldier is wide. To the infantryman, an enemy tank one hundred yards away is conspicuous, not only because of its size but also because of the serious danger it represents. To a pilot preparing to attack the tank from a mile and a half away, it is a large dot; and if its camouflage matches its surroundings, it may not even be visible. Conversely, the pilot sees features that the soldier cannot detect. If the soldier wants the pilot to attack a red-roofed barn, the pilot may see many such barns in his field of view. Further, the battle area is often a confusing, dusty, and smoky bedlam involving forces frantically maneuvering for advantage. The opposing lines are rarely well defined. One cannot say that one will attack everyone facing in a certain direction, for in the heat of battle, both forces often face in all directions. The battle can be so chaotic that those directing the planes—ground or airborne controllers who know the battle area—are themselves confused about who is friend or foe.

However, merging the air and ground perspectives to confirm target location is of life-and-death importance, for the pilot cannot fire until the target is properly identified. Otherwise, one risks shooting the wrong target or, tragically, the friendly troops who expect support. Therefore, good communication is necessary, and though one assumes that airmen and soldiers communicate easily with one another, this is not always so. CAS history features instances where two different military organizations procured incompatible radios. Visual signals exist, but all CAS participants must agree upon their meaning, and disaster looms for the ground troops whose signals become known to their enemy. Nowadays, laser designators and data link devices can instantly transfer target information to airplanes' fire control computers, promising quick target designation. Precision-guided weapons offer unprecedented accuracy. However, these items heap more technology upon an already complex technological undertaking. What happens if the equipment breaks, or there is operator error, or the participants have incompatible equipment or none at all?

Communications also require a system for ordering air strikes, and this is no simple matter either. The U.S. Marine Corps air task structure is relatively streamlined, given that it is a self-contained operation in a small service that often substitutes aircraft for artillery during amphibious operations. But the U.S. Air Force–U.S. Army air task process involves two services that must anticipate large-scale fights in which the Army may already have ample artillery. Further, the two services share a long, turbulent past. Questions arise about the appropriate weapons and minimum level for requesting authority. The Army's own firepower—artillery, attack helicopters, or tanks— may be able to solve the tactical problem; in any case, one cannot send planes hither and yon in response to every platoon leader's calls. Thus, the U.S. Army battalion is usually the lowest unit level that can request CAS, but even this far up, the request makes a long trip up the operational Army chain of command to determine its priority and whether Army firepower can support it instead. If an air support request gets this far, it goes to the Air Force command chain, which determines whether any planes can answer it. If so, one hopes they can get to the battle quickly enough to make a difference. This complex process requires proper staffing and constant active use to work.

Perspective-related difficulties extend to several other war-fighting considerations. There are many soldiers and relatively few airmen, and the soldier's training pales next to the time and cost required to produce a competent tactical pilot. Though Army weapons can be expensive, they do not closely match the average Air Force fighter's cost. During a 1990s interservice debate over force structure, one Army general exclaimed: "If you spend $10 billion on an army you modernize a whole corps. Ten billion dollars on an aircraft programme is nothing."[2] Thus, air leaders often seem more cautious about using their pilots and planes than ground commanders are about using their soldiers and weapons.

Distance from actual ground combat creates another issue. Airmen are concerned about enemy ground forces, but they are not nearly as interested as soldiers, who are in constant direct contact with enemy guns. To the ground troops, the immediate objective and nearby enemy forces are of preeminent importance, and the troops move only in coordination with other friendly ground units.

The pilots see themselves as able to move quickly and relatively freely over terrain, but their restrictions also may seem irrelevant to the ground troops. Pilots do not like flying a machine that is helpless against enemy fighters. Dense antiaircraft fire can appear in areas out of the soldiers' ken. Fuel considerations and pilot fatigue bring back to earth planes with the best endurance. They cannot remain in the battle area indefinitely while the ground folks sort themselves out. Weather that is a nuisance to the soldiers can ground the airmen or prevent them from executing CAS.

Thus, the airmen's operational objectives and priorities differ from those of the soldiers. Even when they were part of the Army, air leaders placed close air support in third priority behind air superiority and interdiction. The mission precedence remains, for the airmen cite historical examples when asserting that air support is difficult, even impossible, to accomplish without air superiority. They also point out that interdiction against enemy forces in the rear area often yields better results both for the air effort and the ground campaign's longer term. Planes flying repeated interdiction attacks against known targets accomplish more missions over time than those remaining on ground CAS alert or flying about seeking permission to hit CAS targets. The airmen may even say that strategic targets deep in the enemy's homeland rate a higher priority effort than the enemy army.

Return for effort is important because many experts consider CAS the most dangerous airpower mission. A fighter patrol may yield no combat if the enemy's planes do not appear or cannot be engaged. An interdiction mission can be very dangerous if there is fighter opposition and dense belts of antiaircraft fire, but its route of flight often passes over unarmed people, even if they are enemy. CAS, however, is always a mission against an armed group of people. Along with the dedicated antiaircraft batteries, nearly all of the enemy—perhaps thousands of troops in a small area—are armed. They may possess only small arms and may not know how to track an airplane, but as one ventures near the target, the metallic maelstrom from all of the aimed and barrage fire can down one's plane, no matter how fast or sophisticated it is. Quite often, CAS pilots must fly near the target in order to identify and then to hit it, and though air and ground force perspectives can vary, CAS pilots' compassion for beleaguered soldiers often leads the fliers to take lethal risks on the troops' behalf.

CAS obviously takes some practice and inclination if it is to be done well, and this raises another problem: both sides must be interested in the mission. The Air Force may say that budget and force structure limitations will restrict CAS training, relegating air support to an ad hoc "emergency procedure," as some airmen have described it. The soldiers may be confident in their own ability to beat the enemy. They may even see Air Force support as a distraction from, or in competition with, their own war-fighting doctrine and weapons procurement.

This last consideration is important. Though the historical record shows that almost any plane can fly CAS—World War II fighters like the P-47 and later transports-turned-gunships like the AC-130 handled it quite well—only a certain kind of plane best meets its varied demands. The plane must maneuver well, and though speed is an asset for responsiveness and survivability, the plane must fly slowly enough for the pilot to see and hit the target. Since finding the target can take time and the soldiers may want multiple attacks, the plane needs good fuel endurance (also called loiter capability) and abundant firepower. It needs durability to withstand the small arms hits that inevitably accompany this mission. Though advanced avionics gear can aid target acquisition, the plane should be a relatively simple, easily maintained design and one that—given the threats it encounters—can be easily repaired. Too much technological complexity increases cost and maintenance demands, especially in combat, and the plane may have to operate from primitive bases near the front. One wants to minimize combat losses in any event, but one also does not want expensive planes in this dangerous arena.

Perhaps more important, an Air Force possessing dedicated CAS planes also guarantees that it has flying units that prepare for such a difficult mission. Further, the pilots will be focused upon doing the mission well. The necessary characteristics will probably prevent the plane from being used as an air superiority fighter, or even an effective deep interdiction plane, since the traditional aerodynamic trade-offs for such a design are a lower top speed and operating altitude. However, Air Force leaders may not want such a plane because it represents a permanent asset diversion from other, more important tasks. Army commanders may not want it either if it competes with their weapons in the defense budget, or if its inflexibility obstructs other war-fighting goals.

The U.S. Air Force CAS Plane

Thus, the purpose-built U.S. Air Force CAS plane, like the CAS mission, lived on a boundary between organizations and war-fighting styles. Given the mission's difficulty and requirement for interservice interest, it should be no surprise that whether as part of the U.S. Army or as an independent arm, the U.S. Air Force did not have such a plane for much of its early existence. It only procured a CAS plane, the A-10, when certain factors combined to make it do so.

The factors inhibiting CAS plane procurement were many, but if there was any sure foundation for the mission and the plane, it was America's war-fighting style and foreign policy needs. The wealthy nation could utilize abundant firepower to achieve tactical advantage and prevent needless sacrifice of its citizen-soldiers. This imperative arose repeatedly in combat, for many wars did not require rigid adherence to Air Force combat priorities. Indeed, in many of the war-fighting situations that the United States encountered, CAS was the most important, and at times the only, combat mission for air units. Combat was the catalyst for the factors that drove the service to build a dedicated CAS plane.

Accordingly, the CAS observer cannot look at the A-10 without considering the historical currents that created it and buffeted its existence. Both A-10 opponents and supporters have appealed to history to justify their positions. Opponents have pointed out that the service derived its doctrine from historical lessons demonstrating air support's relative lack of worth. Supporters have criticized a service that seemed too intent upon fighting a European, high-technology, stratospheric Trafalgar while ignoring the demands of wars that actually occurred.

Therefore, this book reviews the full background of the U.S. Air Force CAS plane, beginning with the CAS mission's birth during World War I and the fortunes of the U.S. Army's airmen before and during World War II. They faced difficulties in developing CAS, a fight within the Army to prove the air arm's independent worth, and promising bomber plane technology. Thus, they embraced other missions, especially strategic bombing. The exigencies of World War II brought them back to flying air support, and they came to fly it well, but the experience supported their impression that fighters could handle CAS when necessary.

This book then recounts the newly created Air Force's experience with CAS, as well as factors that led the service to procure a plane specifically built for the mission. In this period, conflicts such as Korea reemphasized the worth of tactical missions, but Air Force leaders further embraced strategic bombing as part of U.S. nuclear-oriented military policy. The Army's discontent with neglect of CAS helped fuel its pursuit of a new technology, helicopters, as a means of conducting its own air support. The Vietnam War's earliest years highlighted Air Force CAS deficiencies, which, combined with the Army's development of an advanced attack helicopter, forced the service in 1966 to commence construction of a dedicated CAS plane in order to protect its mission prerogatives.

The book then concentrates upon this plane's turbulent existence. After ten years of doctrinal changes, further technological developments, and debates involving the military services and Congress, in 1976 Air Force leaders fielded the A-10. They initially strove to prove the plane's worth, but Army doctrinal changes and the development of a new tactical fighter, the F-16, turned them against it. This ignited in the 1980s a bureaucratic, political, and defense media fight against those who valued the mission and its plane. Political action, budgetary constraints, and success in Desert Storm revived its fortunes. The story concludes with an epilogue about the A-10's later performance, which has favored its continued service.

Aims

The story told in this book goes beyond an airplane's life and retraces the intertwined historical currents and players that affected it. It is about technology's promise and its realities. It addresses U.S. foreign policy and warfighting style, and how both of these made a difficult airpower mission compelling to the U.S. military. Concerning the military, the book recounts much of the history of the U.S. Air Force—an organization that one would expect to embrace all airpower missions and that had made interservice agreements pledging that it would provide CAS. With regard to interservice agreements, the book tells of the often ambivalent relationship between the

U.S. Air Force and the U.S. Army. This history covers American politics—specifically, the U.S. Congress. Any history of U.S. weapons procurement must include this diverse and seemingly fickle body whose budgetary power controls military technology. Technology introduces more actors, for aircraft and weapons developments affected people's opinions about the mission.

The narrative follows some existing historiographical paths and also blazes new ones. In relating the creation and later fortunes of the U.S. Air Force CAS plane, it describes aircraft technology development as a difficult and even unpredictable process. It emulates several important historical works, such as Frederic Bergerson's *The Army Gets an Air Force*, Charles Bright's *Jet Makers*, Michael Brown's *Flying Blind*, and Stephen Rosen's *Winning the Next War*. These works recount the problems involved in making any complex technological device or process work properly. Some also examine how organizations embraced certain aircraft and missions, and they explain how and why, and from where, particular planes and missions acquired influential advocates. Concerning patronage, they focus either within a military service or upon more diverse sources. The story of the U.S. Air Force CAS plane requires the latter approach because Air Force leaders understood the importance of securing outside support for their plans. The plane's constituency comprised an eclectic assortment of people and institutions.

This history of the CAS plane joins Ruth Schwartz Cowan's *More Work for Mother* and Edward Tenner's *Why Things Bite Back* in examining irony in technological development. In this context, "irony" is the unintended consequences or uses of technology—how a device's intended uses and effects often differ from those embraced during its actual operation. Irony also involves what could be called performance niches in technological progress. If the newest inventions are intended to facilitate some human action, devices considered obsolete often remain useful in their own way. An everyday example is people's use of the telephone, an older technology, to communicate, instead of the newer e-mail technology. Given modern tactical jet capabilities, the A-10 was a step backward in terms of performance, even though the nature of CAS dictated this. Further, the strictures and challenges encountered in its development in many ways made the A-10 a better plane. Finally, irony involves what some call myth debunking. In this case, several common assumptions about the A-10's past did not survive scrutiny. One might say that this is not irony so much as straight historical research, but technological developments often create situations that defy assumptions.

More sparsely tracked terrain in the field of history of technology involves analyzing the intentions and results of official studies—congressional hearings, investigations, flight competitions, and weapons tests. Official documents often constitute the bedrock of both news media and academic histories, and though I do not embrace many postmodernists' claim that there are no true facts in history, Mark Twain's quip about "lies, damned lies, and statistics" sometimes applies to such inquiries. Any historian must use care when citing them, especially when dealing with a complex and frequently emotional topic such as CAS. John Tetsoro Sumida's *In Defense of Naval Superiority* covered something similar when describing the gunsight design rivalries accompanying early twentieth-century British battleship programs. However, compared to Britain's battleships, modern military aviation technology is so complex that even valid premises for analyzing an aircraft's performance must themselves be examined. These cases feature a heady mix of national security, professional status, job security, and money issues. Opinions based upon these concerns usually surface within the plans or results. Many of the studies and tests accompanying the A-10's creation and development, and later controversies, aimed to prove a subjective but honestly held opinion about specific war-fighting requirements—and not to establish universal truths about air support.

Better proof of the worth of a mission or weapon can come from real combat. Both during and after the Cold War, the United States pursued any foreign military involvement it deemed necessary, which entailed a wide variety of war-fighting scenarios. Western Europe remained the nation's highest foreign policy priority, but one could not anticipate fighting only the Soviets in Germany when so many other contingencies existed around the world. Regarding this condition, I follow other historians in describing an "American way of war." This thesis states that American commanders prefer using firepower over sacrificing their troops. Their concern for citizen-soldiers, ample access to resources, and faith in technology make them want, in the words of some U.S. Army observers, to "send the bullet before sending the man." This does not mean that American generals have never conducted brilliant economy-of-force operations or that American troops have not taken lethal risks. Research for this book led back to the Washington, D.C., area, where spare time allowed some Civil War battlefield visits. The Shenandoah Valley is a monument to Stonewall Jackson's dazzling campaigns. The sacrifices made at Antietam's Bloody Lane on 17 September 1862 inspire awe. The American military has done and will do these things. However, its leaders tend—especially in foreign wars where national survival is not threatened—to use abundant firepower rather than waste their soldiers' lives.

I believe CAS is part of the American way of war because it is a firepower solution to the soldiers' battlefield problems. In spite of its difficulties, CAS has been used by the U.S. military more than by the armed services of any other country. Indeed, other nations see American CAS as a demonstration of the nation's wealth.

Although the U.S. Air Force historically provided much of that air support, I agree with those who believe that the service was often quick to shun the mission but was then forced by circumstances to re-embrace it. Ironically, in its emphasis upon strategic bombing, the service pursued a firepower solution on a grand scale. As for tactical aviation, the Air Force believed that its mission priorities of air superiority, interdiction, and then CAS were supported by historical examples and a common-sense appraisal of requirements against an opponent with a strong air defense—more specifically, Europe during World War II and the Cold War. Thus, Air Force leaders worried about committing budgetary and procurement resources to a low-priority mission that implied subservience toward the parent service. However, U.S. foreign policy and American war-fighting style repeatedly drew the service back to situations, such as the Korean and Vietnam Wars, that required the mission.

I also agree with those who assert that CAS requires a special plane to fly it properly. However, the Air Force often preferred that fighters fly CAS, and historical precedent seemed to support that choice. Up through World War II, air arms used a variety of fighters for CAS, including both first-line and obsolete models. They flew the mission as well as or better than dedicated attack planes, which were either twin-engine bombers or single-engine dive-bombers that lacked the fighters' high performance. Thus, given its concern for high-threat air war in Europe, the service appeared justified in choosing planes that could meet two mission objectives instead of poor performers that barely met one. After World War II, however, jet fighter performance increasingly became incompatible with that required for CAS; and the wars that the United States actually fought drove the Air Force to use more suitable planes. The creation and continued use of the A-10 is proof.

Unlike some other observers, I do not believe that Air Force resistance was either single-handed, unanimous, or consistent. The U.S Army Air Corps may have emphasized strategic bombing, but it retained dedicated battlefield attack planes longer than any other air service. Once committed to the A-10 in the 1970s, Air Force leaders worked to ensure its success in spite of opposition, even when it came from within their own ranks. Contrary to conventional wisdom, the Army's attitude toward CAS and its planes was sometimes ambivalent. Army leaders appreciated air support; indeed, some Air Force officers observed that the Army would always complain that any amount of CAS was not enough. But tight peacetime budgets sometimes forced the Army to neglect practicing CAS procedures. In the late 1960s, Army leaders were unenthusiastic about the Air Force's CAS plane project because it challenged their own advanced attack helicopter venture. The Army's doctrinal changes of the 1980s also undermined its commitment to Air Force CAS planes.

Given this particular scope and tack, this account breaks with other excellent CAS-related histories because they either did not cover the U.S. Air Force CAS plane story in depth or chose different viewpoints. Peter Smith's *Close Air Support* was a short survey history that understandably did not concentrate upon any one item. Richard Hallion's acclaimed CAS study, *Strike from the Sky*, focused upon air support through World War II, especially that war's European theater. Writing at the climax of the the 1980s CAS plane debate, Hallion issued close air support maxims that endorsed the Air Force predilection for fighter CAS in high-tech conventional wars. Benjamin Franklin Cooling edited a first-rate CAS history anthology for the Office of Air Force History called *Case Studies in the Development of Close Air Support*. The essays in the volume, however, covered selected wars and scarcely discussed CAS issues of the U.S. Air Force in the aftermath of the Vietnam War. Ian Gooderson's *Air Power at the Battlefront* rigorously analyzed the British and U.S. European air support effectiveness in World War II, focusing upon that theater and its lessons.

As for the U.S. Air Force CAS plane itself, the available good studies did not tackle the plane's history in its entirety. U.S. Army Maj. Charles Kirkpatrick's in-service study, *The Army and the A-10*, ended its scope at 1971, but relayed the Army's ambivalence about the Air Force CAS plane project at that time. Two Air Force Systems Command histories, *The A-X Specialized Close Air Support Aircraft: Origins and Concept Phase, 1961–1970*, and George Watson's *A-10 Close Air Support Aircraft: From Development to Production, 1970–1976*, well addressed the technical details of the A-10's creation, but did not deal with larger issues. Aviation writer Bill Sweetman's *Modern Fighting Aircraft: A-10* described design details and addressed some larger influences. However, some of his observations were incorrect, and the book's scope ended with the early 1980s. Ken Neubeck's *A-10 Warthog* contained many historical facts for such a small book, but not enough overall and very little about what made the Air Force purchase such a plane. Finally, William Smallwood's *Warthog* used extensive interviews with Desert Storm A-10 pilots to render an excellent account of the A-10's Gulf War performance.

In all, I have written a much more sweeping study of U.S. Air Force close air support and its CAS plane. I tie together what these fine histories partially addressed. I mean for my work to serve as a historical guide, not only for Air Force and Army readers, but also to others who are curious about the history of military technology and the difficulties and triumphs surrounding the complex CAS mission. Though readers are sure to form their own opinions from studying this often contentious issue, I believe that, on balance, the mission and the plane are worthwhile defense investments.

Caveats

As mentioned earlier, *close air support* (or air support or CAS) is a difficult term to define. I use the following definition of the term as "an air attack against enemy forces near enough to the battlefield to require air and ground party coordination before the aircrew expends ordnance." Even this definition can invite carping. What does "near enough to the battlefield" mean? It changes due to friendly force dispositions, terrain, size of the battlefield, ground speed of enemy forces, and other fluctuating conditions. Another perspective is that CAS is not deep interdiction, but how far behind the lines is "deep" interdiction? At this, I venture one last explanation, knowing that even this will not satisfy the most exacting reader: CAS is flown against enemy ground forces that are directly fighting friendly troops or are near enough to be a factor in a battle's outcome. In any event, it requires coordination between air and ground forces to ensure that the airmen accomplish the mission without hitting civilians or friendly troops who may be nearby.

Concerning military operations and weapons procurement, I stopped short of assessing the specific motives and fortunes of all the military aircraft companies mentioned. The one exception will be the A-10's builder, the Fairchild-Republic Corporation, and even here the coverage remains relatively superficial. The others will be mentioned insofar as they influence, or are part of, the narrative. Besides, corporate interest usually manifested itself in the actions of sponsoring politicians, and this work will cover those.

I will address the command, control, and communication (C3) procedures for CAS only where they are germane to the narrative. Critics may say that this ignores the essential part of CAS, for the expeditious direction of planes to the proper target requires good C3 above all. But my premise for this work is that if there are no dedicated planes to perform this important but difficult process, then it will not be practiced enough to be effective, if it is practiced at all. The Air Force proved this repeatedly to its own chagrin. The CAS C3 process and technology is a very complex topic that rates separate treatment at length. In the meantime, the history of the relationship between the U.S. Air Force and the dedicated CAS plane is story enough.

THE ORIGINS OF AIR SUPPORT

Military people quickly saw that airplanes might secure battle-field tactical advantage by providing extra firepower. In 1911 the Italian military became the first to make use of airplanes in combat, against Turkish troops. After World War I began, improvements in airplanes and armaments made possible more noteworthy air support efforts. Because of differences in situation and national war-fighting style, those efforts yielded mixed results and first raised the perennially contentious issues surrounding CAS and CAS planes.

Air support showed its promise in World War I. If nothing else, it adversely affected enemy troop morale, especially among inexperienced soldiers. In the German Western Front offensive of 1918 and the ensuing Allied counteroffensive, both sides harassed retreating units and disrupted attacks. The Germans concentrated their air support missions to achieve the greatest demoralization and destruction. By the end of the war, the British used airplanes to strike artillery positions that impeded their tank assaults. In September 1918, American Brig. Gen. William "Billy" Mitchell orchestrated a large-scale air support operation as part of the U.S. offensive at Saint-Mihiel. At the same time, British airplanes achieved spectacular successes on the Italian, Macedonian, and Palestinian Fronts, as they caught and decimated large enemy forces retreating through narrow passes.

However, drawbacks arose that were endemic to the mission and specific to various combat situations. Even at the slow speed of a World War I biplane, hitting the proper target was still difficult. One discerns no distinguishing features of men standing in an open field beyond a range of two thousand feet. Large vehicles, such as tanks, are hard to recognize beyond five thousand feet—closer if camouflage or obscurants such as dust are factors. These were distances at which airplanes often either commenced their attacks or delivered ordnance. Further, combat stress combined with distance and aircraft speed to make a seemingly obvious visual feature inconspicuous to the pilot. If friendly ground positions shifted without the airmen's knowledge, they sometimes mistakenly shot their own troops or did not fire at all.[1]

Since the wireless radio was an undeveloped technology, military services normally did not use it to help the soldiers and airmen confirm target particulars. Most mission details were coordinated before flight. After takeoff, pilots tried such rudimentary signal techniques as revving engines or dropping messages, while the troops used visual signals such as flares. However, as British pilot Arthur Lee recalled about one mission against intermixed forces: "Pockets of our infantry, cut off in the sudden advance, were firing lights of every color as S.O.S. signals. . . . It was [still] difficult to distinguish the Boche from our troops as . . . our people were scattered."[2]

Air support often incurred unforgettable air combat casualties, especially on the Western Front, where British aircraft losses sometimes reached as high as 30 percent. The airmen's desire to help the soldiers, the need for positive identification, and weapons limitations together spurred a predilection for very low-level attacks, which put the airmen well within range of enemy ground fire. Lee wrote: "You may see trenches, but you can [only] see who's in them by flying low enough. . . . If in doing this you go down to 100 feet or lower, you're an easy target and dozens of guns will be turned on you."[3] Enemy troop resistance increased as soldiers lost their fear of air attacks and fired back; although Western Front fliers also encountered enemy fighters and dedicated antiaircraft artillery (AAA) batteries, intense small arms fire in the battle area apparently accounted for most air support losses.

The loss rate issue warrants further consideration. British fliers on the Western Front became outspoken critics of the mission's cost versus return. However, their air support usually consisted of roving fighter patrols that located and attacked enemy positions, thus risking longer exposure to gunners. They did not procure armored attack planes until the end of the war, so that meaningful comparative results could not be obtained.[4]

Unlike the British, who assigned fighter pilots to the mission, the Germans provided special training for aircrews to fly coordinated mass air support strikes. Their targets were easily recognizable enemy forces either directly confronting German troops or located just behind the lines. The Germans also initiated air-ground wireless radio contact to assist air support attacks, but struggled with its combat implementation. Further, they modified fast, two-seat escort fighters to fly the mission as armored, dedicated air support craft. By the end of the war they had introduced an all-metal, two-seat monoplane, the Junkers-1, for air support work. This approach yielded good results. German air assaults helped reverse initial British successes in the Battle of Cambrai (1917), a battle that also featured a serious British air support effort. Additionally, air support contributed to the Germans' early success in the 1918 Western Front offensive.[5] However, Germany's air support loss rates were not as well reported, and existing histories either were unclear about them or only cited the RAF statistics. Like the RAF's use, late in the war, of different tactics and special planes, this situation prevented meaningful determination of the value of using dedicated planes. Also, the more numerous Allied air forces seized air superiority in mid-1918, thus hindering German air support operations and further limiting assessments.[6]

Interwar Fortunes

The U.S. Army Air Service (the name used for the Air Force until 1926) flew some limited attack sorties during U.S. involvement with the Mexican Revolution. Because of the late U.S. entry into World War I, the Air Service saw comparatively little action in that war. Still, Air Service leaders favored the Germans' wartime approach to air support, preferring to use specially built, armored, two-seat planes for the mission. They retained a dedicated air support squadron and bought dedicated attack planes up through the 1930s. Meanwhile, other services pursued different paths regarding this mission.[7]

Neither France nor Italy had a strong enough air force to rate a noteworthy air support program. As for Britain, Western Front air veteran John Slessor emphasized in his influential study, *Air Power and Armies*, that *"the aeroplane is not a battle-field weapon"* (italics in the original). Slessor and his service preferred interdiction and believed that only emergencies such as enemy breakthroughs justified frontline attack missions. To the RAF, purpose-built attack planes were as wasteful as the mission itself. Fighters could accomplish CAS when needed—and when not busy achieving air superiority, which had been a Western Front concern. England's air support successes in other World War I theaters, as well as in its postwar colonial air control program, did nothing to change the RAF's attitude. RAF leaders considered these to be poor models for conventional war planning, viewing them as farflung, relatively minor campaigns.[8]

Contrary to popular belief, the Germans did not fully embrace close air support. They preferred interdiction as a more productive use of air power, and contemplated flying CAS only when the combat situation dictated it. Influenced by their experience in the Spanish Civil War, they commenced development of a dedicated air support plane. This was the Heinkel-129 and not the Ju-87 Stuka dive-bomber, which attained such notoriety in World War II air support efforts. The Germans intended to use the Stuka for interdiction, while fighter planes would fly CAS after securing air superiority.[9]

The U.S. Air Service eventually followed the European tack, as the U.S. Army Air Corps (the service's name between 1926 and 1941) drifted away from the mission. The Air Corps joined other air services in believing that air support would not justify the losses it would entail. Also, Air Corps leaders faced tight budgets, a struggle for institutional independence, faith in strategic bombing, and technological factors.

The U.S. Congress reduced military spending severely after World War I, and this especially hurt the Army's airmen. Flying is very expensive, and to ensure credible air strength, the airmen required money for technological research, airplane procurement, pilot training, maintenance, and a host of support functions. Air Service officers like Billy Mitchell believed that air power would someday have a decisive impact upon warfare, and resented the

Army leadership's failure to invest in its potential. Possibly influenced by aviation theorists, such as Giulio Douhet in Italy—and, like Douhet, influenced by the battlefield carnage of World War I—Mitchell increasingly asserted that airplanes could end wars with less overall bloodshed by bombing the enemy's homeland. He advocated a separate flying service, and advertised the potential of air power through publicity stunts like his bombing of mothballed battleships in 1921. In 1925, however, his increasingly outspoken opinions earned him a court-martial.

Because Mitchell sacrificed his career for air power and Air Service independence, his experience and his views had a lasting impact on the thinking of Air Force leaders. Low budgets continued to inhibit air power development, and Mitchell's successors followed his lead in seizing every opportunity to sell their force's capabilities. They organized well-publicized tactical exercises, humanitarian efforts, international flights, and an abortive attempt to fly the nation's mail. They also worked with Congress to maintain political and bureaucratic momentum. The airmen slowly gained status within the Army through the resulting series of hearings and review boards. They also acquired a keen awareness of political and publicity matters—not to mention antipathy toward any notion of subservience to the Army. When the independent Air Force later encountered contentious issues, such as CAS, its leaders used a sophisticated public relations apparatus to support their position.[10]

During the interwar years strategic bombing became a high priority, overshadowing not only CAS but all other tactical missions as well. The Air Corps Tactical School (ACTS) helped formulate the branch's war-fighting doctrine; like Mitchell, its faculty members supported long-range strategic bombing as a separate means to victory and as a demonstration of the independent worth of the Army Air Corps. In their enthusiasm for this unproven, technology-dominated approach to war, the ACTS faculty tended to disregard missions such as pursuit and air support. Although strategic bombing was as yet untested, other developments suggested that it might work as predicted. The B-9 and B-10 bombers of the early 1930s flew faster than contemporary U.S. fighters. Further, no radar existed at the time to give defensive fighters an edge. In the late 1930s the rugged, well-armed B-17 with its Norden bombsight offered the chance to strike an industrial nation's vital centers while enduring relatively few casualties.

Technology not only pulled the Army Air Corps toward strategic bombing, but also pushed it away from CAS. In spite of the new emphasis, however, the Air Corps produced ground support planes throughout the interwar years. Air Corps leaders pursued an attack plane design ideal that featured at least two aircrew (one gunner for rear-attack protection), protective armor, and heavy armament. Opinions about the best air support type shifted in the 1930s. Out of concern about battle area threats, the Air Corps opted for single-engine, fighterlike planes possessing both armor and speed, such as the A-12 and A-17. By the end of the decade, increasing aircraft speeds made these planes too slow and also raised questions about the best CAS weapons delivery method. The Army Air Corps did not opt for dive bombers as the U.S. Navy and German Luftwaffe did. Instead, as its air support doctrine shifted toward interdiction, its leaders procured light bombers such as the A-20. These twin-engine planes carried more bombs than the two-seat fighter attack planes.[11]

Relations with Army ground forces also reflected a growing lack of interest in CAS. During the interwar years, no nation's military establishment fully developed the radio communications that could have facilitated the mission, and the U.S. Army's soldiers and airmen were no exception. They did not work closely together during exercises. As World War II commenced in Europe, the Germans' spectacular Blitzkrieg offensives sparked a reappraisal, but prewar exercises still revealed serious deficiencies in air-ground cooperation.[12]

Air Support in World War II

In World War II the Germans did not at first fly true CAS, but instead flew preplanned attacks against Army targets behind the lines. This was due to their prewar doctrinal setup as well as their limited development of ground-air radio procedures. As the war continued, however, they amply demonstrated air support possibilities and further developed close air support procedures. After experiencing some embarrassments early in the war, the RAF improved its air support system, but remained skeptical of air support overall.

The U.S. Army Air Force (the name changed in 1941) worked slowly toward flying CAS. In the war against Germany, its first tactical flying opportunities came in Tunisia in 1943. After a rocky start, the airmen beat the Germans and also achieved a doctrinal victory.

The Army's air support instruction, Field Manual 31-35, assigned control of tactical air units—light bombers, fighters, and such—to a theater commander's subordinate ground commanders. Many ground commanders believed good air support occurred when they saw their planes attacking targets, and they were very disappointed if enemy planes attacked them. Stuka dive bombers caused only 5 percent of U.S. combat casualties in the European theater, but U.S. troops ranked the Stuka fifth among nine German weapons they considered most threatening.[13] Airmen resented Army control because ground commanders often would not allow them to mass their valuable assets to achieve air superiority—a goal whose attainment would reduce or even eliminate enemy air raids while facilitating their own air support attacks.

Unfortunately, a snarled command structure, logistical problems, and Army misuse of air assets sapped the Army Air Force's effort in the early days of the African campaign. Piecemeal usage cost the Army Air Force heavy losses, and German planes attacked U.S. troops anyway. The airmen achieved their tactical war-fighting preferences through an agreement made at the January 1943 Casablanca Conference and also as a consequence of the U.S. Army's embarrassing February defeat at the Battle of Kasserine Pass. The creation in mid-1943 of FM 100-20, *Command and Employment of Air Power*, ratified a longstanding Air Force goal by declaring that an air commander would control all theater air units and serve equally with a ground commander under the theater commander. The field manual also established the Army Air Force's mission priorities: first, air superiority, then interdiction, and last, close air support. These changes permitted a concerted air campaign in Tunisia against German air bases and supply lines, which greatly reduced the Germans' air activity and isolated their ground units.

Close air support did not occur often either in Tunisia or in the ensuing Sicilian campaign. Procedures and communications equipment for executing it were still not perfected. When the Army Air Force flew the mission, it used fighters, because the light bombers that the airmen purchased for Army air support were, as intended, more suitable for interdiction.[14]

The lack of CAS and the publication of FM 100-20 were matters of concern to many ground commanders, who viewed the new doctrinal regulation with "dismay." High-echelon control characterized how ground forces used their heavy artillery, and air superiority reduced or even eliminated enemy air attacks, but the soldiers still wondered if the airmen would abandon air support. One Army assessment opined, "To the soldier on the ground . . . [FM 100-20] nailed the CAS doctrinal door shut." Further, FM 100-20 seemed too rash, given the U.S. forces' fumbling start; its tenets might not apply in other situations. The Tunisian air support situation led Assistant Secretary of War John McCloy to investigate it in spring 1943, before the release of FM 100-20. His conclusions reflected the rising concerns among ground commanders and their citizen-soldier troops: "It is my firm belief that the air forces are not interested in this type work [close air support]. . . . What I cannot see is why we do not develop this auxiliary to the infantry attack even if it is less important than strategic bombing. It may be the wrong use of planes if you have to choose between the two but to say that airpower is so impractical that it cannot be used for immediate help of the infantry is nonsense and displays a failure to realize the air's full potential."[15]

When Army advance stalled in Italy, airmen and soldiers had time to improve CAS operations. Applying a lesson learned in Tunisia, ground and air leaders collocated their headquarters and set up a coordinated request system. Air leaders assigned radio-equipped, pilot-led liaison teams to ground units to provide expertise in directing aircraft onto targets. CAS-related problems still arose. U.S. troops occasionally botched visual signals, while pilots and liaison parties struggled with target briefing procedures. Tragic results sometimes ensued, with airplanes hitting their own troops or not responding quickly enough to urgent requests. Even so, CAS demonstrated its importance in Italy. The Army Air Force's 1944 interdiction campaign, code-named Strangle, weakened German forces but did not induce the collapse that its proponents had anticipated. Meanwhile, CAS contributed to several tactical successes, and earned the appreciation of the U.S. commander in Italy, Gen. Mark Clark.[16]

During the U.S. sweep across France in 1944, Army Air Force air support achieved its greatest effect. Air supremacy in France and an abundance of planes and pilots meant that the U.S. Ninth Air Force could devote itself to tactical operations. Two pragmatic and energetic officers, Brig. Gen. Otto Paul "O. P." Weyland and Brig. Gen. Elwood "Pete" Quesada, commanded the force's Tactical Air Commands (TACs). The TACs worked with a specific U.S. Army; Quesada's Ninth TAC supported Gen. Omar Bradley's Fifth Army, and Weyland's Nineteenth TAC supported Gen.

George Patton's Third Army. Bradley regarded Quesada as "unlike most airmen who viewed ground support as a bothersome diversion." Patton sometimes entrusted to Weyland's TAC the unprecedented responsibility for covering his flanks.[17]

Both TAC commanders ensured proper coordination between the various levels of their respective air and ground commands. Neither rigidly adhered to FM 100-20 or other air power precepts; instead, they adjusted command setups and mission priorities to match the situation. Quesada put a radio and a liaison pilot in at least one of the tanks in the leading armored units and thus enabled his fighters to fly top cover for the advancing armored columns. This measure also quelled "simmering complaints" about friendly fire and "laborious" communications procedures. Quesada pushed his fighter pilots, some of whom were at first reluctant to forsake aerial combat's glories, to learn and appreciate air-to-ground weapons delivery in support of the Army. Weyland said of his pilots' devotion to supporting Patton's troops: "My kids feel that this is their Army."[18]

One reason why fighters became the prime air support aircraft in this campaign was because bombers confirmed their incompatibility. Twin-engine light bombers performed the interdiction mission for which they were designed, but they were vulnerable to air defenses and required at least forty-eight hours' advance notification for a mission. Thus they mattered little to a tank unit commander facing a sudden strong counterattack. As for heavy bombers, the Army Air Force preferred using them as deep strike strategic weapons. Air leaders made an exception when they acquiesced in the Army's desire to use the heavies in facilitating a breakout from Normandy. The bombers were an important breakout factor, but their inflexible attack procedures directly contributed to one of the war's most infamous fratricide incidents.[19]

Freed from air superiority tasks, and possessing better tactical performance and smaller logistics requirements, fighters were much more adept CAS machines. As the Americans' Normandy breakout forced the Wehrmacht into a maneuver war requiring open vehicular travel, German soldiers encountered swarming fighters, which attacked everything that moved. This happened mainly within the approximately twenty miles of the front lines that made up the close air support arena. Former Ninth Air Force fighter pilot Robert Brulle later wrote about the fighters' effect upon German ground forces: "We knew they called us 'Jabos' . . . and that the cry 'Achtung Jabos' created an instant reaction. The foot slogger hit the dirt . . . the driver and passengers bailed out of moving vehicles . . . and even armored vehicles raced for cover. Captured German soldiers claimed that the Jabos were 'the most terrifying weapon on the Western Front—they are Eisenhower's secret weapon.'" U.S. ground officers observed that German resistance was less fierce after air attacks, while the morale of U.S. troops soared. Perhaps the most prominent example of CAS effectiveness came when one German army commander insisted upon surrendering his unit to O. P. Weyland.[20]

Later studies showed that fighter air support had the greatest impact upon German forces through disruption more than outright destruction. This finding raised questions about close air support's destructive effect, and had implications for later doctrinal and weapons procurement debates. Detailed analysis contradicted fighter pilots' claims for many destroyed tanks at such places as Mortain and the Falaise Pocket. Bombs had to hit very near a tank to destroy it—though napalm had considerable psychological effect upon all German troops. Rockets required a direct hit, and they could be hard to shoot accurately. Strafing with 50-caliber machine guns did not pierce a German tank's heavily armored forward turret area, but it could at least damage the engine and treads. Ricochets could potentially penetrate a tank's vulnerable underside. Although fighters did not destroy as many tanks as did artillery or other tanks, all studies agreed that the fighters' weapons wreaked havoc upon unprotected troops and the less-armored vehicles that made up the bulk of any modern conventional army's maneuver units.[21]

As in World War I, accounts of Western Front fighter air support mentioned high loss rates. Fighter CAS losses were high when compared to fighter losses on air superiority missions, but not as high as bomber losses over Germany. Indeed, historians noted that World War II air combat was an attritional fight with high losses in all missions. In his superb analysis of World War II Allied air support in Europe, Ian Gooderson found that fighters suffered fewer losses flying troops-in-contact CAS than when flying armed reconnaissance missions behind the lines. This recalls World War I, and England's problems with air support conducted via roving patrols. Fighter pilots' memoirs, like Robert Brulle's *Angels Zero*, show that CAS was dangerous—especially against well-prepared defenses on static fronts—but that more losses occurred on roving patrols behind the lines.[22]

One Army Air Force commander in Europe, Maj. Gen. Hoyt Vandenberg, complained that artillery should assume the fighters' CAS load in France, but as Gooderson noted, CAS costs were irrelevant to any soldier facing a dangerous task against a tough opponent. Air-dropped bombs were larger and packed more explosive than most army artillery shells. Further, CAS offered flexible direct fire support against specific targets instead of the indirect barrage fire associated with heavy artillery. CAS often had to serve as the Army's artillery when rapid advance outran artillery support, or when terrain impeded armor and heavy artillery movement. Gooderson observed that ground commanders used CAS not only to deal with emergencies but also to exploit advantages. The U.S. citizen-soldier Army relied upon any available firepower to achieve battlefield success. As one Wehrmacht soldier recalled, "the GIs let the high explosives do the hard work."[23]

Lack of artillery was one condition that made CAS important in many Pacific theater battles. Shipboard guns could provide firepower during amphibious landings, and interdiction of enemy seaborne supplies was very important in isolating Japanese-held islands. However, CAS served as direct firepower support in reducing enemy strongholds, especially since the rugged jungle terrain inland often precluded transporting artillery or tanks. CAS eased the soldiers' situation in meeting determined Japanese resistance in island battles, such as those of New Guinea, Bougainville, Peleliu, Luzon, and Okinawa. Also, once air superiority was attained, U.S. leaders lavishly apportioned CAS missions; in one case, five squadrons attacked a well-fortified Japanese *platoon.* Still, CAS development required work in this theater as well. The jungle lacked recognizable landmarks, and air-ground radio communications were inadequate at first. Thus, most air support missions hit targets just behind the lines, while confused visual signals and friendly fire incidents sometimes marred CAS attempts. By 1944, both Army Air Force and Marine aviators had set up better radio communications, established signals and ground control parties, and practiced the mission enough to render excellent CAS in the battles fought late in the war. A variety of planes initially flew the mission in the Pacific. The Navy and Marines relied upon the Dauntless dive bomber, while the Army Air Force used obsolete fighters and various bomber types. Eventually, all services mostly used fighters for CAS.[24]

The evolution of fighter CAS in both major war theaters provided apparent justification for later Air Force claims that such planes and their pilots could handle the mission when not performing other duties. Wartime fighter commander and later U.S. Air Force TAC Commander William Momyer opined, "What I've always said is that if you can train a guy to do air-to-air, with a minimum of additional training he can do the air-to-ground mission. . . . Then you have that flexibility to be able to shift back and forth in war." The Army Air Force's star air support fighter was Republic Aviation's P-47 Thunderbolt, a rugged radial-engine brute that could also beat German fighters. Pilots praised its durability, a characteristic that probably contributed to keeping CAS aircraft losses in Europe lower than expected.[25]

Conversely, Air Force leaders saw any dedicated attack plane as a vulnerable, limited-capability asset that could be a death trap in certain situations. Beyond their concern for mission execution was the desire not to inflict disadvantages upon their aviators. A bomber commander spoke for all airmen who ever faced a superior enemy fighter force when he recalled his thoughts on a mission that suffered high losses: "I suppose this feeling of being caught in a hopeless situation is far from new. . . . I think of the Middle Ages. I see myself strolling across an open plain with a group of friends. Suddenly we are beset by many scoundrels on horseback. . . . We cannot run, we cannot dodge, we cannot hide—the plain has no growth, no rocks, no holes. And it seems endless. There is no way out, then or now." In his acclaimed CAS survey Richard Hallion asserted that World War II showed fighters to be the better and safer CAS planes. In one section, he even listed the number of fighters that successfully flew World War II CAS and compared these to bomber and attack plane designs—while highlighting these other planes' vulnerabilities.[26]

However, the fortunes of World War II fighter CAS planes rate a more detailed review. Success in air combat relied upon subjective factors as much as performance absolutes; as Air Force historian Robert Futrell observed, this global war generated massive documentary resources that could support any lesson, depending upon the situation. Compared to their later, faster jet counterparts, top World War II fighters better met close air support demands because they could operate from unpaved strips, required relatively little maintenance, and were quite maneuverable. Their fuel endurance and load carriage capability still compared favorably with those of purpose-built attack planes. In spite of fighters' multi-role compatibility, the question remained whether fighter pilots could do all of their assigned tasks well. Even

Momyer had reservations: "But, there is a limitation on the number of weapons that you expect him [the fighter pilot] to deliver. Obviously, you can't have him proficient across the board." Quesada's biographer, Thomas Hughes, observed that the fighter pilots flew air support well because Quesada made them do it well. Techniques for attacking German vehicles required practice to be effective, and the Americans had to set up a special school to teach fighter pilots how to fly ground attack.[27]

Second-rank or obsolete fighters, such as the Hurricane, P-39, and P-40, served as CAS machines, which demonstrated that air combat prowess need not be the criterion for CAS plane selection. As the airmen asserted, contested airspace made it difficult to accomplish any mission, regardless of which plane flew it. Thus, obsolete fighters and dedicated attack planes flew CAS in situations where the enemy could not mount an air threat or where their own dedicated air superiority fighters protected them.

The Dauntless dive bomber used by the Navy and the Marines thrived through several early Pacific Ocean carrier battles that featured Japanese fighter opposition. Nevertheless, air war historians criticized its air combat vulnerabilities, especially the losses incurred when the Army Air Force used Dauntlesses on sustained air campaigns against heavily defended Japanese targets. The Dauntless was no fighter, and its carrier battle opposition was sporadic compared to what the Army Air Force faced. Still, less adverse conditions and good Navy and Marine fighter protection allowed it to become a CAS workhorse until other attack planes—and fighters—began replacing it. The U.S. Marine Corps used Dauntlesses for CAS through 1945. (The Free French also used Dauntlesses during the Allied sweep across France and into Germany.)[28]

More notorious was the Germans' Stuka. The slow dive bomber ravaged Allied Army and refugee columns during Germany's early victories, but RAF fighters forced its withdrawal from the Battle of Britain. Many observers vilified its vulnerabilities and cited it as an example of the Luftwaffe's fatal attraction to air support and the clunker planes designed to fly it. Though the Stuka needed improvements, or even a replacement, other war theaters allowed it to be successful in various roles. Its losses were still high at times, and the Stuka hero Hans Rudel survived several shootdowns. Loss rates and disruption increased for all missions and planes when an air arm lost air superiority, as the Luftwaffe did as the war progressed. The Stuka conducted

Eastern Front air support nearly until the end of the war, thanks to good tactics and to the fact that fighters normally avoided the battle area and its considerable dangers.[29]

Remarkably, the Russians' Ilyushin-2 Shturmovik attack plane received little of the opprobrium heaped upon the Stuka. The Shturmovik was heavily armored, carried a respectable weapons load, and for much of the war was the Soviets' weapon of choice for targets just beyond artillery range. Given its performance specifications, it fared as poorly as the Stuka against fighter opposition. Its success was due to its ruggedness, fighter escort, and the Russians' ability to produce and deploy it en masse.[30]

At the close of the war, any debate over CAS plane preferences would have to wait. World War II ended after B-29 strategic bombers dropped atomic bombs on Japan. Though bombers had significant impact upon the outcome of the European war, the strategic bombing campaign had encountered high losses and other problems that undid the bomber advocates' prewar claims about singlehandedly winning wars. That the atomic bombs induced the surrender of the Japanese, who had fanatically defended various small Pacific islands throughout the war, reinforced the belief of the Army Air Force bomber men that they possessed the ultimate firepower trump card. Besides, the early atomic bombs were big, so only heavy bombers were capable of carrying them. In succeeding years, strategic bombing dominated military thought, U.S. foreign policy, and the newly independent Air Force. Close air support became a neglected mission for a while.

3 AIR SUPPORT IN THE ATOMIC ERA

The relative importance of close air support or any other war-fighting mission depended upon the conflicts that the nation might encounter while conducting its foreign policy. However, Americans have been ambivalent—even fickle—about foreign affairs, thus creating a potential for conflict anywhere and in any situation. For example, during a time of relative isolation for the United States, the 1870s, U.S. Marines stormed forts in a far-off place called Korea. Beyond the world wars that engage people's memory, by 1945 U.S. troops had fought other battles in China, Cuba, the Dominican Republic, Haiti, Mexico, Nicaragua, North Africa, and the Philippines. As might be expected, the reasons varied widely: protecting American interests, serving humanitarian concerns, or teaching foreigners to elect good leaders.

Foreign affairs ambiguities persisted after World War II. As the Cold War escalated in the late 1940s, the United States assumed the lead in thwarting communist expansion, a role that entailed global responsibilities. At the same time, the country reduced its defense establishment from thirteen million uniformed citizens to just over one million; and Congress rejected President Harry S. Truman's proposal for universal military training.

What would be the means for exercising world influence? Air Force bombers and the atomic bomb provided an American firepower solution. In 1947 the President's Air Policy Commission stressed the importance of air power as a deterrent to any future aggression. Hanging over the commission's deliberation was not only the atomic bomb, but also Pearl Harbor. The board warned that the United States could no longer afford to lose the first battle in a war. A congressional inquiry drew similar conclusions. As one historian put it, "No matter how ill-conceived the Combined Bomber Offensive against Germany or the strategic offensive in the Pacific had been, it all became irrelevant to the American public, Congress, and the Air Force. Overnight, strategic bombardment had gone from thousand-plane missions . . . to a single bomber with a single bomb." Further, given Americans' desire to face overwhelming Soviet military strength in Europe without a large standing army, the bomber and its nuclear cargo dominated current national war plans.[1]

The Air Force and Its Bomber Men

The new role of the bomber was a boon for the U.S. Air Force, which gained institutional independence in 1947. Its leaders embraced the nation's charge, for they believed that the recent war had confirmed their earlier claims for air power. The leaders were either former bomber men or officers who supported the bomber mission. As part of a service built upon advanced technology, many also were engineers who believed that technology could overcome the age-old vexations of warfare. A postwar scientific study led by scientist Theodor von Karman urged the Air Force to exploit the Germans' aerospace advances because these were not only feasible, but also would render current airplanes obsolete. Hap Arnold asserted that "for twenty years the air force was built around pilots. . . . The next twenty years is going to be built around scientists." Von Karman's findings spurred a chase for an invincible jet bomber.[2]

These new circumstances affected Air Force leaders' attitude toward other services. Some airmen publicly dismissed most Navy warships, including aircraft carriers, as obsolete craft that were helpless against nuclear bombers. Such comments inspired animosity among Navy leaders, who feared an Air

Force–Army alliance against them in the increasingly intense interservice struggles over the ever decreasing postwar defense budget. Further, Navy fliers remembered that Billy Mitchell had demanded that only one service possess all warplanes.[3]

Ill feeling climaxed in 1949 as Defense Secretary Louis Johnson canceled a Navy supercarrier in favor of the Air Force B-36 bomber (this was not the jet bomber, but an interim propeller-and-jet model). Navy criticism of Air Force bombers threatened the service's most cherished mission, and Air Force leaders accordingly escalated their claims on its behalf. In spite of high bomber losses on unescorted raids against Nazi Germany, they asserted that bombers could successfully strike targets deep in Russia. As one general later observed, "We . . . don't advertise our limitations to demagogic politicians, and we certainly don't advertise our limitations when we're talking to members of the press, who are looking to . . . denigrate the speaker or his service." Thus, debate-inspired claims highlighted the service's public relations efforts, which its first secretary, Stuart Symington, cultivated. Symington built a sophisticated public relations program that used sympathetic writers and scholars to support the Air Force. He also ordered Air Force leaders to perfect their already impressive briefing and staff research procedures to present a professional, unified front. The Navy failed to reverse Johnson's decision, and the Air Force image soared while the Navy's clumsy public relations antics earned its efforts the title "Revolt of the Admirals."[4]

The Air Force also neglected to support its parent service, the Army. However, with only occasional rumblings, Army leaders accepted Air Force priorities. The atomic bomb dazzled many ground leaders, who were now unsure how future wars would be fought. With regard to tactical operations, World War II made Army leaders understand that achieving the air superiority mission helped the ground war as well. Some soldiers even agreed with Air Force control of a theater's air assets, though others feared the service would then commit entirely to strategic bombing. U.S. Army Chief of Staff Dwight Eisenhower told Congress that a campaign commander wanted one air power representative instead of several selfish subordinate ground commanders who were neither as qualified nor as able to concentrate on air matters. Thus, the soldiers supported the Air Force's independence, though Eisenhower made the creation of a tactical air arm within the Air Force a condition of Army acceptance.[5]

Officially, the Air Force remained committed to CAS. The 1948 Key West Agreement that tried to clarify interservice relationships also carried an Air Force promise to accomplish the mission. (Its definition of CAS survived ensuing years and controversies: "The attack by aircraft of hostile ground or naval targets which are so close to friendly forces as to require detailed integration of each air mission with the fire and movement of those forces.") A 1949 Air Force–Army accord confirmed the Air Force's obligation by restricting the Army's own planes to small liaison aircraft. The services worked together to establish an air task apparatus that featured co-equal air and ground commanders serving under a theater commander, along with liaison officers from both services assigned to a Joint Operations Center, and subordinate air and ground commands. This required that any air task request be reviewed up and down both services' chain of command, but the services explicitly rejected the Marines' more streamlined method. The Marines used smaller units with lighter ground firepower in shorter duration operations, while Air Force and Army leaders remembered the massive campaigns of Europe in World War II. Even so, with the Air Force committing so much of its resources to strategic bombing, and Army leaders facing severe manpower reductions, neither service gave the air task apparatus proper attention.[6]

The lack of attention carried the expected consequences: liaison positions went unfilled, or were filled by unskilled or apathetic people. Radio links between pilots and ground parties became unreliable. Joint maneuvers between 1947 and 1950 suffered air response delays and instances when lack of communications prevented any CAS. The misuse of the "bomb line" exemplified the situation. This line set a distance in front of the most forward friendly troops that designated where ground commanders required prior coordination for air strikes. In postwar exercises the bomb line became more a strict demarcation between ground fires and air strikes; the two services effectively gave up on CAS.[7]

U.S. nuclear war strategy affected intraservice relationships as well. The new jet bombers compelled maximum commitment. Jet engines were much more complex and expensive than reciprocating engines. They sharply increased aircraft speeds, and presented aerodynamic and structural challenges which the bombers' size and complexity magnified. Fuel-hungry jets required the development of tanker planes to support long-range flights.

Given the special preparation needs for nuclear warfare, bomber units required constant, intensive training in a top-notch organization. After a shaky start in the late 1940s, the U.S. Air Force Strategic Air Command (SAC), assigned to conduct nuclear war, flourished under Gen. Curtis LeMay's aggressive leadership, but it required most of the service's resources.[8]

In contrast, the U.S. Air Force Tactical Air Command (TAC)—the organization whose establishment was Eisenhower's precondition for Air Force independence—found itself in the budgetary and bureaucratic wilderness. In spite of their bomber orientation, Air Force leaders initially wanted a large force that included tactical planes, and they continued development of jet fighters. However, the Truman administration's military budget cuts removed the Army units justifying a large tactical Air Force and encouraged the service to neglect missions other than strategic bombing. In 1948 Air Force leaders subordinated TAC to another major command and reduced its staff to a skeleton force. TAC Commander Pete Quesada resigned his post in frustration. To him, the move violated agreements with the Army and removed the best insurance against that service reentering tactical aviation with its own aircraft. The Navy publicized the situation during the Admirals' Revolt, but nothing changed—a reflection of American war-fighting preferences ensured that.[9]

Even if the Air Force had given more attention to CAS, air support exercises revealed problems with its jet fighters. Wedded to what one aviation historian called the "fighter-bomber" concept, Air Force leaders claimed that F-80 and F-84 jet fighters could also fly ground support missions, just as P-47s had done in World War II. Though the frantic push for ever faster jets would soon render these planes obsolete, their operating speeds were almost double, and their turn radius almost three times, those of the World War II fighters. This made ground target recognition more difficult and hindered the ability to maneuver to attack a suddenly recognized ground target. Higher fuel consumption compounded the problem by restricting the time over target. The Air Force defended its fighter-bombers against Army and congressional criticism, but fighter technology advances forced hard choices about optimum mission performance. One could not have it both ways with the best air superiority and close air support designs. These problems were exposed more glaringly in another war, in that far-off place called Korea.[10]

CAS AND THE KOREAN WAR

U.S. Air Force chief of staff Hoyt Vandenberg later complained to a Senate committee that the Korean War did not conform to Air Force doctrine. Actually, it did not fit anything Americans wanted to do regarding their world role. Nuclear deterrence did not prevent aggression. Korea was a hot, conventional-style war. The Army fought undermanned and outgunned in many early battles, which meant that air support was needed to save the situation. This was the price of global foreign policy: varied war-fighting situations in which enemies would not do what Americans wanted them to do.[11]

However, air power decisively boosted the combat fortunes of the American-led United Nations coalition. Both U.S. and communist leaders cited its importance in stopping two major communist ground offensives. Interdiction bombing denied the North Koreans their airfields and crippled their supply efforts. U.S. Air Force fighters maintained air superiority over the battlefield, allowing U.N. air support planes to fly their missions unhindered. Nonetheless, the Air Force bickered with the other services over CAS.

Most of the early air support missions were interdiction strikes not far behind the vanguard of the North Korean advance. The communists indiscreetly massed their troops and vehicles, creating lucrative targets for attack planes. However, some U.S. Army generals quickly became dissatisfied with U.S. Air Force CAS. The Air Force jet fighter pilots in Korea who initially flew CAS had focused upon air-to-air tactics and knew little about air support. During the initial communist invasion, F-80s and F-84s could not operate from the rough fields within the ever constricting U.N.–held Korean territory. The increased distances from Japanese bases further limited the jets' usefulness, especially considering their already poor loiter time. Since they were air-to-air planes, their bomb-carrying capacity—aggravated by their heavy fuel consumption—was very poor. Radio problems resurfaced with a vengeance, and the jets' faster attack speeds made independent visual confirmation nearly impossible.[12]

Both services shared the blame for some of the CAS foul-ups, given their prewar neglect of the air task system, but other planes' CAS achievements convinced the soldiers that rapid improvements were possible. They preferred the air support services of the Air Force's World War II–vintage P-51 (now designated F-51) Mustang, relegated in Korea to fly ground support. It operated from short strips close to the battlelines, flew over the battlefield longer than the jets, and achieved better target acquisition through its maneuverability. Finally, since the Air Force committed the plane and its pilots to ground support, the pilots strove to perform it well. The real aggravation for some Army leaders came when they directly witnessed or heard of outstanding U.S. Marine Corps air support—especially during the Chosin retrograde. The Marines remained dedicated to CAS, and many Marine pilots in Korea had flown the mission in World War II. They and Navy pilots flew radial-engine, propeller-driven F4U Corsairs and A-1 Skyraiders. Both planes were superior to the F-51 for CAS work; the Skyraider was especially well suited, for it could loiter, absorb punishment, and carry a lot of ordnance. The combat comparison between these planes and CAS-deficient F-80s was glaring. Navy Adm. John Thach remembered what one Air Force forward air controller (FAC) told a flight of F-80s arriving on target low on fuel, carrying only two 100-pound bombs per plane, and demanding priority over Navy prop planes already in the area: "Well, take your two little firecrackers and drop them up the road somewhere because I've got something . . . that has a load." U.S. Army leaders in Korea continued their public protests, and their efforts led Army Chief of Staff J. Lawton Collins to formally request greater Army control of CAS as well as development of a dedicated CAS plane.[13]

The CAS controversy occurred early in the war, and for its duration Air Force leaders fought the other services over this issue almost as much as they fought the communists. They tried to gain control of Marine air units, arguing that the air support system would work if all U.N. air units were under one command. From general officers down to pilots, they tried to eliminate the air support system's problems, though they cited economic reasons for not committing enough people to run the air control system. The airmen used their public relations skills and defended their CAS record through a media campaign that included studies by outside experts. They closed ranks against the soldiers, whose opinions were not unanimous about Air Force CAS. Indeed, the Air Force cited instances of praise by Army leaders, and a

1951 interservice CAS study revealed that Army generals relied too much upon air support while not fulfilling their own manning obligations for the air task system. Air Force leaders remained concerned about fighting outnumbered against the Soviets in Europe, and thus defended their multirole fighter force policy for guaranteeing flexibility under combat stress. Finally, the two services resolved divisive issues, such as the Army's desire for CAS planes and its increasing use of a new aviation technology, the helicopter. Two interservice agreements allowed the Army to use these aircraft, but stipulated that the Army could not duplicate Air Force fixed-wing missions. Army aircraft would perform limited roles close to the front lines and its fixed-wing planes would remain small.[14]

After the war, the Air Force continued to bend results to its liking. Its accounts of CAS during the Korean War dismissed Army criticism. Likewise, the airmen and other chroniclers insisted that CAS performance of jets improved with increased mission exposure. (Jet-compatible bases in Korea increased with rising U.N. fortunes, improving jet loiter time.) Further, they asserted that jets carried bigger weapons loads, suffered fewer losses, and could fly other missions if needed. Conversely, the airmen downplayed the F-51's accomplishments by emphasizing its loss rate performing CAS. Some of these claims failed to stand up to scrutiny. Fighter jet pilots required diversion from their other missions to master CAS, a difficult task given the characteristics of their planes. Jet multimission capability did not necessarily include air superiority. Air leaders committed F-86 interceptor jets to that mission, while F-80s and F-84s could not defeat the MiG-15s they occasionally encountered. The in-line cylinder engine of the F-51 used a coolant system that was vulnerable to small-arms fire. Had the rugged P-47 still been operational, the loss rate might have been lower.[15]

Concerning vulnerabilities, Air Force historians included CAS losses in Korea as yet another example of the mission's relatively high cost. However, a later historical study of CAS in the Korean War stated that no conclusion was possible because flight records did not differentiate between air support and interdiction missions. Also, to reduce casualties, air leaders ordered *all* pilots to release their weapons at higher altitudes to avoid heavy small arms fire—thus belying the Air Force claim that the faster jets somehow outran concentrated antiaircraft fire in their weapons runs.[16]

The lessons that Air Force leaders took from this war did not support maintaining what CAS capability they had established during its duration. One service report stated that Korea's "lavish" CAS was "unlikely to exist in future wars." Instead, they preferred interdiction during the conflict, and as the ground fighting stalemated, they launched various interdiction campaigns that they hoped would force communist acceptance of U.N. terms. When the raids failed to achieve this aim, air leaders complained about the idiosyncrasies of the Korean situation as compared to World War II—though they chose to forget that interdiction had failed to achieve decisive effects under roughly similar conditions in Italy during World War II. Indeed, Air Force secretary Thomas Finletter declared that Korea was "a special case, and airpower can learn little from there about its future role in United States foreign policy." In keeping with national defense priorities as well as its own opinion of what constituted real air war, the Air Force refocused upon the Soviets in Europe.[17]

The "New Look" and the Air Force's Separate Path

Most Americans thought the Korean War was abnormal. The Army seemed to throw a "fit of pique" as it discounted "victory" as a limited war objective in the 1954 version of its doctrine manual. Even Gen. Matthew Ridgway, a later critic of U.S. nuclear-centered defense policy, asserted in 1953 that "war, if it comes again, will be total in character." Americans preferred the general war and unconditional victory of the World War II variety, and scorned the uncomfortable limited fights that were the price of globalism. The Eisenhower administration accordingly developed a defense policy called the "New Look."[18]

The New Look reduced U.S. ground forces and placed heavier accent upon SAC's big bombers. It was supposed to deter aggression at any level through the threat of nuclear annihilation. Though one might argue that the New Look drove the Air Force's predilection for bombers, the policy suited its leaders' preferences quite well. As Vandenberg's biographer put it, the general "consistently and single-mindedly sought to establish strategic air power as the cornerstone of American defense policy." Nuclear bombing raids deep in enemy territory were supposed to be the ultimate deterrent or, failing that, combat trump card. Further, Air Force leaders assumed that preparations for total war against the Soviet Union in Europe would also cover readiness for any other warfare scenario. SAC Commander and later U.S. Air Force Chief of Staff Curtis LeMay even suggested that the Air Force give its tactical planes to the Army.[19]

Obviously, tactical aviation faced a struggle within the Air Force under the New Look policy. The director of Air Force intelligence declared that "in this fast-moving age we no longer can build non-nuclear forces at the expense of our atomic strike and defense units." The Korean War returned TAC to independent command status, but pro-SAC national and Air Force priorities forced TAC's new commander, O. P. Weyland, to secure nuclear capability for his planes or face bureaucratic extinction. Weyland was concerned about the Air Force's nuclear bomber tack, and occasionally expressed his reservations. However, he still "struck a Faustian bargain with the atomic Mephistopheles"; all accounts of TAC in the 1950s mention its attempt to become a small-scale SAC. Gen. John McConnell, who later figured prominently in the birth of the A-10, noted that "we did not even start doing anything about tactical aviation until about 1961 or 1962."[20]

This attitude duly affected tactical jet design. The Air Force's nuclear strike trend combined with the pursuit of higher speeds to produce the "Century Series" fighters. Nicknamed thus because their designations ran in sequence from F-100 to F-106 (there was no operational F-103 model), the planes achieved supersonic speeds at the cost of maneuverability, fuel efficiency, and weapons-carrying capacity. The F-101, F-102, F-104, and F-106 were designed as dedicated fighters or interceptors. As for the F-100 and F-105, the Air Force asserted that they were fighter-bombers even though their CAS performance capabilities were even less than those of the F-80 and F-84 of Korea. This was no problem for TAC, which wanted the fighter-bombers to carry small nuclear bombs on high-speed, one-pass attacks. Indeed, the F-105 represented a more concerted TAC effort to enter the nuclear strike business, since it was designed to perform supersonic, low-level, nuclear bomb runs.[21]

Yet the F-100 and F-105 could be made to do almost anything, as their later performance in Vietnam attested. Their pilots could achieve acceptable competence in a diverse mission like CAS, but adaptation was difficult because the planes had poor cockpit visibility, and their high attack speeds and poor maneuverability led to relatively distant weapons release ranges. To provide safe escape from both the ground and the blast effects of his own ordnance during dive pullout, an F-100 pilot had to release bombs five thousand feet above ground level (AGL) in a 45-degree dive bomb delivery. Such an attack procedure had several consequences. It limited the pilot's ability to gain and maintain sight of a ground target, and it raised the weather visibility and cloud ceiling minimums required for visual attacks. High speeds and longer attack ranges also placed a high premium on precise aiming and correct release parameters—a task made more daunting by the primitive, fixed-gunsight aiming apparatus these planes carried (fire control avionics had not kept pace with aerodynamic progress). Finally, a Century Series jet's pitifully poor fuel efficiency at low altitudes forced the pilot to find the target quickly. Given ground combat conditions and the performance limitations, this was very difficult.[22]

The New Look policy's emphasis upon the nuclear mission meant that the Air Force's 1950s-era fighter-bomber pilots were not fully competent in tactical missions. They sporadically practiced conventional air-to-ground weapons delivery and CAS, and the lack of proficiency appeared in poor weapons scores and exercise results. Contingency operations, such as the 1958 Lebanon crisis, highlighted the deficiency. After that affair, one Air Force officer observed that "there is considerable doubt . . . as to the conventional combat capability of the F-100 units. Only a few of the F-100 pilots had strafed; none had shot rockets or delivered conventional bombs." Pierre Sprey, an expert on fighter design who influenced Air Force fighter procurement in the 1960s and 1970s, pointed out that if one could not expect a plane to accomplish multiple mission requirements equally well, then one could not demand that a pilot do so.[23]

The Air Force desire for fast, multiple-role, jet fighters contrasted with the trend in naval aviation. Though naval leaders also pursued the nuclear mission during the New Look era, they still recognized that the varied combat situations they encountered required more specialization in aircraft design. As the Air Force fighter-bomber design philosophy encountered difficulties in the 1960s, Navy planes would serve as stopgap solutions.[24]

Army Views and Actions

The Air Force was not shy about expressing its narrow war-fighting view to the Army. An Air Force general told *Army* magazine correspondent Robert

Asprey that in future wars with communist nations, "we are going to need all the fast aircraft we have," and that CAS was "a maximum waste of firepower." Still other air leaders told the *Army* reporter that the F-100 was perfect for CAS or that they accepted its CAS mission weaknesses in exchange for its aerial combat viability. Asprey's *Army* article appeared in 1961, and in it he expressed the Army's increasing exasperation with Air Force attitudes.[25]

It was not that the Army completely disagreed with the Air Force notion of war-fighting. One reason for the Army's disorganized complaints in Korea was that its commanders still appreciated the air superiority umbrella as well as the air assistance during the communist offensives. Also, one expected a nuclear superpower to keep its strategic strike force in top condition. The Army was in the nuclear game itself, as it purchased nuclear artillery shells and rockets during the 1950s. Thus, in spite of his criticisms, Asprey opined that "the free world owes SAC enormous gratitude." Further, some soldiers accepted Air Force arrogance due to the airmen's pre–World War II travails and wartime vindication. Finally, the Army's own Eurocentric war-fighting orientation led its leaders to accept the Air Force rationale.[26]

However, the soldiers believed that the Air Force had strayed too far, especially regarding close air support and the planes the airmen claimed could perform it. Though he was not an Air Force pilot, Asprey could still discern the fighter-bombers' CAS performance difficulties. He and the generals saw them amply demonstrated during joint-service exercises, which confirmed the impressions that the Army had poorly articulated during the Korean War. Hamilton Howze, who became one of the Army's top aviation leaders, recalled how a two-thousand-foot AGL broken cloud ceiling (relatively low clouds, but not overcast) with unlimited visibility underneath had stopped one exercise air attack. Other attack demonstrations he witnessed featured familiar problems: no ordnance expenditure because the pilots could not see the well-marked targets, or missed shots because of the fast jet's weapons delivery challenges. Resenting the way the Air Force almost literally flew away from its CAS duties, Howze and others looked within their own service for a solution.[27]

Control of air assets was the primary Army motivation for obtaining its own air service. The Air Force did not care to be at an Army commander's beck and call, and had set up what one Army leader referred to as an unresponsive monopoly holding company for tactical air support. The Army used a budding aviation technology, the helicopter, to encroach upon that holding company's turf. At times, Army leaders also launched sporadic forays into the fixed-wing air support business.[28]

Helicopters first saw combat use late in World War II and then served well as flexible small logistics aircraft in Korea. In one instance they resupplied a defense against an attack upon an exposed American flank. In 1954 the charismatic U.S. Army general James Gavin wrote an article for *Harper's* magazine entitled "Cavalry, and I Don't Mean Horses." Gavin envisioned helicopters moving troops rapidly—something he and others considered necessary in modern war where weapons of mass destruction could annihilate massed, stationary Army forces. He also believed that a sky cavalry could provide rapid strike and reconnaissance just as horse cavalry had done in earlier times. Referring to the U.N. debacle in Korea in late 1950, he thought a helicopter-borne force might have detected Chinese forces and slowed their offensive. But the sky cavalry needed some form of armed escort to suppress enemy gunners. The Air Force would not do this, so the Army explored its own air support options early after the Korean War.[29]

Army leaders were also concerned about Russian tanks encountered in a European war, and they explored using the helicopter as an antitank weapon. (They also considered adapting Cessna T-37 jet trainers for observation and reconnaissance duties, but probably had something more ambitious in mind.) Initial helicopter weapons tests were conducted in 1955 and yielded the obvious result that more work was needed. The Army aviation school commander at Fort Rucker, Alabama, Brig. Gen. Carl Hutton, and his Combat Developments Office chief, Col. Jay Vanderpool, set to with gusto. Vanderpool urged aviation contractors to submit bids and conduct tests at Fort Rucker. The Army aviators knew they were probably violating Air Force–Army agreements, so they operated informally in what one scholar called a bureaucratic insurgency. This testing method was almost literally hit and miss, as when they learned while flying that machine-gun firing shattered certain plastic canopies. But they got results, and when the Army commenced the first of its large-scale sky cavalry experiments in 1956, machine-guns in transport helicopter doors were commonplace. The enthusiastic aviators began conducting helicopter firepower demonstrations for other Army units.[30]

Air Force senior leaders were aware of these events and complained to Defense Secretary Charles Wilson. The secretary already did not like the Army's attitude. During this time, Army Chief of Staff Matthew Ridgway had resigned his post over defense policy differences, and there had been a "Colonels' Revolt" involving senior officers who criticized the New Look for sacrificing Army combat readiness. In an autobiography written shortly after his retirement, Ridgway warned the Air Force that if it would not field dedicated CAS planes, the Army would develop its own. Responding in a 1956 memorandum, Wilson sharply clarified—some Army men believed he curtailed—Army flying activities. He extended the Army airplane weight restriction to any special short-field capable plane, and he limited Army helicopter size to less than twenty thousand pounds. He warned the Army to limit its aviation research and to rely upon the Air Force. However, Wilson left important loopholes that the Army men would later liberally use. He stipulated that the Defense Department could waive weight restrictions for certain aircraft models, and his memorandum left some restrictions on in-house aircraft development open to interpretation. Furthermore, Army carping about the terms led Wilson to warn the Air Force to meet its various obligations to the Army.[31]

The Air Force had no intention of honoring Wilson's policies, and neither did the Army. Technological advances had forced too wide a divergence in their war-fighting approaches, as Colonel Vanderpool observed: "Each advancement our air force made, separated it further in speed and distance from the Army. The necessary differences in [the] mission of the air force and the Army left a partial vacuum between [the] ground and the fringes of space." Joint talks between the services to update CAS procedures broke down in 1957 over disagreements about greater Army control. The Army obtained waivers from the defense secretary for larger planes, such as the AC-1 Caribou transport and the OV-1 Mohawk. (The Air Force still blocked the Army from using the Mohawk—a twin-turboprop plane ostensibly designed for observation—for CAS.)[32]

The Army continued cementing its hold upon helicopter development. Aviation commanders qualified other Army leaders as pilots and thus created an influential clientele within the service. The Army aviators desired no separate aviation branch. This prevented a repeat of the old Army Air Corps sense of separateness and gave other service branches, such as armor, a stake in helicopter development. The aviators continued to stage firepower demonstrations for their fellow troops, as well as for corporate, government, and political leaders. They also cultivated a political following among members of Congress.[33]

One technological development that boosted Army helicopter fortunes was Bell Helicopter company's creation of the jet-powered UH-1 Iroquois in the late 1950s. Nicknamed the "Huey," or "Slick," it could go faster and carry heavier loads than piston-engine helicopters. The heavier loads included weapons, which companies such as General Electric agreed to build. Army leaders still moved carefully to avoid another challenge from the Air Force—even sneaking in Navy weapons experts to demonstrate how to use rocket launchers and such.[34]

The Army's attitude and larger events put the Air Force on the defensive as the decade ended. A growing chorus of critics took aim at New Look policy problems. Defense intellectuals, such as Bernard Brodie and Henry Kissinger, pointed out the strategic bankruptcy of a policy that made massive nuclear assault the answer to every world crisis. Army generals, such as Matthew Ridgway and Maxwell Taylor, retired in frustration and then made similar observations. All critics either directly or indirectly attacked the Air Force, which was so closely associated with the nuclear war outlook. Air Force tactical difficulties in real-world events, such as the 1958 Lebanon crisis, supported their views. More ominous for the Air Force was the Soviets' shootdown, using a radar-guided missile, of a high-flying American U-2 spy plane in 1960. This implied that SAC's high-altitude, supersonic B-70 prototype was itself already obsolete.[35]

Thus the Air Force, which had embraced the New Look emphasis upon nuclear warfare, found itself under attack as the 1960s began. Thanks to defense secretary waivers, the Army developed armed helicopters and purchased planes that violated the spirit of previous agreements. It continued to explore the combat capabilities of some T-37s it obtained. Late in 1960 the Army's Aircraft Requirements Board approved purchase of the UH-1, which later figured so prominently in Army aviation. Indeed, the Army already possessed over five thousand aircraft at the end of the 1950s.[36]

If the Air Force thought that it would receive respite with the new presidential administration, it was mistaken. President John F. Kennedy and his advisers—one of whom was Maxwell Taylor—wanted to shift defense policy away from overreliance upon nuclear warfare. Further, the new administration was not afraid to fight the kind of wars for which the Air Force was unprepared.

4   VIETNAM AND THE A-7 CHOICE

"Let every nation know, whether it wishes us well or ill, that we shall pay any price, bear any burden, meet any hardship, support any friend, oppose any foe to assure the survival and success of liberty." Possessing a dynamic worldview and strong anticommunist outlook, President John F. Kennedy backed these words from his inaugural speech with actions. Three months after taking office, he failed in his attempt to overthrow Cuban leader Fidel Castro at the Bay of Pigs. Later, he stood firm against Soviet moves in Berlin and Cuba. He also took a stand against communism farther afield, in Laos and Vietnam.

These situations reinforced the Kennedy administration's position, as expressed in the "Flexible Response" defense policy, that the American military should prepare for conventional as much as nuclear war. Additionally, the administration wanted more civilian control of military affairs. The brilliant but arrogant defense secretary Robert McNamara and his assistants in the Office of the Secretary of Defense (OSD) led this reorientation from Eisenhower's course. Dubbed "Whiz Kids," the assistants included Alain Enthoven, who headed OSD's Systems Analysis Office, Harold Brown, and John Foster. McNamara and the Whiz Kids believed that highly planned statistical analysis, called systems analysis, could solve most any problem.[1]

The OSD team obtained mixed results applying their approach to the military. Systems analysis required excellent methodology, which in turn demanded understanding not only statistics but, more important, the subject and one's own biases. They sometimes failed in the latter two tasks—shortcomings highlighted by their own lack of military experience and boasts that they could handle the military better than its own uniformed leaders. As one Whiz Kid recalled, "They could make a case for anything . . . [but] they made a case for a lot of bad things." They streamlined the military budgeting and procurement process, but the Vietnam War was the better known result of their preference for quantification over military science.[2]

The same applied to military aviation. Given the increasingly obvious deficiencies of Eisenhower's New Look policy, the McNamara OSD team favored a more conventional tactical Air Force. Indeed, Brown and Foster figured prominently in the development of the Air Force CAS plane. However, the Whiz Kids often based their decisions on economics and statistics rather than knowledge of air combat and warplanes. McNamara ordered that the F-111 be built to excel at most Air Force and Navy tactical missions—an aerodynamically impossible task. After the Air Force purchased the Navy's F-4 fighter and temporarily redesignated it the F-110, he thought they were two different planes and publicly discussed their relative merits.[3]

Air Force leaders could not mount an effective defense against the Whiz Kid onslaught. They clung tenaciously to their strategic bombers in spite of serious need for tactical reemphasis. They insisted that their forces must prepare only for worst-case, high-technology wars and not squander effort elsewhere. Also, given their own faith in technological solutions, some were too willing to abandon military science for systems analysis.[4]

McNamara, the Services, and Air Support Aircraft

McNamara quickly revealed his conventional warfare tack by reducing Air Force funding for strategic bombers. As for CAS, he and the Whiz Kids wanted a dedicated plane, but their desire to improve both conventional warfare capability and economic efficiency intertwined with other factors to create a tortuous path toward its purchase. The process began in 1961, when McNamara tasked all services to study CAS plane development. Air Force leaders answered by insisting upon multirole designs, with an emphasis upon air superiority. Indeed, the Air Force liked McNamara's F-111 program for this reason. The Navy, however, rejected the F-111 because it wanted a specialized plane and also believed that the F-111 was unsuitable for carrier operations.[5]

In spite of the Air Force response, McNamara insisted upon a dedicated CAS plane because "aircraft which are optimized for 'air superiority' missions are not fully effective in an air support role." He wanted such a plane to replace the Air Force F-105 and Navy A-4, but his Whiz Kids explored making the relatively inexpensive, agile A-4 the F-105's interim replacement. The move highlighted the pronounced differences between the two services over tactical plane preferences, as Air Force leaders emphatically rejected the A-4. In the words of one Air Force tactical issues staffer, Gordon Graham, "We hadn't bought an attack plane since World War II. The general doctrine . . . is that those aircraft are not the kinds of machines that would survive in a sophisticated environment." However, the little attack plane gave the Navy, the Marines, and other countries' air arms few problems in the following decades because it answered their specific tactical needs. Also, it carried a remarkably large amount of bombs and rocket pods externally, and naval aviators understood that such external stores prevented supersonic flight anyway.[6]

To guide reluctant Air Force leaders toward abandoning the F-105, OSD staff conducted a cost-effectiveness study to determine the best attack plane from a large list of candidate jets. On the simplistic basis that the Navy F-4 interceptor could carry more ordnance over longer distances, they concluded that it was the best CAS plane. The Air Force approved this choice because the F-4 was a fast fighter. The service also demonstrated its mastery of the convenient use of systems analysis when TAC staff produced a study condemning the A-4 and praising the F-4. McNamara revealed his strong predilection for economic solutions when he approved the F-4 as the F-105's interim replacement in late 1961. He liked its bomb-carriage efficiency, its dual-service utility, and its mission flexibility. Although a preference for mission flexibility contradicted his previously stated CAS plane philosophy, McNamara stated that the F-4 was still a better attack plane than the F-105. This was not saying much, for one pilot opined that maneuvering an F-4 was "akin to driving a dump truck through an obstacle course."[7]

Meanwhile, the Navy agreed with OSD's desire for an A-4 replacement. Though the A-4 would serve the Navy for many more years, its fuel endurance and load-carriage capacity were not what the service ultimately wanted. In one of those cases where systems analysis and good military sense coincided, research confirmed naval aviators' belief that a supersonic design did not serve them well tactically. Beyond the external ordnance problem, speed hurt fuel endurance, especially if an attack jet flew low to evade enemy defenses. Thus, the Navy chose the Ling-Temco-Vought (LTV) A-7, which well met its goals for range and weapons. The Air Force still needed a permanent F-105 replacement, and the A-7 was a likely candidate—but only after McNamara and OSD spent a few years trying to reconcile the airmen's and soldiers' views of air support.[8]

The Flexible Response policy meant that "life in the army took on new meaning as soldiers once again felt appreciated." It also encouraged Army leaders to pursue their aviation goals, which included research on armed helicopters and even light tactical fighter jets. As a result, the Air Force now faced both McNamara's OSD and a looming Army incursion into the CAS mission. Air Force leaders, however, would not retreat passively. A January 1961 Pentagon conference revealed the rising tension, as both Air Force and Army chiefs of staff discussed a possible interservice fight if the Army pressed for its own attack aircraft. Facing tactical unit cuts that were the legacy of the Eisenhower defense budget, Air Force leaders initially tried to retain these outfits and quell Army discontent through Operation Menu. This involved dedicating to Army support eleven tactical squadrons equipped with planes chosen by the Army from an Air Force list of candidates. The Army resisted the offer because it did not like the airplane choices and because it had to fund the operation. In fact, Army Chief of Staff George Decker issued a response in spring 1961 that was quoted in some Army histories: "The Army's requirement is to have close air support where we need it, when we need it, and under a system of operational control which makes it responsive to army needs."[9]

After the Kennedy administration reversed the Air Force tactical unit drawdown, Army leaders continued the tough talk about air support. The Army general serving as chairman of the Joint Chiefs of Staff (JCS), Lyman Lemnitzer, recommended that a dedicated CAS plane be built. However, perhaps due to the Kennedy-driven expansion of Air Force tactical units, concern about a nasty interservice fight, and desire for their own air support aircraft, Army leaders backed away from specifying their desires any further.

Even as he approved the F-105's interim replacement, McNamara continued prodding the Air Force about a CAS plane. In a November 1961 response, Air Force Secretary Eugene Zuckert stated that the soldiers were satisfied with his service's airplane choices. Perhaps provoked by this reply, Army Secretary Elvis Stahr detailed desired CAS plane characteristics: short-takeoff-and-land (STOL) ability, maneuverability, endurance, all-weather capability, and good communications gear. Stahr also disagreed with Zuckert's proposed CAS force size. Both issues generated an ongoing formal exchange of views, called Project Crossfeed. Its one positive result was that it helped lead McNamara to form a joint Air Force–Army unit, Strike Command, which was supposed to improve such cooperative ventures as CAS.[10]

Army aviation activists pushed beyond this debate when they drafted a memorandum that a sympathetic McNamara agreed to sign as his own. Published in April 1962, it was a virtual ultimatum to Army leaders to take a "bold new look" at their concepts of aviation and mobility. Army leaders quickly assembled a pro-aviation study board led by Maj. Gen. Hamilton Howze. The board produced a report that August.[11]

One board member's admonition summarized the approach that the board took: "Go big. Don't ask for battalions, ask for divisions." Howze later wrote that resentment of Air Force neglect helped fuel an around-the-clock effort that produced a sweeping final report filling over one hundred foot-lockers. It primarily addressed coordinating agencies, but it reflected the aviators' air cavalry concept of large units traveling mostly by helicopter. Of course, these units would require protection as they advanced, and Howze claimed that he still wanted Air Force support. However, his board spent much time testing helicopter weapons employment, especially in the anti-tank role, and it challenged previous Air Force demands by recommending that the OV-1 be used for CAS.[12]

The ambitious nature of the report shook even McNamara, who generally praised it while criticizing how it downplayed the logistics and fire support he expected the Air Force to provide. The Air Force was more hostile, for the Howze Board threatened tactical prerogatives that it took for granted. Howze had let Air Force officers observe his proceedings, which helped that service quickly respond with a rival board chaired by Lt. Gen. Gabriel Disosway. This board's report appeared barely a month after the Howze board's brief, and it attacked the Army effort as that service's attempt to usurp Air Force missions. Concerning CAS, Disosway and the Air Force leadership believed that neither helicopters nor the OV-1 would survive in a high-threat environment. Disosway and company even used some McNamara-style systems analysis to assert that the F-4 carried more ordnance faster than Army aircraft.[13]

Facing an open fight with the Air Force, Army generals moved circumspectly as the Kennedy administration asked for even more studies and exercises. Helicopter technology was promising but unproven. There were other CAS factors to consider, such as command-and-control arrangements, which could possibly alleviate interservice friction. Responding to a President's Science Advisory Committee inquiry about CAS, Army leaders stated that they agreed with Air Force priorities, while preferring attack plane performance characteristics similar to those given by Secretary Stahr. Meanwhile, they created the 11th Air Assault Division (AAD) to develop the Howze Board's helicopter concept. It was well that they moved as they did, for the Air Force chief of staff, old bomber man Curtis LeMay, attacked the Howze Board and helicopters while asserting his service's ability to meet Army needs. LeMay also increased the size of the flying unit that he had formed to satisfy Kennedy's interest in counter-insurgency wars, such as the one brewing in Southeast Asia.[14]

In early 1963 McNamara tried to enforce interservice cooperation through a memorandum directing better coordination of the new airmobile tactics and Air Force logistics and fire support. The services' response was to convene competing close air support boards, each of which produced reports supporting CAS plane designs that fit their own concepts. Again, the Army refrained from escalating the confrontation, even if its signals were sometimes contradictory. Army Chief of Staff Earl Wheeler asked Army Secretary Cyrus Vance to approve development of a dedicated attack helicopter; and though Vance rejected the request, he exhorted his service to study even more technologically advanced models. Wheeler's deputy chief of staff for operations, Harold Johnson, actually sanctioned a study to determine if CAS was worth all the fuss and effort. The study concluded that U.S. ground commanders would want it, no matter what the profit-and-loss analysis might show. Meanwhile, Wheeler vetoed his board's suggestion that the Army buy close air support planes. Also, in a semantic attempt to avoid reminding the Air Force of the support implications of helicopter fire, he ordered that it be called "direct support."[15]

Still, competition and interservice sparring continued, with occasional calls for peace. Strike Command–sanctioned exercises revealed strengths and weaknesses in Air Force support operations. Air Force prowess in reconnaissance and tactical logistics led McNamara to cancel further Army purchases of OV-1s and Caribou transports. However, fast jet air support still failed to impress the Army, as F-4s could not escort slow transport aircraft. The service also relearned command-and-control lessons and promised a more streamlined process and more forward air controllers. In May 1964 Maxwell Taylor, who was now the chairman of the Joint Chiefs of Staff, tried to ease tensions by means of an interservice agreement that would divide service aircraft by design and function rather than by weight. All sides rejected the proposition. Zuckert and LeMay demanded that the Air Force assume responsibility for all joint-service air operations. The Army also adopted an intransigent attitude, even though the Strike Command leader, U.S. Army Gen. Paul Adams, criticized his service for a too zealous and parochial view of airmobile operations. There was even talk of disbanding the 11th AAD, though another Strike Command exercise was planned. Neither event happened, for in summer 1965 the 11th AAD—now redesignated the 1st Cavalry Division (Airmobile)—deployed into a situation that might determine its worth better than any test. This was the escalating war in Vietnam.[16]

Vietnam and Air Support

U.S. involvement in Vietnam highlighted the interservice maneuvering over all forms of air support. The Air Force had deployed its counter-insurgency unit to Vietnam, and friction arose as Air Force leaders suspected the Army of using its aircraft there to usurp Air Force roles. LeMay at one time challenged Army Chief of Staff Johnson: "You fly one of these damned Huey's and I'll fly an F-105 and we'll see who survives. I'll shoot you down and scatter your peashooter all over the . . . ground." Johnson coolly replied that he would be happy to accept.[17]

However, even before the 1965 escalation, the war brought conditions that did not fit either service's preconceptions. Since 1961 the Air Force counter-insurgency unit had had to use ex-trainers and World War II–vintage attack planes, such as the T-28 and B-26, to meet the war's special demands. It had even had to borrow spotter planes from the Army. Revealing their persistent focus on big conventional wars, Air Force leaders started arming the T-28s with Sidewinder air-to-air missiles even though there was no air threat. Senior U.S. commanders later canceled the effort, which was just as well, for the T-28s and B-26s were too old and started experiencing catastrophic structural failures during combat maneuvering. The Air Force increasingly used an old Navy plane with impeccable air support credentials, the A-1 Skyraider.[18]

Army helicopters had also been in Vietnam since 1961, and though Army leaders had not considered counter-insurgency combat when they conceived airmobile operations, their helicopters had well supported the unexpected fights that occurred in normally inaccessible places. However, the aircraft were vulnerable against even light defenses, as Air Force leaders had claimed. Compared to warplanes, a helicopter was slow and relatively unmaneuverable. Its main rotor, rear stabilizer rotor, and engine-rotor gearbox could not operate with anything but very light damage. Helicopter pilots' cocky attitude aggravated their lack of combat experience—a situation made worse as communist gunners acquired proficiency and better guns. As a result, they suffered embarrassing losses in such fights as the Battles of Ap Bac (1963) and Binh Gia (1965). Aviation units modified logistics craft to be armed escorts, but even the jet-powered Huey could not keep up with its charges if it carried many weapons. Commanders then faced a dilemma of either lightly arming the escorts, which meant reduced firepower, or heavily arming them, which entailed poor performance. The 1st Air Cavalry's first serious fight with North Vietnamese regular troops occurred at the Battle of Ia Drang (1965). Revealing helicopters' inability to deliver heavy fire support, the Cavalry's commander praised Air Force planes and not the modified transport helicopters for giving his men critical firepower protection. Increasingly, Army aviators wanted a dedicated attack helicopter, which they hoped would solve the problem.[19]

Up through 1965 the airmen and soldiers each wanted to fight the war in their own way, and CAS snafus were frequent. The situation changed as combat exigencies demonstrated the value of cooperation. The Air Force had thought that the 1965 escalation might allow its fast jets to show that they could win the war. "A squadron of F-100s over here could puncture the balloon of the skeptics," boasted one staff officer. Unfortunately, the airmen's preferred interdiction campaign could not stanch the communist logistic flow to the south because of political restrictions, the rugged jungle terrain, and a too numerous and determined enemy whose guerrilla campaign required relatively limited resources. However, the Air Force and Army developed an outstanding air task system to handle the numerous air support needs in South Vietnam; some observers later commented that the Army came to rely too much upon it.[20]

As part of the superb CAS effort, supersonic jets such as the F-4, F-100, and F-105 earned their share of praise. With their great speed they answered tasks quickly enough, but the problems that had been identified in the 1950s arose, and now were aggravated by the region's tropical climate. The ability to fly closely and slowly enough to see the target, to work safely in poor weather, to carry sufficient ordnance, and to remain over the battle area were all limited. Col. Jack Broughton, a legendary F-105 pilot, summarized the handicaps when he described a four-plane formation of F-105s flying a road reconnaissance ("road recce") mission in bad weather with a full weapons load: "Sitting there with a heavy load like that, trying to road recce, and looking for a native village was a disaster. The machines were just not built for it. . . . They had no maneuverability and the fuel on all four airplanes was going like crazy as they snaked down the road, giving it the old college try to find that village. . . . Dave [one of the pilots] was hurting. He was running out of fuel and he couldn't get high enough to put the bombs anyplace worthwhile, even if he had found someplace worthwhile." Broughton's observations were not unique. The supersonic jet pilots had other missions to fly. Their high-speed and usually high-altitude deliveries gave them little opportunity to understand completely what was happening on the ground. Sometimes they even felt that CAS missions were useless. One asked, "What's the target? A piece of jungle out here. . . . you don't know what you're hitting; [you're] making toothpicks."[21]

Given these problems, even many of those who favored the supersonic jets conceded that the propeller-driven A-1 was the CAS star. Maj. Richard Griffin was an Air Force forward air controller (FAC) who flew CAS spotter planes in Vietnam during the mid-1960s. He acknowledged the greater abundance of fast jets, particularly the F-100, but he still preferred working with A-1s:

> I think that anybody in Vietnam that flew as a FAC probably has somewhat the same thought. If it wasn't the best, it was probably the most versatile. . . . Number one is that the A-1 would carry an awful lot of crap . . . it had all kinds of bombs and everything hanging on it. . . . Another thing . . . that you always ask for early when you've got fighters on target, say: 'Well . . . how long can you stay here?' And invariably the guys in the A-1's would always come up with something smart . . . 'about an hour longer than you can.' Well, you know, most of the time in the F-100's or F-4's . . . they talked in minutes.

Major Griffin also made a common-sense observation about jet-versus-Skyraider attack speeds: "If you're coming out on a target at 450 knots you, obviously, have to release higher than you do if you're coming in at 250." When he directed a jet pilot's eyes to a target, he "had to talk in terms of rivers and mountains." But with A-1 pilots, "instead of having to talk in large geographic features, you can talk in very much detail . . . [so] he can come right in."[22]

Col. Eugene Deatrick, an A-1 squadron commander in Vietnam, made similar comments while pointing out further benefits. Since he ran a dedicated air support outfit, Deatrick insisted that his pilots exchange visits with the Army troops they protected. He also had his pilots take the soldiers on flights over their own battle areas. As a result, soldiers and pilots fine-tuned procedures involving communications, optimum weapons, visual signaling, and ground unit escape routes from a camp. Deatrick later described the soldiers' reaction to their orientation flights: "Those Air Force people aren't as blind as we think they are. They've got some problems and what we've got to do is find a better way to identify ourselves from the ground." The pilots gained empathy for the soldiers they supported, but Deatrick observed that this was not the case with the F-100 units at the time. Not only did they lack Army-compatible radios, like those of the A-1, but also their pilots did not exchange visits as assiduously as Deatrick's men. "I doubt if they had even seen each other," he quipped. That and the jets' higher speeds meant that a jet pilot was often "sitting up there completely in the dark just circling. All he could see was a FAC [in a spotter plane] flying around. Had no idea what was going on. He might see some . . . smoke go up . . . and still had no idea what it meant."[23]

The A-1 scored in other ways as well. Even from steep dives, its low speed and maneuverability allowed the pilot to execute his attack relatively close to the ground, further improving accuracy. The plane routinely delivered weapons within 50 meters of hard-pressed friendly forces, in accordance with their often desperate requests. One variation on air support that the A-1 flew was armed escort for search-and-rescue helicopters. An A-1 pilot who led such missions, Maj. James Costin, stated that one reason for choosing the Skyraider over fast jet fighters was its accuracy hitting potential captors pursuing the downed aviator. "On at least two occasions," he said of the fast jets, "I saw them drop on the survivor. . . . They're too high . . . and they just weren't that accurate." A-1s earned respect for attacking targets in poor weather unsuitable to jets, and both Army ground units and helicopter pilots praised the plane for sticking around in foul weather to provide life-saving heavy fire support. Colonel Deatrick's fliers made sure this happened by devising procedures for working beneath low clouds to support each camp. The A-1's poor-weather success raises the question of its night capability. Though it did not have the radar or infrared equipment carried by designated night/all-weather (N/AW) planes, it flew many night missions. The pilots improvised tactics to prevent mid-air collisions, identify the target via flares, and, above all, avoid being shot down. Interestingly, A-10 pilots later used many of the same tactics in Desert Storm.[24]

A final impressive quality of the Skyraider was its ability to survive in a hostile environment. Survivability involves two characteristics: ability to avoid being hit and ability to survive being hit. Critics and even A-1 pilots observed that the plane was too slow to evade antiaircraft fire, especially the new SAMs that North Vietnam obtained from the Soviet Union. Even so, A-1s flew into certain parts of North Vietnam until the early 1970s, and Navy A-1s even shot down two MiG-17 jet fighters whose pilots had become careless. The plane's most impressive survival attribute, however, was its ruggedness, because A-1s took many AAA hits in their low and slow attacks. Costin said the A-1 was "the type of aircraft which could take a good hard hit and a lot of them and not get hurt too bad." Deatrick recalled returning to base with his engine shot up and one landing gear shot away. His unit lost only one plane to ground fire, and even that one managed to belly-land back at base. Another squadron commander asserted that jet aircraft could not take the punishment that he saw A-1s endure, mentioning one so badly shot it looked like a "sieve." The available A-1 pilot oral histories reveal that losses occurred when pilots became too aggressive or used bad tactics.[25]

Concerns of weather capability and survivability are also important considering the Air Force's dedicated CAS counterpart to the A-1, and later the A-10—the gunship. Modifying transports such as the C-47 and C-130 into bristling gun platforms was another Vietnam-related story of how tactical common sense prevailed over rigid adherence to Air Force doctrine. These lumbering planes had much better loiter capability than even an A-1, and their firepower, which included Gatling guns, cannon, and lots of ammunition, was both lethal and ample. They made many impressive saves of beleaguered units up through 1972. However, they were slower, much less maneuverable, and less rugged than the A-1, and any serious antiaircraft opposition restricted their operations. Their weapons delivery altitudes and lack of maneuverability precluded work beneath low clouds. Beyond South Vietnam CAS, the Air Force used gunships to interdict communist supply lines in Laos. They flew night missions in the dry season in areas prepared by other planes, such as the A-1. Still, like the A-1 and later the A-10, they answered certain needs within the U.S. air power spectrum.[26]

The Pike Committee's Examination of Vietnam CAS

The A-1's performance was one part of the Air Force CAS effort that received congressional attention in 1965. With the services still working out CAS procedures in combat conditions, the legislators heard many complaints. However, they considered the Air Force most to blame, given its long, well-known, and intentional neglect of the mission. Starting a trend whereby ex-Marine congressional representatives and staff served as CAS advocates against the Air Force, Rep. Daniel Flood was a conduit for Army complaints about Air Force CAS snafus. Rep. Otis Pike, who had flown Marine dive-bomber CAS in World War II and wanted more focus upon unconventional warfare, later said that he also received complaints from lower ranking Army officers. (Senior Army leaders apparently did not want an interservice fight played out in Congress and told him everything was fine.) He gladly chaired a subcommittee that convened for seven days in early autumn 1965 to investigate Vietnam CAS deficiencies.[27]

The committee members addressed slow air task response times and incompatible radios. They also brought in Marine witnesses so they could show how better CAS could be accomplished. However, they spent most of the hearings criticizing the Air Force failure to procure an airplane dedicated to meeting its agreed-upon CAS commitment to the Army. They asked all of the witnesses, who were lower ranking Vietnam veterans and senior officers of both services, their opinions of the optimum CAS plane. The three Army Vietnam veterans who testified primarily wanted Air Force planes to respond when needed, though all considered the A-1 an outstanding CAS plane. The three Air Force veterans were pilots who had flown Vietnam CAS missions. The first was an A-1 pilot who praised the A-1's demonstrated CAS strengths but wanted the F-100's speed. Overall, he thought the Skyraider and the jet complemented each other. The other two pilots flew F-100s; one still preferred speed and multirole capability, while the other admitted that the F-100 was not primarily a CAS plane.[28]

Pike exposed Air Force CAS deficiencies better when he and his colleagues interviewed the generals. Perhaps wanting to avoid an interservice fight, former director of Army aviation, Maj. Gen. Delk Oden, resisted most of the committee's leading questions. (He might also have wanted to let the Air Force go its own way so the Army could get its attack helicopter.) But Pike asked him if he knew of any Air Force planes that were really designed to do the mission, to which he had to say no. With that, Pike turned his attention to the Air Force assistant chief of staff for plans and operations, Maj. Gen. Arthur Agan. One Air Force general described Agan as *the* fighter pilot in the Air Force plans directorate at that time; as such, he represented the philosophy that Pike wanted to challenge. As he had done with Oden, Pike forced Agan to admit that the F-100 and F-105 were not designed for CAS, and that the A-1s in the service's Vietnam order of battle were originally Navy planes. Pike asked him, "Can you give me any aircraft that the Air Force has developed since World War II for which the primary mission was close air support?" Agan replied, "Not that way; no, sir." Both agreed that the New Look strategy drove the current warplane design deficiencies, but Agan had no answer for what the Air Force planned for CAS. He insisted that a dedicated CAS plane was fine as long as it could defeat enemy fighters.[29]

Air Force Systems Command chief, Gen. Bernard Schriever, and the Air Force deputy chief of staff for research and development, Lt. Gen. James Ferguson, appeared together, though Schriever answered most of the questions. He reminded the panel that the Air Force's doctrinal and budget concerns bred its desire for multirole tactical planes. Thus, it preferred the F-5, a new, sleek, low-cost, air-to-air fighter, for limited conventional conflicts like Vietnam. However, Pike forced him to admit that the F-5 was not primarily a CAS plane, and he agreed with the panel members' assertion that limited war scenarios were too widely varied to follow a specific doctrinal imperative. Schriever also deviated from the normal Air Force stance when he told the panel that with air superiority guaranteed in many of the post–World War II limited war scenarios, he had concluded that the service should have fielded a dedicated CAS plane. His revelation signaled a change in attitude among some Air Force leaders.[30]

THE A-7 CHOICE

Air leaders' attitudes changed because pressures beset the service from many directions. In 1964, as the A-1 started earning favorable combat reports in Vietnam, McNamara and the OSD considered reopening Skyraider production. They did not do so, because its manufacturing cost would be five times greater than before. In a January 1965 memorandum responding to Secretary Zuckert's budget request to accomplish a CAS plane study, McNamara told Zuckert to assume air superiority in the study's calculations and consider various types of aircraft already in production, including the A-7 and F-5.[31]

McNamara's directive meshed with his Whiz Kids' ongoing efforts. The previous June, the director of defense research and engineering (DDR&E), Harold Brown, had told McNamara that although the Air Force and Army could not agree on the ideal CAS plane, a mixed force of high- and low-cost planes might satisfy both services' needs. The F-111 represented the high end, and other OSD staff pushed the Navy's A-7 as the low-end candidate. As one Whiz Kid, Alain Enthoven, put it, "for the kind of wars the tactical Air Forces were likely to fight . . . the A-7 would simply be substantially better." LeMay dismissed the A-7 as "not much good," and his staff supplied the usual explanation that the A-7 had no air-to-air capability.[32]

Undeterred, the OSD staff ordered the Air Force to study Brown's high-low proposal in summer 1964. Formally titled "Force Options for Tactical Air" and better known by the name of its project officer, Lt. Col. John Bohn, the study was released in early 1965. Again revealing the Air Force skill in systems analysis, the Bohn study took OSD's simplistic analysis model, which used total payload carried for total miles, and added another model featuring the Air Force's traditional concern, heavy air defenses. It accepted the high-low idea, but for the low end it rejected the A-7, which was no fighter, and favored the F-5, which was supersonic and a fair air-to-air machine. Brown was happy that the service at least accepted his concept, and told the Air Force that action would be required soon. However, the A-7 would remain as a low-end candidate to compete with the F-5.[33]

Tactical Air Command would fly the low-end tactical airplane choice, and its leaders detested the A-7. In December 1964 OSD had sent one of its A-7 proponents, Vic Heyman, to determine the depth and nature of TAC resistance. One of the staff officers who received Heyman recalled, "I laughed at him." Heyman later said, "They really considered the A-7 a dog." Regarding Heyman's encounter, fellow OSD staffer Enthoven deemed it symptomatic of the Air Force worldview, which involved "doing their own thing like winning the air battle while somebody else was winning the ground battle."[34]

Enthoven's opinion leaned toward organizational dynamics, which, given Air Force history, has some validity. However, the fighter pilots believed that historical common sense combined with force structure demands to justify their position. The A-7 summoned memories of Europe in World War II and the severe aerial penalties for technological inferiority. TAC's commander at the time, the same Gabriel Disosway who had rebutted the Howze board, cited the German Stuka's sorry fate in the Battle of Britain as an example of why he did not want a dedicated attack plane. He and others believed that such an aircraft risked encountering situations where "someone's going to send it up to shoot MiGs and they're going to get themselves killed." General Agan remembered telling Enthoven about the large fighter escort required for World War II bomber raids and then reminding the OSD staffer that such a force would not be available to protect A-7s in the projected desperate fight with the Soviets. Agan considered Enthoven too wedded to cost-effectiveness, adding that "the 'crime of the time' was that we had to talk to a guy like that."[35]

Feelings were intense enough that one TAC general told the fighter pilot assigned to research the A-7 issue at Air Force Headquarters: "Major, you'd better remember, you're going to want to come back to TAC one of these days. You had better remember that when you're working this thing." But the opposition did not deter Heyman. Believing that systems analysis could overcome all biases and perhaps emboldened by McNamara's directive to Zuckert, he directed TAC in April 1965 to conduct a cost-effectiveness study of CAS plane candidates. TAC was undaunted, and a month later produced a report that condemned the A-7 and praised the F-4.[36]

However, 1965 was a time when such obstinate resistance faced more than persistent OSD bureaucrats. The Vietnam War provided two sources of pressure. First, it created an arena to showcase the A-1's talents while revealing Air Force CAS deficiencies that then became grist for a hostile congressional hearing. Second, and more serious to Air Force mission prerogatives and associated funding, the Army started reviewing contracts for developing its Advanced Aerial Fire Support System (AAFSS) project. Officially designated the AH-56 Cheyenne, it was a super attack helicopter that would provide fire support for Army airmobile formations. In the meantime, Bell Aircraft Company tried to sell the Army an interim attack model.[37]

New Air Force Chief of Staff Gen. John McConnell faced this situation. He agreed with much of LeMay's air power philosophy and apparently did not get along with Army Chief of Staff Harold Johnson, but he was more politically practical than LeMay. Although at least one source opined that this trait contributed to problems with the overall conduct of the Vietnam War, it did help resolve Air Force–Army disputes over such things as CAS. Understanding his own service's preference for the F-5, McConnell quickly deployed twelve of the new fighters to Vietnam for an operational evaluation of their air support capability. He also requested that OSD approve purchasing two F-5 units, but the office demurred, perhaps to see how the fighter fared. The results by autumn 1965 were obvious, given the plane's design. Most pilots liked its sports car handling qualities, and its initial maintainability was also good. But others still preferred the A-1, especially since the F-5 had only a fraction of the Skyraider's fuel endurance, weapons load, and maneuverability. Probably due to their rushed activation, the F-5s also developed mechanical problems as their combat time increased.[38]

Air Force headquarters staff meanwhile had conducted another study to determine whether the F-5, the A-7, or a stripped-down F-4 would be most suitable for CAS. Named for its presiding officer, Col. Howard Fish, the study was a monstrosity based upon a computerized air war featuring several different combat scenarios. Because its complexity left too many premises open to biases, the Fish study achieved no definite conclusion and contributed to disagreement about the best choice. OSD pressed hard for the A-7 as a cost-effective attack plane solution for two services; TAC wanted a good fighter that could do CAS on the side; and some Air Force officers grudgingly favored the A-7 in order to get outsiders off their backs. So intense were opinions that in the study briefing to General McConnell, Colonel Fish presented only the options instead of the traditional recommendation for a single course of action.[39]

General McConnell saw the military and political realities, and on 5 November 1965 he chose to take the A-7. In later discussions, he said that he didn't pay much attention to the computer study:

> [There was] considerable pressure by certain elements in the military and by certain elements in the Congress because they said we had never provided a capability or a specialized plane for close air support of the Army. At the same time, the Army was coming in with a strong close air support proposal which was the AH-56, the advanced helicopter. In order to demonstrate that we did want to give the Army every possible means of close air support—and I know that we can do it better than they can, particularly with the AH-56—we opted for the A-7 in sufficient quantities to provide close air support for the Army in an environment that did not have intensive air defenses.

Later, McConnell summarized his decision: "The thing that was pushing was that we had to get something to give the Army close air support. First, it was our job. Second, if we didn't do it, somebody else was going to do it for us." That "somebody" was the Army and its Cheyenne, with the attendant threat of a lost mission and lost aircraft funding for the Air Force.[40]

The former DDR&E and incoming Air Force secretary, Harold Brown, later spoke of the war-fighting implications of the decision:

> It was perfectly clear by late 1965 and early 1966 that the Air Force was going to be put to the test both by the existence of the Vietnam War and its nature—however representative or unrepresentative they would be of a war somewhere else—and by Congressional interest and by OSD interest. . . . therefore it [the Air Force] had to look at the question of close air support specifically, and not just say as had been part of doctrine . . . within the Air Force for many years, that whatever can fight the air battle can then go ahead and do the Close Air Support role. I think there was coming to be an awareness in the Air Force that, as a result of constraints inherent in limited war, you might not be able to fight the air war.

In other words, some Air Force leaders realized that there were indeed other wars besides the European one with the Soviets.[41]

The Air Force had taken a momentous step in acquiring a subsonic tactical jet dedicated to air-to-ground weapons delivery. Multiple factors led to this decision, and one would think that the A-7 buy settled things. This was not to be, because the forces that promoted and opposed this move were not spent yet. Indeed, Air Force secretary Brown saw even the A-7 as an interim measure toward getting a better CAS plane. The Vietnam War continued, which guaranteed that the Army still wanted attack helicopters for combat. Several Air Force generals fervently believed that the A-7 was an unwanted, inferior plane that Congress and the OSD had forced upon them, but so far that pressure consisted only of public embarrassment in a congressional hearing and continual prodding by Whiz Kid staffers. With the Army proceeding rapidly toward buying attack helicopters, at least General McConnell realized that Congress and the OSD could eventually grant CAS responsibility and funding to the Air Force's parent service. This would be a costly and humiliating loss for the Air Force, given its fight for independence, its status as the nation's premier air arm, and its leaders' desire to control all air combat missions. The service was not by any means done with the CAS plane issue.[42]

The Ju-87 Stuka's sorry fate flying interdiction missions in the Battle of Britain was the lesson that Air Force leaders recalled about dedicated air support planes. It flew air support on the Eastern Front until late in World War II.
Hugh Morgan

The P-47's excellent aerial combat and air support record confirmed many tactical air leaders' belief that fighters could fly both missions and that dedicated attack planes were unnecessary.
Painting by Stan Stokes; reproduction courtesy of Defense Visual Information Center (DVIC).

Although the F-80 was a straight-wing plane, its thin wing, relatively small fuel supply, and engine limited its loiter and weapons carriage capability. In the early stages of the Korean War, it was flown by pilots trained in aerial combat and not air support.
Hugh Morgan

The A-1's superb air support performance in the Korean and Vietnam Wars upstaged the jet fighters and helped influence the A-10's design.
Hugh Morgan

The F-105 represented the ideal 1950s strike fighter-bomber.
DVIC

The AC-130's outstanding loiter and weapons carriage ability has made it a frequent favorite for low-threat air support.
DVIC

As the OSD criticized the Air Force in the early 1960s about its lack of attack planes, many service leaders favored the F-5 fighter as the appropriate answer.
DVIC

The AH-56 Cheyenne was the Army's ambitious attempt to build a rotary-wing air support aircraft. Its development was the biggest factor in the genesis of the A-10.
U.S. Army Aviation Museum

The Air Force picked the Navy's A-7 over the F-5 in its initial attempt to defeat both Vietnam-inspired CAS criticism and the AH-56.
DVIC

The A-9 was Northrop's entry in the A-X flyoff.
National Archives

The A-10 was Fairchild's A-X flyoff entry. These planes are carrying long-range fuel tanks.
DVIC

An A-10 in maneuvering flight. Note the Maverick
missiles underwing.
National Archives

The AV-8 Harrier was the Marines' entry in the CAS plane controversies of
the early 1970s and late 1980s.
DVIC

The man-launched SA-7 was an air
support problem in both the late
stages of the Vietnam War and the
1973 October War.
DVIC

In the October War, the ZSU-23-4 gained a fearsome reputation as a battlefield antiaircraft weapon.
DVIC

Congressional interest made the Piper Enforcer an A-10 competitor in the mid-1970s.
Hugh Morgan

The AH-64 joined advanced ground weapons and the Army's AirLand Battle doctrine as factors in that service's reduced desire for CAS in the 1980s. *DVIC*

An outstanding fighter, the F-16 became the Air Force's controversial choice in the 1980s as the A-10's replacement.
DVIC

Although it bore a passing resemblance to the A-9, the Soviet Su-25 was designed more for interdiction missions.
DVIC

The Mi-24 Hind was the Soviet CAS choice, but it encountered severe anti-aircraft problems after its initial successes in the Afghanistan War. *DVIC*

5

THE A-10'S
IMPROBABLE CONCEPTION

The factors that had driven the Air Force decision to buy the A-7 intensified throughout 1966. In February, Otis Pike's subcommittee released a report that spent half of its fourteen pages scoring the service for its tactical plane selections, stating: "It has never developed one plane for this particular purpose. It is not developing one today. In fact, it insists upon multipurpose aircraft." The committee believed that in the service's "magnificent accomplishments in the wild blue yonder it has tended to ignore the foot soldiers in the dirty brown under." The Skyraider's good work in Vietnam was one item the subcommittee applauded. Eugene Deatrick, then serving his A-1 squadron commander's tour in Vietnam, affirmed the plane's strengths when service leaders had him brief them about its performance.[1]

The Air Force's A-7 selection did not slow the Army's attack helicopter efforts. Vietnam was one factor, but its audacious helicopter design project became an even bigger influence. In the same month that the Air Force announced its A-7 purchase, the Army capped a two-year review of AAFSS contract proposals by selecting Lockheed Aircraft to build the AH-56 Cheyenne. Lockheed's design "exceeded the Army's wildest expectations," as one historian put it. It featured an innovative combination of propulsion and lift features: stubby wings, a main rotor, and a pusher propeller. The rotor assembly itself was revolutionary in that it was more rigidly constructed than conventional models. The result was that the Cheyenne could fly at 210 knots—an unheard-of speed for a helicopter, and nearly as fast as the A-1. The unique aerodynamic design also allegedly enabled it to fly dive-bomb runs, like the A-1 and other fixed-wing planes. But Lockheed designers went further. They loaded the Cheyenne with an impressive weapons array, including a turret-mounted 30-mm gun, a 40-mm grenade launcher, rocket and bomb pylons—all told, an alleged capability to haul 8,000 pounds of ordnance. Such a load carriage, approaching that of the A-1, was outstanding for a helicopter. Finally, Lockheed incorporated a most advanced all-weather avionics package—"more complicated than a B-52"—that included a laser range finder, terrain-following radar, inertial navigation systems, and an autopilot.[2]

Lockheed's daring design eventually helped kill the Cheyenne program. For one thing, the company could not deliver the aircraft soon enough to meet the Army's urgent need for attack helicopters in Vietnam. In the meantime, Army leaders bought an interim attack helicopter that Bell Aircraft built on its own initiative, the AH-1 Cobra. Though not as impressive a design as the Cheyenne, the sleek Cobra could still travel fast enough and carry enough weapons to escort transport helicopters and provide direct fire support.[3]

The Cheyenne and Cobra were unmistakably CAS machines, but could the Air Force's A-7 remove the need for them? By spring 1966 the service saw that the answer was no. The plane made its first flight in September 1965, and subsequent test flights confirmed that its takeoff and landing rolls were very long, and that it lacked maneuverability. Further, the service wanted a substantial upgrade of the A-7's weapons delivery avionics, and this increased the plane's cost and lengthened its development schedule. The Air Force would keep the A-7, but it still had an air support problem.[4]

An Agreement and a CAS Plane Decision

In early spring 1966, service chiefs McConnell and Johnson met secretly to resolve air support differences that the Vietnam War had aggravated. Both apparently feared a war-driven interservice clash that OSD or Congress might settle in an undesirable manner. Vietnam combat confirmed that the Air Force provided airlift support better than the Army's transport planes, and also cemented the Army's commitment to helicopters and rotor-wing fire support. Army aviators still wanted to arm OV-1s, which one OV-1 unit in Vietnam did until Air Force leaders successfully demanded that it disarm. This shows one reason why the chiefs kept their discussions secret. Many Air Force leaders wanted no mission ceded to the Army. Johnson dealt with Army aviation activists who wanted to control not only helicopters but any fixed-wing aviation directly supporting Army activities. They had fought hard to get transport and reconnaissance planes for their service, and did not want to lose them now.[5]

The Johnson-McConnell Agreement, as it became known, followed Maxwell Taylor's earlier suggestion to shift the aviation responsibility criterion from aircraft weight to aircraft type. The Air Force would still provide CAS, but it recognized that Army helicopter missions included fire support. The Army ceded its large, fixed-wing transports to the Air Force. Additionally, the Air Force kept some helicopters for SAR work, and the Army kept some fixed-wing planes for observation and liaison but not fire support. Subordinate officers in both services chafed at the agreement, but it displayed an encouraging pragmatism on the part of both leaders. The agreement was a start in settling overall aircraft-type-versus-mission problems, but like previous interservice accords, it did not address all the specific issues. The Cheyenne still threatened to duplicate the Air Force CAS mission as A-1s flew it in Vietnam. In June 1966 the Army capped its initial development work with Lockheed by presenting a Cheyenne mock-up. Also, Bell Aircraft was about to deliver its first production Cobra to the Army, whose helicopter budget had risen substantially.[6]

The Cheyenne's progress rated a review in June 1966 by the Air Force Doctrine, Concepts, and Objectives Directorate, headed by Maj. Gen. Richard Yudkin. This office studied long-range goals and requirements, and formulated doctrine as it pertained to service roles and missions. It studied the Cheyenne because the aircraft affected whether or how the Air Force would conduct close air support. As early as 1964, the directorate had an officer, Col. Ray Lancaster, assigned to monitor the Army's armed helicopter development and assess its implications. To Lancaster, the Cheyenne was an expensive "monstrosity" representing nothing less than "the Army wanting to take over the close air support mission." The mock-up unveiling spurred him to support any effort to build a successful rival. Another staff member in the Air Force Doctrine, Concepts, and Objectives Directorate, Col. Avery Kay, agreed. Recalling that the Army probably had a legitimate complaint about Air Force CAS neglect, he added that the Cheyenne program threat drove the service to answer it with a dedicated CAS plane. However, not every Air Force officer understood this. TAC commander Disosway did not hide his contempt for the CAS plane idea: "I was very much opposed to it. I argued with McConnell about it. . . . He said, 'No, we have to do that for the Army.'"[7]

Yudkin had friendly Army contacts, but he did not agree that the Air Force had always neglected CAS, especially in view of what he considered an insatiable Army demand for it. However, his personal opinions or those of other officers were not his concern. He later said that though he anticipated "inevitable controversy" about the required course of action, he had encountered a doctrinal problem that demanded going beyond the usual policy boundaries and "identifying in conceptual terms a required hardware response." Favorable reports about the Skyraider, as well as Vietnam veteran A-1 pilots serving at Air Force headquarters, also contributed to his decision. Yudkin briefed McConnell in August, telling him that though the Army was satisfied with Air Force CAS in Vietnam, current Air Force jet fighters possessed a poor to unsatisfactory capability for helicopter escort and direct fire support. The Army filled the gap with its armed helicopters, and Yudkin recommended that besides paying more attention to CAS matters, the Air Force "should take immediate and positive steps to obtain a specialized close air support aircraft, simpler and cheaper than the A-7, with equal or greater characteristics than the A-1." Yudkin recalled that McConnell took an unusually long time to review the report's recommendations, but on 8 September 1966, the Air Force chief issued a decision paper directing his service to develop and procure a dedicated close air support plane, which would initially be called the A-X—and, after years of development and struggle, the A-10.[8]

For the first time in its existence as a separate service, the Air Force now moved to acquire an airplane dedicated to the CAS mission, something many of its leaders then and later believed that it would not do. This skepticism

could be seen in Disosway's attitude, as well as in a 1989 *Air Force Times* article in which former Air Force Vice Chief of Staff Robert Mathis dismissed
the A-10 as an "aircraft designed for the last war and pushed through by a
coalition of lawmakers from the Northeast and influential staff members."
His impression of the plane's genesis resembled what many service leaders
had believed about the A-7—that ignorant or malignant civilians had forced
it upon a properly unwilling service. The same article, written during a later
effort by the service to divest itself of the A-10, mentioned that Air Force
leaders had resisted the plane. In oral history interviews, some used the terms
"stupid plane" and "ridiculous" to describe it. Therefore, it is important to
review the reasons behind its creation and development.[9]

The Air Force's central motivation was to counteract the doctrinal and
budget threat posed by the Army's Cheyenne project. As Robert Dixon, a
1970s TAC commander who promoted closer Army ties, observed, the central issue was who would control CAS air assets. Pierre Sprey, another official intimately familiar with the A-X project, asserted, "The A-10 was
invented strictly to defeat the Cheyenne." Perhaps lampooning the incipi-
ent rivalry behind the CAS plane's birth, a cartoon in a 1968 issue of *Armed
Forces Journal* showed a winged tank sitting behind a "Tactical Air Command" sign. An Air Force general glares at the craft, while a subordinate says
to him, "No sir General it won't fly, but it will sure scare hell out of the
Army!" Some Army officers saw the CAS plane simply as an Air Force move
to preempt their Cheyenne; one charged that "the A-10 was illegitimate in
conception."[10]

Other factors also helped create the dedicated CAS plane. The man whom
many praised as vital to the plane's conception, Richard Yudkin, said of the
Cheyenne-only theory of the A-10's birth: "You make a mistake if you over-
simplify it that way." Some of the Air Force fixation upon action by OSD and
Congress was valid, because both pressured the Air Force to pay more attention to CAS. They also encouraged the Army to challenge the Air Force for
control of air support after years of the airmen's perceived and real neglect.
Technology played a role, in that the helicopter provided opportunities for
the Army to fill an apparent void in air support.[11]

The Vietnam War was a major catalyst for all the factors involved. It
accelerated the Army's armed helicopter development. It inspired congressional hearings that criticized Air Force CAS plane choices. As a low-intensity

ground war with no associated air superiority struggle, Vietnam was not the Air Force's kind of war. It highlighted the A-1's strengths, while revealing deficiencies in the F-5 jet fighter that was the CAS plane favorite of many tactical air leaders. Kennedy's Flexible Response defense policy shifted emphasis away from Eisenhower's pro–strategic bombing stance, but the Air Force resisted this tack until the war exposed weaknesses in how it waged nonnuclear combat. Thus, the war caused the Air Force to depart from the strategic bombing–oriented war-fighting doctrine that it had embraced since its 1947 birth. General McConnell stated in a 1968 speech that the service needed to develop "new equipment, new applications, and new tactics" to deal with conventional wars.[12]

One "new equipment" item would be the A-X. Although McConnell did not specifically explain why he bought the A-X, the reasons were probably similar to those he gave for buying the A-7. The A-7 was the precedent for the A-X, but not its predecessor. In congressional testimony, McConnell pointed out the A-7 shortcomings that helped drive the campaign for a dedicated CAS plane. In a later interview, Air Force Secretary Harold Brown confirmed that he and McConnell agreed about sincerely redressing Air Force CAS deficiencies, and that they both saw the A-7 as an interim gesture on the way toward a more capable CAS plane design.[13]

DESIGN CONCEPT WORK

The Air Force Headquarters department associated with aircraft design, the Directorate of Research and Development, Operational Requirements, and Developments Plans (office symbol AFRDQ) assumed responsibility for the new project. Though McConnell expected full compliance with his decision, the A-X was both doctrinally important and the object of considerable in-service contempt. For these reasons, Yudkin and his staff were allowed to involve themselves in the project as much as they could. To determine what deficiencies the A-X had to redeem, they produced a chart showing how various Air Force fighters' air support performance matched that of the Cheyenne. The F-4 and F-105 failed utterly, while the A-7 lacked maneuverability. The officers needed further expertise, and Yudkin brought in more A-1 Vietnam veterans and a young OSD staffer who expressed interest in the A-X project, Pierre Sprey.[14]

Sprey was an aerodynamic engineer whose short stint working at Grumman Aircraft showed him that "it would be twenty years before they let me design an aileron." After meeting one of McNamara's Whiz Kids at a symposium Sprey became one himself, working tactical air combat matters on the staff of the assistant secretary of defense for systems analysis. After finishing a controversial study criticizing World War II air interdiction, Sprey became an advocate of close air support because it was "something that actually makes a difference in warfare." As a brilliant and energetic participant who helped ensure that the plane's design remained practical, he influenced more than an airplane aileron's construction.[15]

Early concept work yielded a Requirements Action Directive (RAD), which AFRDQ sent to the U.S. Air Force Systems Command in December 1966. The RAD was only the first of many A-X development directives, but it set the program's tone and reflected the design team's influence. The A-X would support troops in close combat, escort helicopters, and fly armed reconnaissance of enemy positions. It had to be "simple, lightweight, reliable, highly survivable and capable of operating from medium-length semi-prepared airstrips with a high utilization rate. It must be able to carry a large payload of mixed ordnance and deliver it accurately. It must have sufficient low-altitude range and loiter capability, airspeed range, and aerial agility to perform the entire spectrum of close air support missions." Beyond the general guidance, the RAD's specific performance details revealed a plane like the A-1 but with improved speed, durability, and weapons carriage. It was supposed to operate from four-thousand-foot runways, carry at least an 8,000-pound weapons load, fly 300 knots, maneuver as slow as 120 knots, turn in a tight, one-thousand-foot radius, and loiter at least two hours. It would carry bombs, rockets, and a gun, and had to sustain hits from 14.5-mm shells as well as the shoulder-launched, heat-seeking SAMs appearing on the scene.[16]

The A-X project's blending of A-1 strengths with performance improvements led one officer familiar with the project to describe its finished product, the A-10, as a "jet-age A-1." It also led later critics to claim that the plane was too much a Vietnam War design. Although the A-10 incorporated many lessons derived from Vietnam, the concept designers produced an attack plane that could handle—as the Army intended for Cheyenne—a wider air support spectrum.[17]

Testimony to this came in the preliminary design's large gun. Several sources claimed that the Israeli Air Force's success against Arab tanks in the June 1967 Six-Day War sparked the desire for a large, rapid-firing gun, but some of Yudkin's staff members and Sprey recall that other factors inspired it. The A-X had to fly CAS in other environments besides the guerrilla war in Vietnam, and a tank-killing gun could enable the plane to help the U.S. Army defeat a massed Soviet tank assault in Europe. Also, the Cheyenne would carry anti-armor weapons, and Sprey asserted that the big-gun idea had to succeed or the Air Force would not "have a way of fending off Cheyenne." Other influences were German aircraft engineer Hans Multhopp, who had recently made an unsuccessful proposal to Martin Company for a big-gun CAS plane, and Stuka pilot Hans Rudel, who had fired single 37-mm cannon shots to destroy Soviet tanks in World War II. (As for the influence of the Six-Day War, the gun proposal appeared before it occurred; postwar analysis of Arab tank hulks revealed that the claims of Israeli aviators were inflated; and some of the sources had other motivations for their claim.)[18]

In early 1967 Systems Command commenced work upon the next big step in the development process, the Concept Formulation Package (CFP). Interestingly, the command's F-X Systems Project Office (SPO), which would produce the F-15, assumed responsibility for this phase. The F-X was the service's response to the failure of "multimission" fighters, such as the F-111 and F-105, to meet aerial combat demands. Also, Air Force fighter pilots believed that dedicated attack planes required a strong dedicated air-to-air fighter to protect them. However, one would have expected a separate A-X SPO, even if air superiority was a higher doctrinal priority than CAS. The setup perhaps reflected a desire to coordinate the two projects, but it more likely indicated that though the Air Force wanted the A-X to protect a mission, its development would proceed only as fast as mandated by the special conditions surrounding its birth. The service had changed its tune about air support, but not by too much. Sprey recalled that the A-1 pilots with whom he worked used a collection of mimeographed sheets as their tactics manual. To Sprey, this showed "how much hind tit these guys were getting; they couldn't even get a hard cover to put on a manual!"[19]

By March 1968 the working group had a CFP ready for OSD to approve and commence the construction contract phase. Perhaps reflecting a

Cheyenne-induced sense of urgency—the helicopter first flew in September 1967 and McNamara approved a Cheyenne production run in January 1968—the CFP directed an A-X initial operational capability (IOC) date of December 1970. This required *concurrency*, a term that signified accomplishing normally sequential developmental tasks simultaneously. This approach risked major snarls if something went wrong, and the Air Force had indeed encountered such situations previously. But due to doctrinal exigencies and the fact that the A-X was a simple, cheap plane, it was willing to take the risk. However, in a process that Sprey called "bloodshed," Air Force Headquarters staff modified the CFP to ensure certain aircraft characteristics. As a result, the final version did not appear until two months later—a critical loss of time. The finalized CFP described A-X CAS in the same terms as the RAD, but for the first time, the document specified four design guidelines for A-X construction: lethality, simplicity, survivability, and responsiveness.[20]

Lethality. In the CFP, lethality signified not only increasing the A-X's underwing pylons to ten and its ordnance capacity to 16,000 pounds, but also incorporating the antitank gun. The Whiz Kid emphasis upon maximizing ordnance weight per mile reinforced the push for extra bomb and rocket carriage. To ensure firepower, the CFP specified a 30-mm gun with a six-thousand-rounds-per-minute rate of fire and a twelve-hundred-round magazine. Lest one think that the designers shortchanged future CAS pilots with a fire rate six times what the magazine held, they accurately grasped gun employment concepts. The longest anticipated burst length per pass was two seconds maximum—enough to hurl about seventy rounds at a target and ensure at least ten passes. The round's constitution, velocity, and weight had to guarantee that this burst length would kill a tank, and thus the traditional tungsten-carbide antitank rounds would not do. Armed with such ammunition, Stuka pilot Rudel had to fire from a range of six hundred feet at a weak spot in the rear of Soviet tank turrets to destroy them. The concept designers accepted a short firing range, but preferred four thousand feet to allow steep dive recovery, given the A-X's higher speeds (compared to an A-1 or Stuka). They discovered that depleted uranium was appropriately dense and could ignite like magnesium upon high-velocity impact.[21]

Depleted uranium and other gun design issues were tough to resolve because of physical concerns and bureaucratic rivalries. Slightly different gun displacements either lacked the required round strength, like the 25-mm, or induced unacceptable airplane design penalties, as with the 35-mm. The staff also studied using a recoilless rifle, but its accuracy and muzzle velocity were poor. The Gun and Rocket Division at Systems Command's Eglin Air Force Base Armaments Laboratory objected to the 30-mm proposal. They suggested modifying the 20-mm gun used in Air Force fighters, but this weapon was incapable of piercing tank armor. The Eglin lab also did not like using depleted uranium, though current studies showed it to have no unsafe radioactive properties. The U.S. Army Ballistics Research Laboratory at Aberdeen Proving Ground, Maryland, revealed the growing friction between the A-X and Cheyenne projects by challenging the effectiveness of the 30-mm gun. This was the first of several clashes between aggressive CAS plane advocates and the respective Air Force and Army armaments laboratories.[22]

Simplicity. To match the A-1's simplicity and undercut the Cheyenne's cost, the Air Force made sure that the A-X did not carry the abundance of avionics goodies that had escalated other planes' price tags. For example, it would not have an all-weather radar like the F-111, or a sophisticated inertial fire control system like the A-7. Nor would it have even a fraction of the Cheyenne's avionics gear. Instead, it carried sufficient radios to talk to ground troops, a basic optical sight for weapons delivery, and a laser spot tracker for detecting a forward observer's laser designation. The design did allow inclusion of cockpit instruments to accommodate either the television-guided Maverick missile or Forward Looking Infrared (FLIR) night vision apparatus. In all, the CFP projected a per-aircraft cost of just over $1 million, which was well short of the Cheyenne's looming estimate of $6 million apiece.[23]

Survivability. The CFP demanded extensive and unprecedented survivability features, reflecting the impact of aircraft battle damage studies as well as concern about the CAS environment's lethality. Eighty percent of aircraft combat losses occurred due to fire and loss of control, so the A-X would have flame-retardant foam in its self-sealing fuel tanks, its armor-protected engines would be well separated from the fuel tanks and each other, and its gun magazine would have blow-out doors. Analysis of U.S. jet fighter losses in Vietnam revealed that flight control hydraulic systems were in vulnerable spots, or next to engine hot sections where a leak could ignite the flammable hydraulic fluid. Worse, no back-up system existed in case of total hydraulic failure, making the planes' flight controls unmanageable. Many F-105 combat losses occurred due to small hits on these vulnerable components; accordingly, the A-X would have protected hydraulic lines, dual flight controls, control-jamming protection features, and a manual flight control system. There would also be a "bathtub" of heavy armor to protect the cockpit from small arms fire. The designers also preferred maneuverability to overall speed because maneuvering helped defeat antiaircraft tracking solutions, while flying fast on a straight, predictable course did not. (Because the design staff insisted upon maneuverability over speed, TAC refused to participate in the concept formulation phase.) Finally, even though analysis did not conclusively prove that two engines improved combat survivability, designers still wanted a twin-engine plane for the common-sense reason that one engine allowed safe return if the other failed, regardless of cause.[24]

The A-10 that was created from this design concept proved its ability to sustain hits, survive, and even return to combat after short maintenance repairs. It was, as Sprey asserted, "the most survivable plane ever built for the close air support mission." However, it was not designed to fly the high-speed dash runs required for deep interdiction missions against the Warsaw Pact. Critics later disparaged the plane for lacking such a capability, but the designers were charged to build a CAS plane, not a strike fighter.[25]

Responsiveness. Sprey recalled that the A-1 Vietnam veteran pilots wanted loiter capability because it allowed more time in the battle area and also determined responsiveness more than aircraft speed:

> The other thing these guys taught me, just banged it home, because it didn't make sense to me at first, was the overwhelming importance of loiter time. . . . that if you want very fast time response. . . . you can't get it with speed, because you're always too far away. If you're on the ground it's too late . . . you can have an airplane go Mach 3 and it's still too late. The only way to get super quick response time is to be in the air, loitering, and checked in with a FAC . . . you need three to five minutes response time . . . come any later and you might as well not have come.

Engine performance was a key factor in determining the fuel endurance required for loiter, and the Air Force's initial sense of project urgency drove the choice for a two-engine turboprop design. A turboprop provided better thrust than a reciprocating engine, beat a turbojet on fuel efficiency, and surpassed the emerging turbofan engine technology on both counts.[26]

The staff designers believed that responsiveness depended not only upon loiter capability but also upon maneuverability, the ability to fly from short, austere fields, and proper communications. A maneuverable plane could quickly turn to re-attack a target. Short-field capability increased basing flexibility and allowed A-X units to be closer to the troops they supported. (It would also better enable the A-X to compete with the Cheyenne.) These requirements created a future problem because the plane needed a thicker wing to carry all of its planned ordnance, allow STOL performance, and possess the fuel efficiency for slow-speed loiter. The wing produced a lot of drag that prevented fast speeds regardless of powerplant strength. Finally, to solve a problem that had inhibited U.S. air-ground coordination in at least two wars, the CFP demanded that the plane carry the appropriate Army-compatible communications gear.[27]

The CFP's later completion date and stipulations, such as inventing a new gun and fitting it into the plane, led the service to slip the A-X's IOC to 1972. Another reason for the slide was that the Cheyenne's development problems were becoming public, which relaxed pressure to build a competitor as soon as possible. It was just as well, for the modified CFP that the service sent to OSD for approval snagged upon some of its particulars.[28]

Dr. John Foster's DDR&E office would approve the CFP and in turn write a Development Concept Paper (DCP) for the defense secretary to approve and then commence the next phase in the long, alphabet soup–laden procurement process—convening a Defense System Acquisition Review Council (DSARC, a project milestone oversight group) to commence contract definition and issue a Request for Proposal (RFP) to be sent to prospective contractors. Foster's CFP review involved mostly routine, analytic questions. For example, could an already operational aircraft meet the requirements; or should concurrent development take place on a more complex A-X that could do night and all-weather CAS? One other question reflected a procurement philosophy change that occurred with Defense Secretary McNamara's departure in March 1968: should competitive prototypes be developed and flown before a particular design was selected? McNamara's belief in systems analysis caused him to favor using written studies and computer modeling to choose the winner. However, the problems surrounding his push to build the F-111 left many people cold. Foster and the Systems Command commander, Gen. James Ferguson, were interested in competition between contractor prototypes—called competitive prototyping or "fly before buy." Though it had not been done in the Air Force since the 1950s, Foster made "fly before buy" a procurement issue by incorporating it as an A-X CFP review question.[29]

DDR&E's review yielded some concerns about the design concept, and in Development Concept Paper (DCP) 23, completed in December 1968, it recommended deferment of approval. Foster thought the plane's survivability specifications for small arms fire (14.5 mm) were insufficient given the heavier automatic weapons, such as 23 mm, available to most Warsaw Pact ground units. Additionally, he wanted to see if anything from the development of the A-7, A-37, and OV-10 could be transferred to the A-X. Finally, Foster was suspicious of the CFP's expansion of the A-X's size and weight. He did not want to repeat the F-111 experience, where attempts to add too much capability nearly produced a monstrous failure. Also, it seemed that the A-X's planned size, range, and payload specifications would probably create a plane similar to the A-7. Foster wanted an A-X, but he also wanted to lock the simplicity and survivability concepts into its construction.[30]

OSD reviewed DCP 23 and approved contract money in the fiscal year 1970 budget—contingent on Air Force review of Foster's concerns. Through early 1969 the service examined these points and made small size and weight reductions to the design. As the year progressed, the Air Force urged OSD to proceed with the contracting phase. The renewed sense of urgency came from outside factors that had rekindled service interest in the A-X's fortunes.[31]

Interservice Rivalry and Other Factors

OSD staff resisted the desire to rush, in part because they wanted more opinions, especially from the Army. As a recipient of Air Force CAS, the Army would be the prime beneficiary of this plane's creation, but the A-X was also the Cheyenne's competitor. Interservice actions during this period did not indicate harmony about CAS aircraft procurement. A 1967 Army report revealed that its commanders in Vietnam liked Air Force CAS, but preferred aircraft that could guarantee quick response and target acquisition—an indication that they might prefer attack helicopters. Through 1968 the Air Force conducted comparative analyses that described the A-X as a more cost-effective aircraft than the Cheyenne. The OSD considered using its Weapons System Evaluation Group (WSEG) to provide an unbiased means of resolving the issue, but in February 1969 the WSEG responded that it should not be used for this purpose.[32]

In so doing, the WSEG presaged the inconclusive, contentious nature of later air support studies. Any finding would be based upon varying definitions and impressions of CAS, as well as the various players' views of their own present and future needs. A later analysis of Vietnam CAS claimed that only 10 percent of the missions that either the tasking authority or the pilots called "close air support" actually involved troops in direct contact with enemy forces. Furthermore, just after the Vietnam War ended, a survey conducted by a U.S. Army Command and General Staff College student revealed confusion among both services' veteran officers over CAS procedures and definitions.[33]

The Army wanted air support, but it simultaneously played semantic games to secure its attack helicopters' existence, and also backed away from demanding an input in Air Force tactical airplane development. Army leaders continued to use the term *direct fire support* to describe helicopter CAS. When Defense Secretary Clark Clifford asked the Army its opinion of the A-X, its Combat Developments Command (CDC) acknowledged that the A-X and the Cheyenne competed for defense funds and asked if the A-X was a plane designed too much for Vietnam instead of other war environments. Sounding like some of the Air Force's A-X opponents, the CDC's commander (and a helicopter advocate), Lt. Gen. Harry Kinnard, opined that the A-X's dedicated CAS design added little air capability for the financial and force structure sacrifice. He wanted the Air Force to choose tactical aircraft using its traditional doctrinal priorities.[34]

Kinnard's attitude reflected the Army's anxiety about the Cheyenne program, whose troubles endangered Army plans for an armed helicopter. As a result of project cost overruns, one Cheyenne cost twice as much as an Air Force F-5 and the same amount as the F-16 fighter that the Air Force bought in the 1970s. The innovative rotor technology caused problems; one Cheyenne's main rotor smashed into its canopy during a 1968 test flight. In

February 1969 a Cheyenne crashed and killed its pilot during flight tests, and in April the Army ordered Lockheed to cure the program's ills. When Lockheed failed to do so, the Army canceled the Cheyenne contract, but since it wanted an advanced attack helicopter, it later renegotiated with Lockheed for an improved Cheyenne rotor system. Because the rotor was important to the whole design, the Army and Lockheed had to resume the program almost from scratch. Later accounts by Army leaders and historians described an overly ambitious program that failed due to misunderstandings between, and errors by, the Army and Lockheed. The Army abandoned the notion of a simple, inexpensive aircraft in favor of something reflecting the Air Force's penchant for expensive and complex machinery. The service aggravated matters by insisting that Lockheed help recoup its research investment on an unworkable gun by installing it on the Cheyenne. Lockheed had never built a military helicopter before, and the Cheyenne was far too ambitious for a first attempt. Finally, the company encountered problems with its C-5 transport plane project, and channeled most of its attention to that plane.[35]

Thus, the C-5 received a lot of criticism from Congress, which was increasingly irritated over Pentagon aircraft programs. One reason for this was America's disillusionment with the Vietnam War. Another was skyrocketing costs of aircraft as weaponry became ever more complex and inflation gripped the country's economy. A congressional report on military aircraft costs complained that "fighter aircraft now being developed for procurement in the mid 1970s will cost more than five or six times more than comparable aircraft at the beginning of the 1960s." Unfortunately, the early 1970s were a time when all of the services wanted new aircraft in order to remain superior to Russian planes.[36]

Thus, the CAS aircraft argument came at a bad time, and congressional tempers flared during hearings in summer 1969 that addressed the A-7 project. The Air Force failed to describe the A-7's specific worth well enough, and also could not show how the program would avoid cost overruns. For this reason, the lawmakers canceled the plane's funding and restored it only after the service completed corrective action. As for the A-X, in 1968 the Senate Armed Services Committee had questioned the need for it, when the military already had A-1, A-6, and A-7 attack jets. In September 1969 the House Armed Services Committee suspended A-X funding because some members did not think that a turboprop design represented enough of a technological advance to warrant support. After all, the Air Force had many low-technology planes, such as the A-37, OV-10, and gunships, for air support. Other committee members, such as antiwar liberals in the Democratic Study Group, thought the plane represented an OSD promotion of limited wars like the unpopular one in Vietnam.

Congressional pressure spurred the Air Force to make a stand for the A-X when the Cheyenne's threatened demise might have led the service to neglect it. In an *Armed Forces Journal International* article about the A-X budget suspension, Air Force officials rebutted favorable reports about the OV-10's CAS potential by saying that the OV-10 lacked the A-X's ruggedness and lethality. In another article in the same journal they reminded congressional skeptics that the A-X was the successor to the popular A-1. Both articles cited the alleged Israeli fighter success against Arab tanks in the Six-Day War as proof that the A-X's big gun made it a valuable CAS asset in armored warfare environments, such as Europe. The Air Force Association (AFA) was the service's civilian advocacy and lobbying group, and in January 1970 its *Air Force/Space Digest* magazine presented the salient A-X characteristics given in the CFP and DCP 23, and told readers that A-X was "persuasively cost-effective" in many ways. The service also elevated the A-X project's stature. The A-X project had remained stuck within the F-X SPO, and as the Cheyenne encountered problems, Systems Command removed people from the A-X to work on other programs. Normally, an officer of at least colonel's rank supervised a new Air Force aircraft project, but at the time of Congress's budget suspension, a major was running the A-X project. By April 1970 Col. James Hildebrandt was the chief of the newly created and independent A-X SPO.[37]

Colonel Hildebrandt's office remained small, but this was due to a combined OSD and Air Force response to the congressional budget axe. DDR&E and Systems Command had already expressed interest in competitive prototyping, and now concern about overall defense costs and developmental fiascos led both a blue-ribbon defense policy panel and Deputy Defense Secretary David Packard to implement it as policy. Following the new practice, and probably spurred by Congress's September 1969 A-X budget suspension, the new Air Force secretary, Robert Seamans, announced on 10 October 1969 an austere approach called "parallel undocumented development." As later promulgated in the A-X's DCP 23A (successor to

DCP 23), this developmental style featured a minimally manned SPO that monitored selected contractors' efforts to produce a design that best met Air Force specifications. DCP 23A expressed the hope that parallel undocumented development would prevent concurrency, minimize development time, and, above all, reduce cost.[38]

Parallel undocumented development reflected the Air Force's doctrinally consistent budget priorities. In congressional testimony at the time, the new Air Force Chief of Staff John Ryan told the legislators that the F-X air superiority fighter had precedence, followed by the B-1 strategic bomber, then the Airborne Warning and Control System (AWACS) plane, and finally the A-X. Neither the F-X, the B-1, nor the AWACS had a prototype flyoff competition. Some observers pointed out that the contractor risk involved in building these more expensive and complex planes for a winner-take-all competition was simply too much. (However, there were competitions for their various component parts.) Another expert averred that the Air Force's desire to create the ultimate air superiority machine led it to make the F-15 the best plane money could buy, literally. Conversely, the A-X represented a reaction against high technology and cost. It was supposed to be an upgraded A-1, the propeller plane that seemed to disprove the need for high speed and complex fire control avionics in the CAS arena. Given doctrinal priorities, an Army CAS aircraft threat, and congressional budget limits, it is not surprising that Seamans, Ryan, and company used the A-X to reintroduce prototype competition.[39]

Congress had also quickly endorsed the Army's move to suspend funds for the Cheyenne, and insisted upon comparing the A-X and the Cheyenne in order to determine the best CAS aircraft to buy. In response, the OSD reviewed the two programs in summer and fall 1969, and brought staff members from both services to discuss their respective air support aircraft. These actions enhanced the Army's wariness on the issue, and in the OSD meetings Army leaders echoed General Kinnard's opinions as they asked that the Air Force follow its traditional doctrine of preferences. They also questioned whether the Air Force could afford the A-X. The senior Army representative, force development chief of staff (and former Army aviation advocate and leader), Gen. Robert Williams, told the group that if the Air Force really wanted to support the Army, it should upgrade its attack jets to accomplish night and all-weather CAS. When queried about the Air Force's preferences, the senior Air Force representative, Vice Chief of Staff John Meyer, said that given the demonstrated lack of congressional fiscal support, he would choose air superiority fighter projects.[40]

An Army representative at the meetings found Meyer's statement "rather remarkable," given congressional scrutiny and the services' high-stakes competition over this mission. But as his chief of staff had done before Congress, Meyer simply voiced his service's doctrinal priorities under the pressure of an either-or choice. Other Air Force leaders, however, were only too happy to scrub the A-X. When the Army canceled the Cheyenne contract, General Disosway of TAC approached Ryan and told him he "could get rid" of the A-X. Ryan, who wanted to protect his service's CAS charter, resisted such a precipitate move. Given the rising budgetary and emotional currents surrounding air support, the Air Force and Army secretaries intervened at this time with an agreement that helped secure the A-X's future versus the Cheyenne.[41]

"Two service secretaries entered this arena where angels should fear to tread," Robert Seamans later wrote of his CAS aircraft agreement with Army Secretary Stanley Resor. Congressional heat about CAS aircraft funding drove David Packard in January 1970 to ask both men to resolve their services' CAS aircraft differences. They recognized that interservice rivalry threatened to ruin what they believed were two aircraft that complemented each other in the CAS mission. Seamans had visited combat units in Vietnam and wanted the Air Force to build better CAS planes; while he believed that helicopters provided valuable support, he felt that they were too vulnerable to perform CAS alone. Sources indicate that Resor needed an Air Force guarantee of support for the advanced armed helicopter now that the Cheyenne was in trouble.[42]

Their March 1970 agreement specified that the fundamental differences between the two aircraft made them suitable weapons for different types of air support. The Cheyenne possessed a helicopter's flexibility in maneuver, take-off, and landing that made it better at such operations as helicopter escort, urban CAS, and highly fluid, close-quarters ground combat situations. The A-X was faster, flew farther and for a longer time, and carried greater firepower than the helicopter. (Seamans also believed it was far more rugged than a helicopter, though this was a contentious issue and was left out of the agreement.) The secretaries also envisioned situations in which the two aircraft would work together to provide mutual support in finding targets, avoiding and suppressing air defenses, and providing combined fire support. They did not agree on everything, and one scholar criticized them for not achieving agreement on fundamental doctrinal issues as well as aircraft suitability for every air support task. Given the intense feelings at the time—not to mention external pressure from Congress—this was a tall order. The secretaries may only have been able to "agree to disagree" on some issues, but their proposal to keep both programs alive was important to the future of the CAS plane and the advanced attack helicopter.

The agreement enraged uniformed subordinates in both services. Seamans had the following recollection of his pact with Resor: "I later heard that Stan had said, 'Bob and I must have done something right. Both of our staffs told us that we had sold them down the river.'" Seamans recalled that General Ryan would not talk to him for a few days out of fear that he "would be too emotional" if he personally spoke to the secretary. Advocates on both sides felt that their respective secretaries reprieved a threat to their own service's aircraft. Intense feelings remained, and one example was General Ryan's reply to a January 1970 assessment of Army CAS needs by Army Chief of Staff Gen. William Westmoreland. Westmoreland reiterated General Kinnard's assertion that the Army's air support priorities matched Air Force doctrinal preferences, and that perhaps the Air Force tactical plane designs "should emphasize multi-mission capabilities." Ryan matched this ironic twist by sharply reminding Westmoreland: "We believe that only by optimizing an aircraft for close air support can we provide the capabilities outlined in your memorandum. Our experience has shown that a multi-mission design must sacrifice too many of the characteristics desirable in a close air support aircraft in order to meet the requirements of its other roles." Thus, between 1960 and 1970, various developments had brought about a swap in the services' CAS plane philosophies. With the factors that had sparked the A-X's creation still extant—and others arising—Air Force leaders pressed on enthusiastically through journal articles, SPO reinforcements, and ringing declarations on behalf of dedicated CAS planes.[43]

By spring 1970 the OSD had convened a Defense System Acquisition Review Council (DSARC) on the A-X and had released DCP 23A, which incorporated changes to DCP 23, outside inputs, and the Army's concerns. (The changes included prototype competition and parallel undocumented development.) The new DCP asserted that the A-X would provide CAS in all theaters and not just Vietnam—though it admitted that the plane would need fighter escort in situations where enemy fighters contested air superiority. The document also contained a promise that the Air Force would answer the Army's desire for night/all-weather capability by considering such features for the A-X. Finally, it included provision for a jet-powered plane.[44]

The last two items rate further discussion. Especially in troops-in-contact (TIC) battle situations, visual identification is the surest method to achieve correct target acquisition. The current artificial means for determining the target, such as radar and infrared, were not reliable—or at least required enough advance planning and increased pilot workload to make them unsuitable for the fluid circumstances of battlefield air support. Yet the Army always wanted, and the Air Force accordingly searched for, technology that would allow fixed-wing TIC CAS at night or in clouds, in any defense threat situation, and at any altitude over any type of terrain. Placing such technology— radar, inertial navigation systems, FLIR—into a plane that was fourth on the service's funding priority list raised concerns over cost. If one required proof of the development or cost pitfalls in night/all-weather technology, one had only to look at the F-111 and Cheyenne projects. Air Force leaders would examine the possibilities, but they also rightfully insisted that visual recognition was important to CAS. Thus the A-X, like the A-1 before it, would possess the maneuverability to work visually in weather poor enough to prohibit other planes from doing so.[45]

Though there is no written proof that DCP 23A wanted a jet-propelled A-X because of the recent congressional rebuff of a turboprop version, the available sources do mention that turbofan engine technology had caught up while the design process encountered delays. The first stage of a turbofan's compressor blades extended beyond the engine's diameter at the front to direct air along its exterior, thereby generating extra thrust and also reducing operating temperatures. The intention was to create greater thrust than a turboprop with less fuel consumption than a turbojet, but the early models were not impressive. Meanwhile, several aerospace engine companies had developed the technology so they could break into the airliner market, where fuel-efficient engines provided financial benefits. By 1968 the Air Force noticed two new turbofan candidates that promised to meet the A-X's thrust and fuel specifications. General Electric had won a Navy contract to put its

TF34 into an antisubmarine patrol plane, the S-3. Meanwhile, Lycoming rebuilt its T55 turbojet into a turbofan engine to compete for airline business. DCP 23A included them as likely A-X powerplants.[46]

Initial Contract Work, Gun Development, and More Congressional Pressure

In May 1970 the Air Force finally released its A-X RFP to twelve contractors for construction bids. Even though the document ran to more than one hundred pages, it was short and concise compared to RFPs for other aircraft. By August the Air Force stated that it would conduct further review of six returned proposals, and after a December meeting of the A-X DSARC, it picked two companies, Fairchild-Hiller (later to be called Fairchild-Republic) and Northrop, to build prototypes and compete in a government-run fly-off. The initial buy would be for six hundred aircraft with a twenty-six-month competitive development period and an IOC of 1975.[47]

The economic philosophy that drove competitive prototyping and parallel undocumented development included other items that later made the A-10's life interesting. The Air Force wanted this project to survive while it pursued big-ticket weapons, and the RFP accordingly introduced another new weapons policy, design-to-cost. This required that both the contractor and the government adhere to the aircraft's initial cost estimates—accounting for later currency inflation—in all development actions. Failure to meet the cost goals at various points would incur denial to proceed with the next phase. Given Air Force budget aims, the A-X CFP's simplicity goal, and a Congress keen to enforce military fiscal restraint, the A-X project's design-to-cost adherence was stringent and would deny the plane some sensible improvements.[48]

Design-to-cost was intertwined with another issue, the A-X's gun. In June 1970, as the service worked on contractual issues for the plane, it also sent to DDR&E its DCP 103 concerning gun design. DCP 103's direction that the 30-mm gun be built only for the A-X married the plane and its gun in fortune as well as in fact—something guaranteed by the RFP's demand that the plane meet both design-to-cost restrictions and weapons-effectiveness goals. The marriage of economy and lethality would be difficult, for DCP 103 demanded that the gun be effective against Soviet T-55 tanks (then the most numerous model), and that designers use the best shell. DCP 103 based gun selection upon competition between the two best prototypes.[49]

The marriage had to work. Congress let the Air Force continue with A-X competitive prototyping, but it still questioned the need for the plane, given budget constraints and the existence of other aircraft that flew CAS. Throughout the early months of 1970, the House Appropriations Committee pressed the OSD and the Air Force to justify their push for several planes that could do the same mission. The legislators wondered why the U.S. military could not use either the OV-10 or Cheyenne. Secretary Seamans and General Ryan repeated their findings that the A-X surpassed both aircraft in most CAS plane performance categories. Additionally, they asserted their service's CAS prerogatives, according to past interservice agreements. In April 1970 the House Appropriations Committee chairman, Rep. George Mahon, asked Dr. Foster if it was "essential" to support both the A-X and Cheyenne "in view of the fiscal problems which appear to be ahead." Foster replied that the choice was still undetermined, since neither machine was developed enough to make comparisons. Mahon also asked why the A-7 could not be used, and Foster replied that modifying it to fly the CAS mission as the A-X would was too difficult. Though favorable to defense needs, Mahon was a fiscal conservative from Texas, the home state of the A-7's maker, Ling-Temco-Vought (LTV). His questions on the A-7's behalf foreshadowed the partisan politics that would affect the A-X.[50]

As if there were not enough attack planes already, the U.S. Marine Corps wanted to buy a jet that it felt was optimum for its war-fighting environment, Britain's Hawker-Siddely AV-8A Harrier. Since 1969 the Marines had conducted their own bureaucratic insurgency by procuring a small number of these first-ever vertical/short-takeoff-and-land (V/STOL) tactical jets, ostensibly for evaluation. They considered the Cheyenne for their CAS needs, but the helicopter's complexity and escalating cost cooled their interest. Though the Harrier cost roughly the same as a Cheyenne, it was an already developed design. Not only did it possess the Cheyenne's vertical-take-off-and-land (VTOL) capability—though with serious weapons carriage and loiter penalties—but it was also a tactical jet. As such, it fit the Marines' need for a jet providing firepower and some aerial protection during amphibious operations in which carrier decks and very short fields were often the only available bases. By 1970 the Marine Corps caught Congress's attention due to its sudden request to buy many more of the foreign-built jets.[51]

The Marine Corps move was the last straw for Congress, whose commentary on its fiscal year 1971 budget intoned, "There is a serious question as to whether or not future Defense budgets can support the development and/or procurement of three separate aircraft weapons systems . . . to perform essentially the same mission." The Senate Armed Services Committee supported its House counterpart in the same report, condemning the services for a history of "parochialism" concerning the close air support mission. In spite of its complaints, Congress allowed Fairchild and Northrop to continue their work because the legislators liked the idea of a low-cost plane developed from active contractor competition. However, in October 1970 the House Appropriations Committee directed the OSD to "reevaluate the roles and missions and aircraft relative to close air support," and determine which of these planes was most suitable. The Air Force CAS plane's fate now lay with the OSD and Congress.[52]

6

CAS AIRCRAFT ON TRIAL

In late 1970, the OSD's initial study of the services' CAS aircraft desires went to Congress, which dismissed it as too "perfunctory," according to one account. In response, Deputy Defense Secretary David Packard directed a deeper study of interservice CAS differences as they pertained to choosing a CAS plane. Packard's directive may have been a response not only to the legislators' rebuff but also to Vietnam War events that threatened to escalate the CAS plane issue into a public interservice fight.[1]

In February 1971 a South Vietnamese Army advance into Laos, Operation Lam Son 719, stalled and then turned into a rout. U.S. ground forces could not participate because Congress forbade them to do so after a public uproar over the 1970 Cambodia incursion. Still, U.S. aircraft provided most of Lam Son 719's air support; and though the South Vietnamese Army's poor showing stirred national media debate about South Vietnam's military viability, the air combat results stimulated the lesser known CAS debate.

U.S. Army helicopters supporting the South Vietnamese encountered far tougher antiaircraft defenses—radar-controlled AAA in some instances—than previously seen. Their normal low-altitude flight paths (less than two thousand feet but not skimming the ground) made them easy targets for communist gunners. Additionally, South Vietnamese soldiers at the landing zones seemed more intent upon boarding the helicopters and leaving Laos than in providing suppression fire for the helicopters as they approached. Army helicopter advocates tried to belittle the battle's implications for helicopter combat survivability. They pointed out that after some initial surprises, they changed altitudes and flight routes to avoid further losses. Further, they argued that casualties were not as high as they seemed because a stricken helicopter could autorotate to a safe landing and be retrieved later (an autorotation occurs when a helicopter pilot with a failed engine modulates the main rotor blades' pitch for a controllable descent).[2]

Air Force leaders and other critics countered that Army commanders often used wreckage to claim a retrieved helicopter, and accused the soldiers of doctoring the loss figures. One day before Lam Son had even ended, the Air Force completed an official history of the battle that criticized Army helicopter vulnerability while praising fixed-wing jets for destroying communist tanks and providing what antiaircraft suppression there was. The Air Force maintained that though helicopters might survive against weak air defenses, they would not do so in Europe—where the military increasingly redirected its attention as the Vietnam War de-escalated.[3]

Perhaps due to a combination of Lam Son and the upcoming OSD review, in March 1971 the Army unilaterally rescinded Department of Defense (DoD) Directive 5160.22, which had denied to the Army and reserved to the Air Force the right to conduct CAS. Another public interservice brawl over a mission and its resources threatened to break out, but though there was sniping over Lam Son, senior leaders in both services did not want such a fight. It recalled the 1949 Admirals' Revolt and a more recent command squabble during the 1968 Tet Offensive. While Tet raged, the commander of the U.S. forces, U.S. Army Gen. William Westmoreland, clashed with the Marines over their refusal to place their tactical air units under centralized Air Force command. The debate waxed so intense that Westmoreland considered resigning over it—and not over the real battle, which caused so much consternation among Americans at home. Fear of such infighting, and Packard's inclusion of senior officers from the affected services in his group, led to greater cooperation in the study project.[4]

Because resolving interservice CAS demands in favor of one service or aircraft was an intimidating task, Air Force and Army leaders had so far opted for halfway measures. However, Packard's study group seemed game for a universal settlement between three services and their A-X, Cheyenne, and Harrier aircraft—or at least for proof that a solution did not exist. Covering nearly three pages, the study's questions addressed four different warfighting environments: Lam Son 719 (including queries about what had really happened there), a Soviet attack in Europe, a Middle East war, and a Korean war. For each case, further questions addressed each aircraft's capabilities, tactics, and weapons against various threats. Finally, the study asked each service how CAS supported each campaign, how it would conduct CAS in that campaign, and how it would use the other services' machine if its own craft was not available.[5]

Along with the four theaters and dozens of specific questions, the study group identified six different CAS sub-missions and eleven different types of CAS targets, each of which invited creation of an aircraft optimally designed for one or a combination of several. The group also established twelve key competitive factors, with attending interpretational problems. "Payload," for example, could be free-falling bombs, which all helicopters—including Cheyenne—had difficulty delivering. Or it could be guided missiles, such as the Maverick or TOW (tube-launched, optically aimed, wire-guided antitank missile), which presented their own target acquisition and guidance challenges. "Responsiveness" could mean aircraft speed that enabled a quicker trip to the battlefield. It could signify V/STOL characteristics, which allowed forward basing, and thus overlapped into the "basing flexibility" factor. Or it could be a simple machine that could fly more sorties over time, which in turn overlapped into the "utilization rate" factor. Concerning the machines' performance against defenses, the group's computer analysis contained variables so complicated that it finally used a scenario featuring a lone aircraft facing only one antiaircraft weapon—a situation that the group admitted was unrealistic.[6]

Packard's June 1971 report did not choose a single aircraft because it found, not surprisingly, that *"close air support is a complex mission"* (italics in the original). Indeed, it stated that analyzing the relative merits of the three aircraft— not to mention those of other, already operational attack machines—was task enough. It recommended that the A-X, Cheyenne, and Harrier should all be developed because each delivered unprecedented air support specific to its service's needs. Likewise, none could duplicate the others' strengths, designed as it was to maximize its own. The A-X could not operate in poorer weather or shorter fields than the Cheyenne, and the Cheyenne lacked the A-X's simplicity, firepower, ruggedness, and loiter strengths. The Harrier's V/STOL capability answered the Marines' need for it to accompany their amphibious operations, but it also inflicted a weapons carriage and loiter handicap. Packard qualified his recommendation by requiring each aircraft to meet its cost and effectiveness goals. He also wanted flight tests of the fully developed A-X and Cheyenne to resolve any specific questions.[7]

Overall, the service representatives concurred with Packard, since he granted their respective aircraft a reprieve and averted interservice war. But to punctuate their strongly held opinions, they filed addenda disputing certain conclusions. The U.S. Air Force TAC commander, Gen. William Momyer, wanted only the A-X to be approved, since CAS was an Air Force mission and the A-X was the optimum plane. U.S. Army force developments chief and helicopter advocate, Gen. Robert Williams, discounted the report's skepticism about the helicopter's relative ability to acquire targets. And though Marine air leader, Maj. Gen. Homer Hill, was his service's representative, the Marine Corps commandant, Gen. Leonard F. Chapman Jr., spoke for him to insist that their service could support Harrier units in the field, no matter how austere the surroundings.[8]

The Senate and the CAS Riddle

One could say, and at least one observer intimated, that Packard and the generals intentionally created a study whose findings suited their beliefs. Its ultra-complex premises precluded determining either the primary CAS service or the best plane for the mission, and thus sustained the legislators' suspicions. Already that March, Senate Armed Services Committee Chairman John Stennis had charged his Tactical Air Power Subcommittee to conduct hearings on the matter. Subcommittee Chairman Howard Cannon seemed confident that his panel could resolve such matters if the military could not: "I feel reasonably certain in my own mind that when it gets down to the wire we are not going to be able to have a Cheyenne, an A-X, an A-7, a Harrier, and all of these other weapons . . . simply from an economic standpoint."[9]

They should have known better. Congress had also ordered the Government Accounting Office (GAO) to examine the mission and the aircraft that flew it. The GAO did not formally publish its report until after the hearings, and though the findings criticized the services' inability to create joint CAS doctrine, the report gave the impression that the mission's complexity and demands intimidated the GAO as much as they had Packard and others. It evenhandedly listed each candidate aircraft's strengths and weaknesses, and confirmed that they met certain needs. It also detailed the issues that made the mission hard to address, and asked for better weapons testing to answer the tricky questions that they raised.[10]

The hearings opened on 22 October 1971 with Senator Cannon announcing that they would resolve not only procurement questions but also disputes over CAS roles and missions. His first witnesses were Packard and the chairman of the Joint Chiefs of Staff, Adm. Thomas H. Moorer. Packard explained his report's conclusions, and Moorer backed him by pointing out that the Navy did not buy only one type of ship for all of its missions. The subcommittee spent the rest of that day revealing most of its members' ignorance of CAS as it pertained to interservice agreements, aircraft capabilities, and different war-fighting scenarios. Some subcommittee members did know the details, but they backed their favorite service's interests. Sen. Stuart Symington, the former Air Force secretary, wanted the services to choose one CAS aircraft, but questioned the Cheyenne's viability and implied that the Marines' requirements did not match the Air Force's and Army's needs. Sen. Strom Thurmond revealed his pro-Army stance when he made the witnesses confirm their support for the Army's attack helicopters.[11]

Following a day of briefings about Soviet weapons in Europe, the senators called Army witnesses—all senior officers, led by Gen. Robert Williams of the Packard study. Senator Cannon told them, "Today, gentlemen, we are going to get down to the nitty-gritty of these close air support hearings. You in the Army are the primary users of close air support." Perhaps the senator was disappointed, for the soldiers supported Packard's findings. They praised Air Force CAS in Vietnam, told the senators that the A-X and Cheyenne were complementary, and refused one senator's suggestion that the Army assume sole control of CAS and its associated aircraft. General Williams explained that fixed-wing tactical aviation was not the Army's specialty, and therefore he wanted the Air Force to fly fixed-wing CAS. Still, the witnesses supported their attack helicopters. They observed that though helicopters could not perform CAS while in the clouds, they could fly the mission in poorer weather than the A-X. The senators asked about Lam Son and recent Army attack helicopter tactics trials staged in Europe, and the officers downplayed the battle losses and replied that they had adjusted their tactics to handle European combat. Their helicopters would not duel with antiaircraft weapons, but would instead use the terrain to hide until they had a shot opportunity. This answer spurred more questions, since the helicopters' prime anti-tank weapon, the TOW, required them to maintain sight with the target until missile impact. The soldiers again emphasized that good tactics—decoys, suppression fire, and the like—would minimize risk.[12]

The Army's testimony further exposed the senators' different opinions of CAS and the various services. Sen. Thomas McIntyre pushed General Williams to admit that the Air Force was unprepared for CAS in Vietnam, and thus should transfer that mission's responsibility to the Army. This led Senator Cannon and Sen. Barry Goldwater, both of whom were Air Force Reserve officers, to rebuke him in a somewhat heated exchange. "I hope you will stay and listen to all the hearings, so you won't make your decisions from a preconceived idea rather than from the testimony that is presented," Cannon chided McIntyre. Meanwhile, Senator Thurmond asked questions that let the soldiers advertise their advanced attack helicopter needs.[13]

The next day the Air Force sent only General Momyer to deal with the senators. Momyer had opposed the A-X's creation, but as an officer who loyally supported his service, he uncompromisingly presented the Air Force's case to the panel. He defended his service's air support record by citing Army commander praise for Air Force CAS in every war since World War II. Momyer claimed that the attack helicopter was a usurper of an Air Force mission, and that the A-X was a proper response. Further, the Cheyenne was too expensive, carried a light ordnance load, and could not face tough defenses—Lam Son 719 was proof of the latter claim. Warning that the Army's aviation technology needs would escalate, he observed, "I am sure that after Cheyenne—an Army aviator is no different than any other aviator; he is going to want to go faster; he is going to want to go farther; and he is going to want to carry more bomb load."[14]

The senators asked Momyer about the Marines' CAS prowess, and he referred to his command of all fixed-wing units in Vietnam to answer that they were no better at CAS than the Air Force. Also, the Marines' desire to control their own air units undermined a properly coordinated air campaign. As for their Harrier, he believed that since the plane paid a high performance price for V/STOL capability, it was not suitable for the Air Force.

Concerning the A-X itself, Momyer praised it as the best CAS plane design, especially for Europe. Fielding questions about Europe's often poor weather, he answered that if the pilot flew slowly enough, the plane could operate outside of clouds nearly all the time. The A-X's gun and Maverick guided missile allowed it to stand off from enemy defenses and hit the target— something that he claimed the Cheyenne could not do. When the senators asked him why he preferred the A-X to the A-7, he told them that the A-X remedied the A-7's deficiencies. Its high maneuverability allowed it to foil enemy fighter attacks, while the A-7 could neither outrun nor out-turn them. Cheaper than the A-7, it also possessed better loiter, weapons carriage, and short field performance.

Three days later there was a long session with Marine Corps and Navy officers. Cannon announced, "Some say the Marines have the finest close air support in the world. We will be anxious to discuss all aspects of this mission with you." With this mandate, the Marines wasted no time promoting the Harrier as most suitable for their amphibious operations. Indeed, they were so keen on the plane that they told the senators they had deactivated other air units to procure it. They agreed that it lacked payload and range, but citing their World War II experience, they believed that air base proximity to the battlefield best determined responsiveness. The Harrier helped make this possible, and the Marines were confident that they could logistically support Harrier units via ships and helicopters. Concerning attack helicopters, they thought that the machines were too slow to be survivable in a high-threat environment, but they still wanted them in order to meet specific fire support scenarios. (The senators also probed another Harrier issue, one that figured in its later history: vulnerability to antiaircraft hits. The Marines dismissed the issue because they believed the plane was fast enough to avoid ground fire.)[15]

Another part of the Marine/Navy session was a lengthy discussion of the A-7's merits. Impressed with the Air Force A-7D's advanced fire-control system, the Navy modified its own A-7s to carry it, creating the A-7E. The Navy used A-7Es in Vietnam, and the witnesses included veteran pilots who strongly endorsed it. They thrilled the senators with their accounts of the A-7E's weapons accuracy and loiter capability, citing FAC praise of its Lam Son 719 air support as proof. Furthermore, their use of the A-7E's ground mapping radar for medium-altitude navigation (ten to twenty thousand feet) seemed to make the A-7 the night/all-weather CAS plane that the Army wanted the Air Force to buy.[16]

The panel followed this line of questioning two days later when it asked Air Force A-7 SPO staff members and tactical unit pilots about the plane. They confirmed the plane's weapons and loiter strengths, as well as the praise it received in Army CAS exercises (the Air Force had not yet sent its own A-7s to Vietnam). They also described their work with a FLIR apparatus that enabled them to navigate and deliver weapons from medium altitudes at night. However, they thought the radar, which could be used for dropping bombs in bad weather or night, was not accurate enough for CAS. Significantly, they told the panel that the need for attack capability at low altitudes in poor weather remained.[17]

With the hearings nearly complete, the panel not only seemed undecided about the fates of the A-X, Cheyenne, and Harrier, but now had virtually introduced a fourth candidate, the A-7. Beyond its fuel economy and weapons strengths, the A-7 was a fully developed aircraft, while the A-X and Cheyenne required much more work. The A-7's rising fortunes also marked the entry of more personal and commonly political considerations, as legislators jockeyed on behalf of pet causes, home state companies, and associated voters. Goldwater said he came to favor the plane after watching a performance demonstration, but he also admitted that one of the Air Force A-7 pilot witnesses was the son of a friend. A more obvious connection involved Sen. John Tower, whose Texas constituency included the A-7's manufacturer, LTV. The Texas congressional delegation had a reputation for securing defense contracts, and though Tower had not attended the proceedings so far, it was perhaps because he did not want to be closely associated with their increasingly pro-A-7 slant.[18]

These other political issues surfaced more openly on the last day of the hearings. Sen. William Proxmire appeared as a witness on behalf of the Members of Congress for Peace through Law (MCPL), a group that opposed defense spending and the U.S.–Soviet arms race. Proxmire represented MCPL apparently due to his public role in Congress's termination of the supersonic transport (SST) budget, but if he thought his charisma would solve the CAS plane question, he erred as much as the subcommittee members. The MCPL praised the A-X as the most economical choice while condemning the other designs, but its argument contained faults. For example, it dismissed the close organizational connection between attack helicopters and the Army units they supported. Proxmire's own testimony encountered problems. The Air Force wanted to increase its authorized tactical wings to accommodate the A-X, but Proxmire aimed to reduce the service's force structure *and* buy the plane. The reason, as revealed in arguments with Senators Cannon and Thurmond, was that CAS yielded tangible results compared to other missions, and the A-X was the most efficient CAS machine. CAS was important, but an effective air campaign also required air superiority to reduce air threats and interdiction to constrict enemy reinforcements. Proxmire's reasoning recalled the McNamara Whiz Kids' economic infatuation and occasional lack of good tactical sense.[19]

Proxmire also attacked the Texas congressional delegation. He recounted its military contract successes and asserted that Texas-based LTV needed more A-7 orders to retain employees. This inspired Senator Tower—who certainly appeared at this session—to demand further explanation. Proxmire retorted, "I think you probably know a great deal more about that than I do for a number of reasons. First, you are on this committee, and second, you are from Texas." Tower reminded Proxmire that Texas aviation companies had also lost recent contract competitions, and then asked, "But what I want to establish is whether or not you are accusing certain of us here on the committee of bringing political pressure to keep the [A-7] production line open?" At this, Proxmire backed away, answering, "Certainly not." Tower crossed up Proxmire in his own logic by pointing out that the Wisconsin senator favored other expensive weapons, such as submarines. As a parting shot, the subcommittee inserted into the hearings record not only the MCPL report but also two *Armed Forces Journal* articles charging Proxmire and MCPL with hypocrisy.[20]

Thus the hearings convened to resolve aircraft choices and overall CAS issues ended in indecision and acrimony. Interestingly, no one questioned the need for CAS, regardless of the difficulties involved in grasping its details. Every military leader, OSD official, and senator—including MCPL—wanted U.S. soldiers to receive the best possible CAS. With that premise established, it should have surprised no one that the Senate rejected none of the candidate machines—after all, each service believed that its plane best accomplished CAS as defined by its respective war-fighting doctrine. The senators seemed befuddled by the varying capabilities, mission definitions, and service agreements. A couple of them qualified their initial bluster about choosing only one machine and accepted at least two. Others took sides, which almost guaranteed the survival of all the CAS machines.

The ensuing April 1972 report reaffirmed the senators' desire for the best CAS possible and recommended further development and testing of all of the aircraft. There were differences, however, concerning how the development and testing should proceed. The senators liked the A-X competitive prototype development process enough that they wanted to see the winner, but reflecting their newfound interest in the A-7, they wanted the A-X fly-off victor to compete against the A-7 in a second flyoff. Perhaps reflecting an awareness that the Army would soon terminate the Cheyenne contract for good—which it did that August—the report's conclusions did not mention Cheyenne by name: "Assuming that questions regarding helicopter vulnerability are resolved successfully . . . there is a valid requirement for a more capable attack helicopter." As for the Harrier, the report recommended no more purchases pending operational evaluation. Finally, the report conceded that the mission and its planes were hard to analyze, and as if to underscore this, it contained addenda by dissenting senators, just as Packard's report had featured military dissent. Senator Symington objected to the A-7/A-X fly-off provisions. Sen. Harold Hughes, a subcommittee member who did not even attend the hearings, demanded that the "Army's right to provide close air support with helicopter gunships should be protected." Senator Goldwater asserted that the Air Force held the sole right to perform CAS. Senators Thurmond and Tower believed that all of the aircraft reflected valid service needs. (Tower did not push the A-7, but he did not need to.) With this rancorous sendoff, the Air Force CAS plane proceeded to the next trial.[21]

Competing CAS Plane Candidates

Meanwhile, Fairchild-Republic (henceforth called Fairchild) and Northrop continued building their A-X candidates, and by spring 1972 they had actually flown prototypes. The two planes were similar in many ways, especially in their survivability features. Both had redundant hydraulic flight control systems, blow-out panels, interchangeable parts (between the two sides of the plane, an uncommon feature for many aircraft at that time), protected fuel tanks, an armored cockpit "tub," and a backup (nonhydraulic) flight control system. Their cockpits were roomy and had large bubble canopies to allow unhindered visibility. They carried only basic radios and navigation systems. They were built to accommodate the 30-mm gun then under initial development, and used as many "off-the-shelf" items as possible. The companies were somewhat similar, in that both had experienced mixed fortunes working with the Air Force. Fairchild's last successful effort was the F-105, while Northrop had more recently built the F-5.[22]

Northrop's A-9 was like nearly all tactical jets in that its two engines were mounted within the fuselage. The engines and intakes were low to the ground because Northrop's designers wanted them at a height that allowed easy maintenance access. The A-9 weighed slightly less than its rival, but its engines, Lycoming F102 turbofans, produced nearly one-fourth less thrust per engine than the A-10's General Electric TF34s. In order to meet the government's specifications for loiter, short field performance, and maneuverability, Northrop had to offset the lack of thrust by extending the span of its thinner wings. The plane also had a selectable flight control subsystem that was supposed to let the pilot adjust his target run-in alignment with his rudders (normally, this technique induced unwanted rolling motion on many jets).[23]

The A-10 was an unconventional design. It had a twin tail, and its two engines sat separately atop the fuselage just aft of midway between nose and tail. The twin tail provided obvious control redundancy, and Fairchild engineers asserted that the engine setup prevented one engine's catastrophic failure from affecting either the other engine or other fuselage components. Though the engines sat high above ground, they required none of the maintenance access or extraction effort required for engines placed within the fuselage. They also had less chance of experiencing foreign object damage (FOD) from objects sucked up by the intakes or thrown up from the ground. Striving for maximum efficiency, Fairchild's designers rejected fully recessed landing gear bays, which increased aircraft weight and size, in favor of wing pods that only partially concealed the gear. In order to meet the RFP's CAS performance guidelines, the plane had a wing that was thicker than the A-9's. The wing sat low on the fuselage but angled upward toward its tips to provide better control response and stable handling. The A-10's wing and engine design allowed more weapons carriage pylons to be installed underneath its wing and fuselage. Unlike the A-9, the A-10 had no sophisticated autopilot system for weapons delivery.[24]

The flyoff occurred at the Air Force's high-desert test facility at Edwards Air Force Base, California, between October and December 1972. When the competition commenced, both aircraft had flown over 150 hours, which allowed both companies time to prepare. This also gave the Edwards test pilots time to familiarize themselves with the planes, for the same pilots were flying both in order to make a better comparative assessment. In the flyoff, each plane flew approximately 140 hours in profiles that tested not only weapons delivery suitability (the planes had to use a 20-mm cannon because the 30-mm gun was not ready), but also handling characteristics and cockpit features. Time and cost constraints did not allow using such tactically realistic items as simulated air defenses, but one close observer wrote that it was still a "very stringent" competitive test for planes at this stage of development. All observers believed that the flyoff achieved its aim of accentuating the competitors' differences. Although the planes did not fly against air defenses, one of the tests involved taking their component parts, such as engines, fuel cells, and wings, and firing Soviet 23-mm rounds at them in a wind tunnel. This "unprecedented" test action was the best simulation, short of actual combat, of how these parts would handle being hit by AAA in flight.[25]

The weapons results were nearly even, with the A-9 slightly more successful in 45-degree dive deliveries and the A-10 performing better in 15-degree dive deliveries. Test personnel cited the A-9's strengths, which included cockpit visibility, maintainability, and weapons accuracy. However, the A-9's rudder input adjustment feature encountered problems, and test pilots found that rudder forces were unacceptably high during manual (hydraulic systems out) flight. The A-10 shared the A-9's strong points, but had different deficiencies that later caused difficulties; for example, the engines performed poorly because of their position on the plane.[26]

The DSARC reviewed the results in January 1973 and chose the A-10. Secretary Seamans said that the A-10 won due to its ease of maintenance, ordnance-carrying capacity, simplicity, and more complete development. (These last two items were important to the A-X program's design-to-cost goals.) Dr. Foster added that the test pilots chose the A-10 as the plane they would prefer to fly in combat. Additionally, the A-10—or at least its parts—fared better in the survivability testing. By March, senior OSD officials had approved the DSARC's recommendation, with a stipulation that the plane meet its cost and performance goals. Specifically, they allowed further development but not full-scale production, and paid Fairchild enough money to build ten prototypes that would be used for developmental milestone tests. Deputy Secretary of Defense Packard highlighted the OSD's cost consciousness when he ordered the Air Force to "thoroughly review the design and eliminate any features not absolutely necessary for the accomplishment of the close air support mission."[27]

Some observers surmised that the A-10's flyoff victory was largely political. Two aviation history authors, Bill Gunston and Bill Sweetman, believed that the results were too even, and thus the determinant was which aircraft company could best stand contract rejection. New York–based Fairchild had not had a successful military contract in some time, and was still smarting from the recent SST cancellation (it had been the largest subcontractor on that project). Another rejection could possibly mean unemployment for many New Yorkers. California-based Northrop was in better financial shape. The Air Force allegedly accepted the A-10 because it did not want heat from the politically powerful New York delegation and because the government did not want to see an established military aircraft maker go out of business.[28]

Those closest to the event assert that the result was based upon the planes' relative merits. Robert Seamans later told this author, "I can assure you it was not political." He added that Sen. Lowell Weicker, in whose state the A-9's Lycoming engines were manufactured, accused the Air Force of bowing to political pressure from New York and Maryland (another state with a Fairchild facility). Seamans arranged a special briefing to show Weicker why the A-10 won. Weicker remained unconvinced, and told Seamans and the other attendees that the A-10 won on the "strength of Agnew and Rockefeller," or words to that effect. (Vice President Spiro Agnew was from Maryland and Nelson Rockefeller was governor of New York.) Seamans told the author that Weicker's behavior left him "apoplectic with rage," and that he immediately left the room. Otis Pike, a New York congressional representative who had chaired the 1965 CAS hearings, believed that New York and California were equally matched in political clout, perhaps even more than the respective planes. In such a case, the decision would fall back to their relative merits.[29]

The A-10 had certain unmistakable advantages. It more closely matched the concept designers' desire for widely separated engines. The engines were more powerful than those in the A-9, and were also a proven design. Service leaders stated that the A-10 was better designed to receive the 30-mm gun, and its low wing setup allowed comparatively ample space for weapons carriage. As one of Fairchild's design executives, Robert Sanator, put it: "We literally sat down and designed a plane around the gun that we had to have." Its more advanced state of development, as well as its greater simplicity and maintainability, were most important to government officials anxious about cost constraints. The test pilots also liked the A-10, and its survivability features stood out in realistic testing. Concerning survivability, TAC would later boast that the plane could lose one engine, half of one tail, two-thirds of one wing, and assorted parts of the fuselage—and still fly.[30]

The A-9's underwing engine placement restricted the amount of ordnance it could carry. Also, the closely set engines risked greater aircraft damage if one were hit or caught fire. The engines' intakes were close to the ground, increasing the likelihood of foreign object ingestion. The A-9's 30-mm gun bay did not allow easier access than that of the A-10. Overall, the plane was too complex, even in its simple prototype configuration, and labored under too many questions about its future development and associated costs.

Congress and Combat, Again

Still to take place was the contest between the A-X flyoff winner and the A-7, and Congress pointedly reminded Air Force leaders about it throughout 1973. They replied that the A-10 was not yet sufficiently developed to make a flyoff meaningful. Instead, they offered studies which, they said, proved that the A-10 was better than the A-7, especially in Europe. Since the Air Force seemed reluctant to comply, the legislators played rough. In late August they cut the service's A-10 prototype buy from ten planes to six and reduced the plane's development funding. Money would be released pending the results of the flyoff. Fearing outright cancellation of the A-10 program, the service began to put together a test plan.[31]

Congressional action was spurred by other items besides the Senate report demanding another flyoff. Budget concerns still drove overall skepticism, and Congress sanctioned a GAO study of the A-10. (The report, issued in March

1974, stated that the plane was not developed enough for costs to be accurately judged.) The Texas congressional delegation's interest in LTV's A-7 bred criticism of the A-10; Representative Mahon quipped: "It seems to present a beautiful target." Apparently, Texas's influence became overt enough that Fairchild president Ed Uhl complained about it in a press conference in late August 1973. "It's time for the New York delegation to get out and tell *our* story," he asserted (italics in the original quotation). One could dismiss Uhl's complaint as so much political noise, but some sources, at least one of them Texan, agreed. Brig. Gen. Thomas McMullen took over the A-10 SPO just after the A-9 fly-off (reflecting some increase in project status, since his predecessor was a colonel), and he later said that politically motivated pressure for a flyoff was evident. Texas Rep. Robert Price, who served on the House Armed Services Committee during this time, recollected that the Texas delegation believed the New York delegation was trying to push the A-10 without the A-7 getting its due. Also, to Price and others, there was a growing opinion that the A-10 was no longer suitable for the modern CAS arena.[32]

The legislators reached this conclusion by considering current wars, which included the fortunes of U.S. tactical air power and the A-7 during America's last year in Vietnam, 1972. North Vietnam's Easter Offensive invasion of South Vietnam that year featured a massive conventional assault and more sophisticated antiaircraft weapons systems. Radar-controlled SA-2 missiles threatened U.S. aircraft in Laos and just south of the demilitarized zone; and communist troops used the shoulder-launched, heat-seeking SA-7 throughout South Vietnam. Their action caught airmen off guard, and air losses mounted, especially among slower moving or less maneuverable aircraft, such as helicopters, the A-1 (now flown almost exclusively by South Vietnamese), the AC-130, and the OV-10. Adjusting altitudes and attack routing helped somewhat. Using decoy flares was another remedy, but the initial losses shook the airmen's complacency about tactical air support missions in the south. The invasion also gave the Air Force a chance to use its A-7D in combat. The plane did well, suffering minimal combat losses in missions over both North and South Vietnam, and impressing everyone with its range, weapons carriage, and accuracy.[33]

In the 1973 October War between Israel and its Arab neighbors, the Israeli Air Force (IAF) encountered serious difficulties in the war's early days while providing CAS to its beleaguered ground forces. IAF commanders disparaged

CAS, and IAF units did not practice it much. Because Israel was a small desert nation facing larger, hostile, but militarily second-rate neighbors, its leaders wanted to avoid lengthy, expensive wars by striking their enemies quickly. Such quick action naturally favored interdiction over close air support. Since the Israeli Army had whipped its Arab opponents in every war since Israel's embattled birth in 1948, CAS was almost irrelevant. For the IAF, the main goal was to seize air superiority, as it had in the 1967 Six-Day War, and then use interdiction strikes to decimate any rear-echelon forces that the Israeli Army might later encounter in its inevitable advance.[34]

The first few days of the October War featured a reversal of priorities. Diplomatic considerations and hubris created by past victories led the Israelis to forego a preemptive strike against a looming Arab attack. The war commenced with massive Arab ground assaults across the Suez Canal and toward the Golan Heights. The Israeli Army barely repelled them, especially on the Golan Front, and CAS became the IAF's most important mission as the soldiers frantically called for air support. The IAF was unready for CAS and also for the Arab army units' air defenses. These included the latest Soviet-made mobile SAMs, such as the SA-6, and mobile radar-guided AAA, such as the ZSU-23. The Israelis had neglected upgrades to their electronic countermeasures (ECMs), apparently due to faith in their own tactics and contempt for the Arabs' technological competency. When Israeli jets flew CAS on the Golan Front, they faced one hundred thousand Arab soldiers with fifteen hundred tanks, over four hundred antiaircraft guns, and at least one hundred SAM batteries. This force occupied a front twenty-five miles wide, which a jet going 450 knots could traverse in about three minutes. Avoiding metal aimed at one's jet is difficult in such close quarters against a densely packed defense, especially when one is not well prepared for it. One Israeli observer said that when the first Israeli jets attacked the Arabs' Golan juggernaut, "simultaneously we saw over fifty ground-to-air missiles in the air at one time. Over fifty on a very, very narrow strip of land." In the first few days, the Israelis lost fifty planes—half of the estimated total Israeli air losses for the two-week war.[35]

The Israelis recovered, taking advantage of Arab mistakes and their own successful, combined air-ground efforts against Arab SAM sites. War's end witnessed Israeli ground forces seizing more Arab territory with their trademark slashing attacks and the IAF crushing Arab ground reinforcements with an effective interdiction campaign. Even so, some ill feeling persisted during and after the war about the IAF's CAS effectiveness. Occasionally, its underdeveloped CAS command-and-control setup led ground units to complain about a lack of air support. Arab defenses made the IAF pilots rush their attacks or release ordnance at longer distances, thus reducing accuracy. Though the IAF played a key role in blunting Arab assaults on both fronts, some observers asserted that its effectiveness owed more to its shock effect against Arab troops than to actual destruction of Arab tanks. However, Israeli leaders dismissed any criticism by stating that CAS would remain a backup mission because of its high cost and relatively low return.

The October War inspired a wealth of commentary about tactical aviation's future. This obviously had an impact upon the A-10's fortunes, as Congress pressed Air Force leaders about the war's lessons as they pertained to the A-10's survivability. After all, it was the Israeli Air Force, the internationally acclaimed master of aerial battle, which had encountered these problems. In March 1974 hearings, Senator Cannon asked Air Force witnesses what they thought of the Israelis' postwar opinion that any plane that flew CAS required high speed, excess thrust to sustain hard maneuvering, and the best antiaircraft countermeasures equipment. The witnesses answered that the American-built A-4s and F-4s that Israel used for ground attacks did not possess the A-10's survivability features. They repeated the CAS adage that too much speed hindered the ability to acquire and attack targets, and they reminded the senators that although the Israelis did not like CAS, they still had to fly it when their army was nearly overrun on the Golan Heights. The witnesses could also have emphasized that A-4s and F-4s were fast jets, and that their speed had not helped them much over the Golan Heights and Suez Canal. Senator Thurmond asked if the A-10 could have survived in such an environment, and why the service needed such a plane instead of the combat-proven A-7. The witnesses replied that the U.S. Air Force's more varied arsenal and capabilities would have provided better defense suppression—in other words, the United States was not Israel.[36]

Preparations for the A-7/A-10 Flyoff

Amid rising skepticism, Air Force leaders expressed confidence that the A-10 would win the impending flyoff. They deemphasized the consequences of the event, because if the A-7 lost it would continue service in the Air National Guard (ANG) and would remain part of the service's wartime force structure, complementing the A-10. The OSD made TAC the lead command for conducting the test, with two OSD agencies, DDR&E and WSEG, providing guidance for planning and evaluation. Given TAC's historical opinion of slow-speed planes, one might expect the test to conclude that neither plane passed scrutiny—and that the CAS plane should be the F-4. At the least, one might think that it would favor the faster and more avionics-laden A-7. For that matter, one could also have expected the Army to echo such choices, given the A-10's perceived threat to its armed helicopter plans.[37]

However, opinions were changing, at least at the top. Ever the good soldier, General Momyer told subordinates who opposed the A-10 that they would support it. Momyer, Air Force Chief of Staff George Brown, and Army leaders improved interservice cooperation. When Momyer retired in October 1973, his successor, Gen. Robert Dixon, resolutely followed the new course. Indeed, Dixon asserted to an interviewer that "we would crash airplanes into the opposing troops to provide close air support, if those were the required circumstances. If you think that is a joke, it ain't a joke to me at all. That may be one of the reasons why the Army and I understand each other." These TAC commanders and the U.S. Army Training and Doctrine Command (TRADOC) commander, Gen. William DePuy, pursued a training and operational dialogue that resulted in the formation, in July 1975, of the Air-Land Forces Application Directorate (ALFA) at Langley AFB, Virginia.[38]

Different factors drove the shift toward interservice cooperation and the Army's new appreciation of a CAS plane. For one thing, more commanders in both services either had joint-service staff backgrounds, such as Dixon, or recalled the excellent air support achieved in Vietnam, such as Army Chief of Staff Creighton Abrams. For another, they did not want to repeat the A-X/Cheyenne competition, whose intensity had nearly driven Congress to dictate unsatisfactory terms to both services. A final reason was that with the United States ending its Vietnam involvement, attention had shifted back to Europe and the Soviets' lopsided numerical superiority, especially in tanks. To the soldiers, the October War proved that modern firepower could overcome the disadvantage. High-technology weapons inflicted high ground and air losses on both sides in that conflict, and the Israelis' modern weapons dexterity was a big factor in their narrow victory over the Arabs' Soviet-style onslaught. With this in mind, TRADOC Commander DePuy started revising Army doctrine to emphasize concentrating fires early at the point of attack, because "the first battle of our next war could well be its last battle." Although the Army still had Cobras after the Cheyenne's loss, these aircraft lacked the desired attack helicopter performance and weapons capabilities—and the Cheyenne replacement process would not produce an operational aircraft until 1985. The soldiers now needed the A-10's dedicated air support firepower; and as the flyoff approached, the Army publicly chose the custommade, durable, and long-loitering tank killer.[39]

Defense Secretary James Schlesinger sympathized with those who wanted a strong tactical Air Force to fight the Soviets. Like the 1960s Whiz Kids, he wanted the Air Force to have a mix of expensive and inexpensive planes, which would guarantee a larger tactical force containing an appropriate number of the best machines. In March 1974 he "struck a deal" with General Brown about force structure. In return for extra tactical air units, he secured from Brown the promise to continue supporting the A-10 and the new low-cost, lightweight fighter program that yielded the F-16. When Brown left that July to chair the Joint Chiefs of Staff, his successor, Gen. David Jones, increased Air Force support for the arrangement.[40]

Meanwhile, the planners worked on a test that would address all the concerns and desires expressed by the Army, Congress, and the Air Force itself. Its objectives were to determine (a) the pilots' ability to acquire and attack targets when flying each plane; (b) each plane's ability to evade modern battlefield air defenses; and (c) each pilot's subjective assessment of the two planes. The last aim answered Senator Cannon's mandate that the test's main goal should be "to take experienced combat pilots and let them fly both airplanes, the A-10 and the A-7D, and then make a judgment as to which airplane they would rather fly in combat."[41]

The flyoff would occur at the Army's Fort Riley, Kansas, maneuver area against armored vehicle arrays that represented either a Soviet battalion-sized attack or a breakthrough situation. (Fort Riley's rolling, wooded terrain matched European conditions.) As the planes attacked the target array, Army air defenses would try to track them with weapons that simulated the most lethal—or at least the most notorious—of those encountered in the

October War. The Hawk missile stood in for the SA-6 (though it was more effective than the Soviet missile); the mobile, radar-guided, antiaircraft Vulcan Gatling gun simulated the ZSU-23; and the shoulder-launched, heat-seeking Redeye served as the SA-7's surrogate. The attack profiles addressed European conditions by asking the test pilots to adjust their maneuvers for four simulated weather situations of decreasing quality: (a) clear skies and unlimited visibility; (b) five-thousand-foot cloud ceiling and five miles visibility; (c) three thousand foot ceiling and three miles visibility; and (d) one-thousand-foot ceiling and two miles visibility. Four Air Force fighter pilots with Vietnam CAS experience in either F-4s or F-100s would fly both competitor planes and make subjective judgments. Interestingly, none had flown the A-1, and though the Air Force's European weather–oriented test conditions favored the A-10, one could also say that the pilots' background somewhat favored the faster A-7.

The Question of Comparative Weapons Effectiveness

Problems arose over one flight test item, comparative weapons strengths, because the 30-mm antitank gun that defined the A-10 had experienced a few development adventures of its own. Following the same procurement process as for the A-X, gun project leaders by June 1971 had announced two competitors for evaluation: Philco-Ford Aeronautics and General Electric (GE). Both companies submitted Gatling gun designs as specified in DCP 103. Into this process came a proposal from DDR&E—apparently influenced by Pierre Sprey—to substitute Switzerland's Oerlikon 304RK 30-mm gun for the Gatling gun designs. The Oerlikon had been rejected earlier due to its low reliability and rate of fire. However, its twin-gun configuration delivered that rate of fire instantaneously, while the Gatlings' rotating barrel clusters required a few seconds to spin up to their maximum rate. The designers had envisioned a two-second burst, and Sprey also knew that the most accurate tracking occurred within this time. Thus, it seemed the Oerlikon guaranteed maximum rounds on target per pass. Those responsible for the gun project did not mind competition, but the introduction of a rejected candidate after they had chosen two other competitors was both an embarrassment and a hindrance.[42]

In June 1973 the gun project's DSARC formally announced that GE's candidate had won. Though some Eglin Armaments Lab members apparently favored Philco-Ford's model as a more advanced and lightweight design, it was not well developed, lacked parts, and performed poorly. Competition rules required the guns to fire seventy thousand rounds. GE's gun fired its full allotment, while the Philco-Ford model jammed repeatedly and finally fired only fifteen thousand rounds. A separate test firing of the Oerlikon revealed similar problems; it repeatedly jammed after firing an average of 842 rounds.[43]

GE's winning gun, designated the GAU-8/A, still required substantial development before it would be a fully functional part of the A-10. Contracting negotiations for and construction of low-cost, effective 30-mm rounds was a looming challenge. Further, the GAU-8/A required flight testing aboard its host aircraft. Initial trials in late 1973 revealed no big problems, but more extensive firings in early 1974 featured a continuous secondary ignition of gun gas in front of the plane's nose during firings. The phenomenon occurred due to the propellant's chemical makeup and to the escape of some of this material through imperfectly shaped shell rings as they exited the barrel. Though the gun was already considered too undeveloped to be used in the actual flyoff, the problems encountered guaranteed that it would not be used.[44]

The Flyoff

The test appeared to satisfy a neophyte's view of a fair, head-to-head, competition. Certainly, it was a "test," as recommended by the Senate and GAO, and the Air Force avoided any action that would publicly taint the findings. However, several restrictions reduced the significance of the results. Besides the gun, none of the A-10's other weapons delivery gear would be available (though the pilots got to fire a gun installed on an Edwards test plane). Thus, all weapons results for both planes relied upon postulated hits derived from recorded flight parameters and the various weapons' performance data. Attacks on the target arrays would be flown by one aircraft, thus negating any tactical advantages derived from multiplane attacks. Obviously, air defense hit claims would have to be simulations derived from tracking data, but the planes would not carry countermeasures gear against the defenses. Given the heavy reliance upon postulations, Senator Cannon's demand for the pilots' subjective assessments now assumed even greater importance.[45]

TAC Commander Dixon knew that the pilots' opinions were critical, but he was also aware of the hard congressional scrutiny directed at his service and the test. Just before the test commenced, Dixon informed the pilots that he would not direct their responses. As he later recalled, one of the pilots expressed concern that the Air Force seemed to want the A-10, and that they had better choose it as well. General Dixon said he answered:

> No, no. You didn't listen, and you don't believe me, but I am going to say it again, and you had better believe me. You are going to do this, and you are going to say what you have to say. That is an order. . . . I assure you if you don't, you are going to suffer the consequences and so am I, because I have assured the Chief of Staff you will, and I have assured Senator Cannon you will. I'm not going to tell you what to say, and if you are waiting for me to tell you, you are going to wait a long . . . time, boys.

To ensure that their operational judgment would remain untainted, the pilots could not talk to each other during the test, and their base location was isolated.[46]

The test commenced in mid-April and ended in mid-May. The missions featured single-plane attacks against the defended target arrays, while the pilots and other observers rated cockpit setup, handling characteristics, weapons delivery, and tactical performance. The results revealed that in simulated free-fall bomb deliveries, both planes did equally well; the A-10's closer employment ranges offset the A-7's weapons avionics advantage. When conducting mock Maverick missile attacks, both planes again did equally well, with the test confirming to the Air Force that the Maverick would potentially make a good standoff weapon. Regarding postulated gun attacks, the test authorities judged the A-10's GAU-8/A a more effective weapon due to its much greater hitting power than the A-7's 20-mm gun. Concerning battlefield threats, the Air Force had derived information from the Israelis' SA-6 war prizes, and thus learned that missile's capabilities. The Hawk missile crews had no problem tracking the two planes, but SA-6 postulations indicated that both planes would defeat the missile if they maneuvered and used countermeasures. Similar testing based upon exploitation revealed that the SA-7 could also be defeated by countermeasures. Good tactics and self-protection devices would help against the ZSU-23 as well, though the A-10 was rated as more likely to be hit due to its slower speed and lower attack altitudes. However, since it was a far more rugged plane, and more easily repaired, the evaluators still considered it a more survivable machine against this weapon.[47]

These results gave the A-10 a slight edge, and the pilot assessments confirmed its victory. The questionnaire that they completed focused upon the planes' ability to handle the various simulated weather conditions. In clear weather with unlimited visibility, the A-7 was the flyoff pilots' unanimous choice. But as the weather scenario reached three-thousand-foot cloud ceiling and three miles visibility, one pilot changed his mind and another rated both planes equal. For the one thousand feet ceiling and two miles visibility scenario, the pilots' unanimous choice became the A-10. They liked the A-7's smoother handling and navigation/weapons delivery avionics (though most conceded that the A-10 was still an undeveloped prototype that could improve), but found that the A-10 better handled degraded weather. Design payoffs also appeared in other characteristics ratings; for example, most of the pilots liked the A-10's cockpit visibility and believed that its maneuverability enabled them to acquire and attack targets quicker.[48]

In June 1974 Air Force leaders brought the flyoff pilots with them to brief the House Armed Services Committee. Those closely associated with the flyoff say that the pilots' testimony was the unofficial flyoff winner announcement. The legislators seemed generally satisfied, though one Texas representative wondered why the A-7's combat record did not count toward the selection. Another member wanted to know what the Air Force thought of the Enforcer, a World War II P-51 fighter derivative that Piper Aircraft was trying to sell as a counter-insurgency and CAS plane. The legislator alleged that it "could do everything these planes could do, only more cheaply," and thus previewed Congress's use of the Enforcer to goad the Air Force when the A-10 encountered development problems. Chairman F. Edward Hebert still seemed interested in the comparative costs of the A-7 and A-10. Overall, the lawmakers remained skeptical of the dedicated CAS plane's worth.[49]

In the meantime, the A-X had become the A-10, after undergoing review by the OSD and GAO, Senate committee hearings, a flyoff between two A-X candidates, and a flyoff competition against the A-7. The agencies and Congress were neither able nor willing to assign the CAS mission to only one service or to select only one plane to fly it. As the Packard report revealed,

U.S. foreign policy created multiple war-fighting scenarios that supported the individual services' unique CAS doctrines and aircraft. As for the aircraft itself, aviation technology had diversified enough that planners could customize machines to be unbeatable in their specific mission. Further, with such a complicated mix of factors, the flyoffs could be arranged to yield a specified result.

The A-7/A-10 flyoff was oriented toward operations in the often poor European weather against armored units, a condition for which the A-10 was designed. Its conclusion ended the overt, direct-competition contests intended to determine the Air Force CAS plane's fate, but it would not end questions about whether the plane should be built. Upcoming and less overt trials would require that the A-10 SPO prove that the plane, its manufacturer, and the Air Force could meet both cost and effectiveness goals. Meeting these challenges would allow the plane to enter the operational inventory as the first dedicated CAS plane in the history of the independent Air Force.

7   DEVELOPMENTAL AND OPERATIONAL CHALLENGES

The A-7 flyoff should have been the final major challenge for the A-10, but a variety of external and internal factors enlivened the last two years before its operational debut. Several legislators continued attacking the plane. Some, such as Rep. Jim Lloyd, doubted its suitability for modern combat. Others, like Rep. George Mahon, resented its defeating a favored plane in flyoff competition. And some, including Rep. Les Aspin and influential congressional staffer Anthony Battista, wanted military procurement cuts and considered the A-10 program, with its apparently escalating costs, a good candidate for elimination.[1] As early as August 1974, a House Armed Services subcommittee requested testimony on behalf of the Piper Enforcer, an even simpler and slower plane than the A-10. The Enforcer's creator was politically influential newspaper publisher David Lindsay, and except for wingtip tanks and a nose elongation due to its turboprop engine, his plane resembled a P-51. In 1970 the Air Force had evaluated it as a candidate in its counterinsurgency and light attack plane project, called PAVE COIN. The service liked the Enforcer, but finally rejected it and other planes in favor of the OV-10 and A-37. Since these two planes had been proven in combat, they had demonstrated their aptitude for PAVE COIN requirements.[2]

In the August congressional hearing, Lindsay claimed that he did not want his plane to compete with the A-10. This was not true. Not only did he frequently attack the A-10, but an *Aviation Week & Space Technology* article cited his contempt for it as well as his desire for an A-10/Enforcer flyoff. Many members of Congress abetted him because of their displeasure with the A-10 program. Before Lindsay gave his briefing, subcommittee Chairman Melvin Price submitted letters from other lawmakers—including one from the full committee Chairman, F. Edward Hebert—criticizing the A-10 for cost overruns and suggesting the Enforcer as a cheaper alternative. The members' leading questions, as well as his own apparent desire for a government contract, led Lindsay to shed his hesitation about publicly endorsing the Enforcer over the A-10.[3]

Indeed, Lindsay became even less reluctant as the representatives invited him and Piper Aircraft President J. Lynn Helms to present their plane's case in 1975, 1977, and 1978. The 1977 hearings alone took up 125 pages of testimony and featured an orchestrated debate over the two planes' relative merits between Lindsay and Helms on one side, and Air Force generals on the other. Beyond the doubts about the A-10's economic or tactical viability were more parochial concerns, such as Sen. Strom Thurmond's political connections with Lindsay. The Enforcer had some qualities that apparently justified public advocacy. It was a cheaper plane. It had exceptional loiter performance and handled better on unpaved runways. It carried a remarkable amount of ordnance for its size, and Lindsay and Helms made many promises about the weapons it could be modified to carry. Thus, some members of Congress considered offering the Enforcer to the Army if the Air Force refused, but the Army backed away, apparently fearing another roles-and-missions fight.[4]

However, if Lindsay and Helms thought that congressional interest and attacks on the A-10 would obtain an Enforcer contract, they were wrong. Neither Congress nor the Air Force ever seriously pursued buying it. The Enforcer might have been good for small-scale, counter-insurgency wars, but it could not compete with the A-10 for CAS prowess in all situations. Air Force congressional witnesses observed that the Enforcer's maximum ordnance load was only a fraction of the A-10's; further studies revealed that ordnance carriage incurred serious performance penalties. Even without any load, the plane was slightly slower than the "slow" A-10. Lindsay and Helms claimed that the Enforcer's small size made it less vulnerable to air defenses, especially radar weapons, but the Air Force countered that its propeller was a natural radar reflector. Air Force researchers had disregarded the P-51's battle damage difficulties in past wars, but still found the Enforcer highly vulnerable. It lacked backup features for flight controls and other systems, and its ceramic armor was not as strong as the A-10's titanium shields. The PAVE COIN evaluation had identified as many developmental problems with the Enforcer as currently vexed the A-10—yet despite two flyoff successes, the A-10's problems still spurred hearings examining another potential competitor, the Enforcer. Air Force witnesses pithily observed that even if they wanted the Enforcer, it might lose to yet another plane in procurement competition.[5]

The A-10 would beat this hurdle, but one had to conclude, as two observers close to the issue did, that the lawmakers did with the Enforcer as its name implied and pushed the Air Force to guarantee the A-10's planned performance and cost. They also created a lot more work than was perhaps necessary for all involved those in this debate. Many of the congressional members' motives were fair neither to true circumstances nor to the players involved. Some played a simplistic unit cost comparison game, while ignoring the further development work required on a machine already lacking the A-10's strengths. Senator Thurmond pushed a political favorite's plane against plain evidence that it would be rejected. Even so, he, Lindsay, and Helms secured an Air Force flight test for the Enforcer in 1984. Again, the service concluded that though the Enforcer was a good machine, it was not as good as the A-10, and there were already enough counter-insurgency planes in the inventory.[6]

A Challenge from Within

The A-10 was nowhere near as expensive as other Air Force planes in procurement in the 1970s, such as the B-1 or F-15, whose respective per-unit costs were something over $60 million and $15 million. Even so, questions about its cost made it vulnerable to initiatives by such people as Lindsay,

Thurmond, and Lloyd. Rep. Robert Giaimo complained in a 1976 hearing, "I remember when this plane was first sold to us . . . one of the real key things about it was the reasonableness of the plane. [Now] It is already up to $3.6 million." This was the rub; depending upon how one did the accounting, A-10 per-unit expense either hovered around the design-to-cost $1.7 million goal or, like other current designs, had encountered serious overruns. The design-to-cost estimate accounted for inflation and other outside influences, but estimates based upon later costs put the A-10 at Giaimo's $3.6 million, and even over $4 million. (Lindsay and Helms used the inflated figures in their congressional testimony.) Determining and then later defending the design-to-cost figure was tough, given political and defense trends, inflation estimates, evolving demands, and varying construction costs. Facing tough congressional questioning about the cost estimate, U.S. Air Force Gen. Alton Slay conceded that "in 1970, we didn't know too much about how to do design-to-cost analysis."[7]

The Air Force stuck to its assertion that if the plane was over its cost goal, the excess amount was slight. Nonetheless, the service could feel the heat. After the A-7 flyoff, Deputy Defense Secretary William Clements approved the DSARC's recommendations for a limited buy of fifty-two planes, but he stipulated that the A-10 project had to conduct various flight handling tests and prove the worth of its GAU-8/A gun. The aircraft handling issues resolved themselves well enough, but proving the gun's performance became the critical factor in the project's survival. This took time, and while gun development continued, the service attended to a more immediate concern, Fairchild's performance.[8]

In summer 1974, with Congress upset over the A-10's cost and David Lindsay hawking the Enforcer, the service scrutinized difficulties at Fairchild. The biggest problem was that the company had not undertaken an aircraft project of this size in some time. Among other corrective actions, the Air Force wanted to move at least some A-10 construction from Long Island, New York, to Fairchild's Hagerstown, Maryland, plant, which the service believed had better production facilities. When Fairchild rejected Air Force queries on this matter, the service ordered an investigation.[9]

Air Force investigators examined the A-10 contractual setup throughout late 1974, and harshly judged Fairchild. Their report criticized company managers for lack of experience with large-scale projects. Failing to oversee operations well enough, they created confusion about lower level responsibility and lost control of the pricing and delivery of subcontractors' goods. Further, the company's machinery and skilled work force were aging. It faced problems with equipment reliability and, as the workers retired, with work force competence.[10]

The Air Force action shook Fairchild throughout late 1974 and early 1975, as the company changed top-level managers and streamlined its channels of responsibility and communications. More manufacturing went to Hagerstown, but not as much as the Air Force had desired. Fearing the loss of constituents' jobs, the New York congressional delegation pressed the Air Force to reexamine its findings, which yielded a conclusion that the move would not be as profitable as had been thought. There were shake-ups within the Air Force as well. The service improved coordination between its A-10 project office and its Fairchild plant representative's office. Brigadier general–select Jay Brill replaced Brigadier General McMullen as head of the A-10 SPO (though this was not considered a punitive action for McMullen). Another transfer to the SPO was a hard-charging fighter pilot, Col. Robert Dilger, who oversaw A-10 armaments issues. Dilger already knew the territory, thanks to his previous job handling gun development projects at Air Force Headquarters. Others already serving at the SPO had important roles in ensuring the GAU-8A's success. Col. Samuel Kishline, the A-10 SPO deputy director, was one of these, and Lt. Col. Ronald Yates, the director of A-10 development test, was another—but Dilger would be the standout.[11]

Interestingly, one report finding criticized the flyoffs and budget constraints. Both the report and Air Force congressional testimony pointed out that congressionally directed flyoffs cost Fairchild time and money for development. In one hearing, Air Force Secretary John McLucas asserted, "The A-10 is the first aircraft to which we applied the design-to-cost concept. We set a limit . . . if people think of wonderful things to do to improve it, we are not interested." Further, these sources and later studies—including the GAO—criticized the service's design-to-cost goals for inducing performance penalties not only upon Fairchild management practices but also upon the plane itself. The A-10 that entered the operational inventory lacked various devices deemed necessary for modern air support. However, because the

A-10 was not a high priority program like the B-1 or F-15, the service considered these shortcomings worthwhile for the time being.[12]

Determining Gun Capability

Clements and the DSARC wanted the GAU 8/A gun to demonstrate reliability, effectiveness, and economy because it not only constituted much of the A-10's attack capability but was also a major part of the A-10 itself. Its complete barrel, engine, and magazine apparatus was bigger than a Volkswagen Beetle car. It was also one reason for the A-10's weight increase and corresponding flight performance decrease that roused congressional criticism.[13]

From late 1974 through 1975, the Air Force and General Electric demonstrated the gun's basic reliability via a fifty-thousand-round test of a GAU-8/A in various environmental conditions, including salt fog, dust, humidity, and extremes of temperature. The service also fired GAU-8As using various types of rounds. In all cases, the gun performed well. However, reliability problems surfaced elsewhere in February 1974, as the initial ammunition lot generated secondary gun gas ignition during aerial firings. This phenomenon occurred in front of the plane's nose and disrupted airflow to the engines, blinded the pilot during night firings, and even scorched the cockpit windscreen on one test flight. The Eglin Lab wanted extensive studies, but the program had neither the time, the money, nor the powerful backing to do this. As a research organization more accustomed to big budgets and different deadlines, Eglin seemed oblivious to the increased costs and development difficulties affecting the A-10 program. Thus, A-10 SPO staffers worked with or around—and, according to some accounts, ran over—Eglin to remix the propellant and create a shell that yielded no combustible gases outside the barrel but retained the required high muzzle velocity. To make certain that secondary ignition did not happen, the A-10 SPO replaced copper shell rings with plastic ones. This not only prevented propellant gas from escaping, but also reduced barrel wear and provided a cheaper source of shell rings. Though there would be occasional problems with bad lots of ammunition in later years, the gun gas fix was tested and found sure enough that Systems Command declared the problem solved in November 1974.[14]

As the Eglin Lab's actions during the gun gas episode revealed, gun-related challenges arose even from agencies that should have helped the weapon. As part of the DoD coordinating group overseeing the GAU-8/A's progress, the Army's Ballistic Research Laboratory (BRL) questioned whether a 30-mm gun really could destroy a tank. After all, if tanks used guns of over 100-mm barrel bore displacement to destroy other tanks, what could a gun of one-fourth that displacement do? This sensible skepticism was accompanied by more parochial motives, such as fear of how a successful 30-mm aerial gun might affect Army tank procurement efforts. Also, A-10 SPO officers wanted to shoot real tanks in gun tests because computer models could not realistically account for the synergistic effect of dozens of rounds hitting a tank loaded with fuel and ammunition. Conversely, the labs wanted controlled firings of single rounds against steel plates. Most weapons evaluations of that time were either pure computer studies or controlled, small-scale tests whose results were extrapolated via computer modeling to yield larger conclusions. This was due to an institutionalized faith in computer analysis and the not-unreasonable fear of runaway test costs. Computer analysis was more vulnerable to manipulation, however, due to the lack of actual experience—and the BRL had a motive for downgrading the GAU-8/A.[15]

The fight over determining the gun's abilities intensified an already existing struggle over ammunition. Since the GAU-8/A was relatively small compared to other antitank cannon, it required the best antitank rounds. Early during concept development, the government determined that depleted uranium contained the desired mass and ability to burn metal upon impact (called "pyrophoric effect"), but outside players still interfered. In 1971 the Packard study insisted upon tungsten and steel rounds. The Air Force tested the various candidate rounds, and in July 1973 repeated its assertion that depleted uranium was the best for the armor-piercing incendiary function. The service also had to show that depleted uranium would not produce harmful radioactive side effects; both Dilger and Kishline recalled several congressional inquiries on this matter. Perhaps to ensure an unbiased judgment, the OSD handled the study, and in April 1974 it reported that the material had "no significant medical and environmental impact." The other issue with this material was ease and cost of production—a problem solved thanks to the initiative of SPO staff. They discovered that Ohio-based Battelle Laboratories could mass-produce depleted uranium shell cores cheaply, and was also willing to support ammunition manufacturers.[16]

The SPO won its battle to shoot tanks with the gun. Against lab resistance, and knowing that the A-10's survival might depend upon it, Kishline and Dilger set up a live-fire demonstration against tanks at the Nellis Air Force Base test ranges in Nevada. It featured A-10s using the special armor-piercing 30-mm rounds in close-range dive attacks against tanks carrying a normal fuel and ammunition load. The results were spectacularly successful, and Dilger, who was the live-fire project manager, made sure that they were filmed. Firing bursts set the tanks on fire and in some cases blew them up. At an autumn 1974 DSARC meeting, Kishline impressed the attendees with these films and emphatically ended any development phase questions about GAU-8/A lethality. In autumn 1975 live firings against T-62 tanks captured by the Israelis in the October War punctuated the gun's success.[17]

The plastic shell rings and the Battelle contract were two gun economy measures, but there were other, more significant actions as well. GE had used Aerojet for ammunition supply, but in June 1973 DDR&E approved full-scale ammunition development using a dual-source contract with Honeywell as the second company. Aerojet contacted congressional allies in early 1974 for help in remaining the sole contractor, but the A-10 SPO successfully insisted upon the new policy. Competition among vendors helped reduce the per-round cost from nearly eighty dollars to around twenty, and when one considers the total amount of ammunition required for both practice and combat, this resulted in spectacular savings and helped assuage congressional budget complaints. The A-10 SPO staff found that aluminum shell casings worked as well as steel or brass, which saved weight as well as money. The SPO also suppressed various expensive and risky shell improvement initiatives by the Eglin lab. Finally, Colonel Dilger spearheaded creation of an automatic loading machine that reduced the time and cost of routine reloadings.[18]

Resolving gun issues was the last major hurdle before the OSD and the Air Force allowed Fairchild to commence full production of operational A-10s. Upon the DSARC's Milestone IIIB recommendation in early 1976, the OSD approved a total buy of six hundred to seven hundred aircraft—though Congress would approve spending on a year-by-year basis. This meant that the plane still had to prove itself, as the later Enforcer hearings attested. There was resistance within the service as well; in 1975 A-10 defenders allegedly spoiled an attempt to place responsibility for the program in some bureaucratic backwater.[19]

This brings up the issue of individual efforts on behalf of the plane. In later years—especially in the 1980s when the military reform movement decried the inverse evils of increased weapons costs and allegedly decreased military quality—reformist writers cited the A-10 and its project officers as paragons of common sense and initiative. Specifically, they presented Bob Dilger as an aggressive innovator singlehandedly challenging an incompetent defense management. Certainly, Dilger was aggressive and innovative; all accounts agree on that point. From his service while at the Pentagon, through the development stage just recounted, and onward to later actions, he substantially contributed to the A-10 gun's success. However, he was not the only individual involved with some of the decisive actions mentioned. Official accounts and personal interviews reveal that others, such as Al Casey, Bob Buchta, Tom Christie, John Foster, Sam Kishline, Thomas McMullen, and Ron Yates, played big roles in guaranteeing an operational A-10.[20]

On 30 March 1976 General Dixon presided over the Langley Air Force Base ceremony that formally welcomed the A-10 into the Air Force operational inventory. The plane was the product of nearly ten years of development work, competition, and refinement. Obstacles lingered, and questions remained about whether it had met its cost goal—though even if one accepted the unflattering estimates, it had not grossly exceeded the limit. The A-10 now had to prove itself outside the developmental world as well.

The Warthog

The Air Force later christened the A-10 the Thunderbolt II, in honor of Republic Aviation's rugged P-47 Thunderbolt, which had provided such good ground support in World War II. The new craft quickly acquired a nickname that stuck: both supporters and detractors among the service's pilots called the plane the "Warthog" (or "Hog") for its ugly appearance compared to other tactical planes. Indeed, one late-1970s commander of a new A-10 wing recalled his wife's reaction upon first seeing the Warthog. She said that she was always confused about identifying other jets he had flown, but she would never have any problem remembering this one. Before too long, jokes about the Warthog's relative lack of speed and technological sophistication appeared:

> What is the greatest threat to the A-10 during low-level flight? Bird strikes from the rear.
> Two tanks are driving down the road, and the commander of one radios the other, "A-10s are attacking us from behind! Step on it, and we'll outrun 'em!"
> What is the speed indicator in an A-10? A calendar.[21]

The acceptance issue was an important one, for the plane had enemies within the Air Force. Gen. John Vogt, at the time commander of the U.S. Air Forces in Europe (USAFE), stated that the plane would not long survive European air combat. Nellis Air Force Base in Nevada was the service's premier fighter base, home of operational fighter testing, Fighter Weapons School (FWS), and the newly created "Red Flag" tactical combat exercise. The Nellis non-Warthog pilots' attitude toward the A-10 was and remains cool, if not contemptuous. Also, one aviation weekly noted that "ninety-five percent of the pilots transitioning to the A-10 do not want to come because of the fighter pilot's traditional value on speed and looks."[22]

However, for the men who flew it, the plane inspired fierce loyalty. The same journal that commented upon pilots' unenthusiastic reactions to an A-10 assignment also noted their dramatic conversion after flying the aircraft. Indeed, many of the pilots who later served as A-10 unit commanders during and after Desert Storm had chosen the A-10 for their first assignment. Calling themselves "Hog drivers," A-10 pilots embraced the Warthog nickname and laughed off the derision. A-10 flight instructors teased pilots transitioning from other jets: "Don't worry, we'll let you put a sack over that mother before you get in and fly it." Addressing the Hog drivers' enthusiasm, one Air Force general asserted that pilots naturally develop a loyalty to their specific machine. This was often true, but there were other reasons for the Warthog pilots' attitudes. The Hog was easy to fly, and its maneuverability not only helped during CAS but also made it hard for ground gunners and fast jet fighters to track the plane successfully. Its ruggedness generated confidence in its survivability. Its simplicity and easy maintainability meant that it flew a lot—something quite attractive to pilots. Simplicity generated affection in another way. To pilots turned off by new jets' emphasis upon complex combinations of autopilots, radar, and computers, the Hog was a return to pure tactical flying. The A-10's primary weapon was not a functionally temperamental radar missile, but a simple gun whose destructive impact bred further confidence. The challenge involved in successfully accomplishing a CAS mission motivated pilots, as they had to evade defenses, maintain formation integrity, and then find and hit fleeting targets.[23]

Like the trench-strafers of World War I, Hog drivers found that helping one's own ground troops bred commitment, since it was a worthy task that yielded direct, tangible results. One Army historian noted that this was perhaps the most important part of the Air Force's A-10 purchase: the Hog created a corps of pilots whose mission outlook consisted only of successfully performing CAS. A saying at the A-10 FWS summed up the attitude: "You can shoot down every MiG the Soviets have, but if you return to base and a Soviet tank commander is eating chow in your snackbar—you've lost the war, Jack." The saying had an edge, and it revealed another source of pride. Other fighter pilots' contempt generated defiant camaraderie among Hog drivers.[24]

Still, even A-10 pilots wanted improvements for their plane. The A-10 test team criticized the lack of chaff/flare dispenser, inertial navigation system (INS), stability augmentation system, cockpit heads-up display, and Maverick TV missile cockpit display. The design-to-cost philosophy meant no purchase of these items until the plane became operational and the pilots proved the need. Except for the INS, the service acquired all of them within the A-10's first five operational years. The chaff/flare modification occurred within the first two years, as a result of increased air defense lethality. The INS modification required European deployments to justify procurement, and the service did not make the change until 1983. The pilots did not want a high-technology avionics platform like the A-7 or F-111, but they came to believe that an INS enhanced CAS mission success. It reduced pilot workload in navigation, area orientation, and target acquisition—especially during low-level attacks. This was especially so in Europe, where defenses were supposed to dictate such an approach. Also, Europe's often flat, nondescript terrain and poor visibility required much greater pilot attention to navigation chores than that needed on American desert weapons ranges with large terrain landmarks and unlimited visibility. One could do the mission without the high technology—after all, pilots had often done so in earlier wars—but the INS facilitated the hard, low-level job.[25]

In late summer 1976 GAO investigators gathered one other major complaint—the plane's lack of engine thrust. It remained unaccomplished, however, because it was far too ambitious and costly. Thrust deficiency was the performance price of a wing designed to achieve desirable CAS performance. Its extra thickness created too much drag, and even the powerful engines of the time could not overcome it without incurring other penalties, such as unacceptably high fuel consumption.[26]

Gaining Wider Acceptance

General Dixon later said that regardless of what some Air Force people thought of the plane, the service still had to make it perform as well as possible. This required time, since more planes had to be built and the pilots required experience to develop tactics. Many in the initial test and instructor cadre either had no CAS experience or had experienced CAS flying fast jets in Vietnam's relatively low-threat combat environment. They now had to develop procedures for using the A-10's unique weapons against the kind of defenses the Israelis faced in the October War, and for this reason active participation in exercises did not commence until 1977. Even then there were only small test units at Edwards AFB and Nellis AFB, a pilot training wing at Davis-Monthan AFB, Arizona, and an operational wing still forming at Myrtle Beach AFB, South Carolina.[27]

Nonetheless, the A-10 community moved quickly enough, and in spring 1977 its training unit, the 355th Tactical Fighter Training Wing, participated in Red Flag. As an ongoing series of Nellis range exercises, Red Flag was an effort sanctioned by General Dixon to ensure that his fighter pilots experienced the most realistic simulation of modern air combat. The spring 1977 Red Flag combined its own operations with those of Army maneuvers at nearby Fort Irwin, California, and it was here that A-10 pilots first practiced the tactics they believed would ensure A-10 success and survival. Since they expected to hit forces at or just behind the front lines, they flew very low and used the terrain to conceal themselves from enemy eyes until the actual attack. The gun and Maverick missiles let them stand off from the target and still hit it. When operating within range of known radar gun emplacements, they jinked hard and spoiled the gunners' tracking solutions. They could turn to face any attacking jet fighter, thus neutralizing somewhat the fighters' speed advantage in an engagement. Finally, the plane proved its easy maintainability as the 355th flew from a desert dry lake and exceeded all reliability expectations.[28]

Though flown in the Nevada desert, these tactics had Europe in mind; and this was where the 355th went in June 1977. There was also a tour of U.S. bases in Asia, but Europe dominated Air Force and defense media attention because it was a hard case. The Air Force pilots had to prove the Warthog here, for throughout the late 1970s European defense journals and military officers questioned its ability to handle European conditions. Fairchild wanted to sell the plane to NATO Air Forces, as General Dynamics had recently done with its F-16s. That same month Fairchild conducted demonstration flights at the prestigious Paris Air Show. While at the show watching a film of Bob Dilger's tank shoot, a Luftwaffe general made comments that illustrated the Air Force's and Fairchild's rough road to European acceptance:

> That's all very well for the Nevada desert; for the clear weather conditions predominating in the Middle East and South-East Asia! [?] But what of Central Europe? If the Russians come, they are unlikely to court suicide by choosing a bright summer's day with visibility to the horizon; they will come at night, exploiting the murkiest weather that their forecasters can predict, and they will travel beneath a sophisticated air defense umbrella. What use will your fair-weather A-10 be then?

Further, the Europeans did not like the Hog's commitment to one mission. With their limited defense budgets, they wanted multi-role fighters. Some questioned how the United States could dedicate some of its already outnumbered Europe defense resources to a mission that they believed yielded too little result for too much cost. Meanwhile, Fairchild probably did not change many minds at the air show when its test pilot crashed his A-10 during a flight demonstration and was killed.[29]

The qualms about worst-case European conditions—low-level operations in bad weather against heavy defenses—reintroduced longstanding questions about all-weather CAS. As already mentioned, the workload required to fly low level completely by reference to instruments, identify and employ weapons against a fleeting target, and evade antiaircraft weapons, was simply overwhelming. Ground troops themselves probably could not see their opponents in the conditions envisioned by the Luftwaffe general and night/all-weather (N/AW) CAS advocates. The A-10 was the best available answer to this problem. The A-7 flyoff had demonstrated that the A-10's slow speed and maneuverability better allowed the pilot to remain in visual conditions when encountering marginal weather. The Soviets might launch an attack on a stormy night, but this condition would not obtain for long.

As for other European concerns about the Hog's abilities, A-10 pilots in Europe aimed to showcase their plane's strengths. Participating in NATO's annual Reforger exercise, which tested U.S.–based forces' ability to help allied units repel a Soviet assault, the Hog impressed NATO troops. Indeed, one defense journal article covering the deployment observed that the plane was hard for gunners to track. The Hog drivers also demonstrated their plane's marginal weather abilities as they flew under cloud ceilings as low as eight hundred feet and visibilities as poor as one and a quarter miles (normal visual flight rules minimums are one thousand feet and three miles). However, some U.S. Army helicopter pilots expressed skepticism about the Hog because they believed that they could provide more dedicated air support in even poorer weather.[30]

The Army and the A-10

Regardless of what their helicopter pilots in Germany thought, Army leaders still liked the A-10. First, it was the Air Force's tangible guarantee of its commitment to air support. Second, their concern about NATO's ability to stop a conventional Warsaw Pact invasion had increased. A cartoon of the time revealed this anxiety as it showed two Soviet Army generals reviewing their tanks' victory parade in Paris, with one asking the other: "By the way, who did win the air superiority battle in the end?" A more substantial indication was the 1978 bestseller, *The Third World War: August 1985*, whose authors described a fictional close NATO victory over the

Warsaw Pact, which occurred thanks to military improvements initiated in the late 1970s. Third, TRADOC Commander DePuy's new firepower-oriented doctrine, called Active Defense, had been published, and budget and development problems vexed the Army's Cheyenne successor, by now identified as Hughes Aircraft Company's AH-64 Apache. All of these factors increased the soldiers' need for the Hog. In this vein, the Air Force produced a study showing that the A-10 would be a critical item in stopping the projected invasion.[31]

Thus, the interservice cooperation established in the early 1970s continued. A 1975 House directive to identify duplication of A-10 and attack helicopter functions produced no interservice bickering, and instead generated a formal agreement that the CAS plane and armed helicopter were complementary. TAC Commander Dixon and TRADOC Commander DePuy appeared together in 1976 at classified hearings before both the House and Senate Armed Services Committees, and demonstrated their unity in the face of some occasionally provocative questions. When asked about their respective machines' survivability and effectiveness, the generals did not argue about which aircraft was best. Instead, they touted their weapons' distinctive contributions and emphasized joint service efforts. In the Senate hearings, Barry Goldwater probed for any further bickering about air support and favored aircraft, and Dixon and DePuy assured him there was none. Dixon observed, "I think that we have grown up. I think we understand each other. I think the overwhelming size of what we have to do takes first priority with us." Though there were occasional problems, the cooperative atmosphere continued under Dixon's and DePuy's respective successors, Gens. Wilbur Creech and Donn Starry.[32]

The A-10 embodied these joint service ideals. Even while still familiarizing themselves with the Hog, A-10 pilots flew with Army Cobra attack helicopters during summer 1977 exercises at Fort Benning, Georgia. They discovered that if they coordinated their attacks against simulated air defenses, their kill claims rose while their own postulated losses decreased. The result rated a brief mention by a couple of defense journals as well as increased attention from both services' leaders. One of the journals observed that this tactic could answer "a pressing need to demonstrate to a skeptical Congress that the armed forces need a variety of attack aircraft."[33]

Follow-up exercises called JAWS (Joint Attack Weapons Systems) verified the tactic. Conducted at Fort Hunter-Liggett, California, the JAWS exercise was as "free-play" as possible, and it required that Cobra attack helicopters and A-10s help stop a simulated armored attack accompanied by air defense weapons. The Cobra formation leader acted as a "battle captain" who roved about coordinating ground, helicopter, and A-10 firepower against enemy forces. Flying very low, maneuvering hard, and using Hunter-Liggett's rolling terrain to their advantage, the Cobras and A-10s individually fared well evading defenses while striking their targets. Their combined tactics massed firepower against critical areas, deceived and surprised enemy air defenders, and allowed each element to cover the other during its attack. Better still, exercise judges ruled that coordinated attacks improved "kill ratios"—the grim proportion of air losses and enemy losses—by "multiples of three, four, or more." Enthusiastic leaders in both services sanctioned even more exercises.[34]

They had a good opportunity with an operation that had congressional and OSD interest, the Tactical Aircraft Effectiveness and Survivability in Close Air Support Anti-Armor Operations Joint Test and Evaluation (TASVAL). The CAS debates of the early 1970s had produced requests for realistic testing, and the OSD had ordered an evaluation to resolve continued "uncertainties" surrounding CAS aircraft. Congressional skepticism remained about tactical air support in general and attack aircraft in particular. TASVAL was half complete in June 1979 when the GAO criticized both Army and Air Force aircraft—specifically the A-10 and the Army's attack helicopters—for alleged weaknesses pertaining to European warfare. These included survivability against modern air defenses and N/AW capability. The GAO recommended canceling both aircraft if they could not improve.[35]

TASVAL thus had ambitious objectives. It aimed to determine kill ratios versus various air defense combinations. It would also evaluate combined tactics involving Cobras and A-10s, as well as weather effects upon operations (nearly impossible to do, given the test location's mostly fair weather). More ominous, given outside attention, was that TASVAL had to determine which tactics, weapons, and aircraft generated the best kill ratios. TASVAL was conducted at Fort Hunter-Liggett from April through September 1979. It included two armored battalions opposing one another, with the "enemy" battalion accompanied by a full complement of simulated

Soviet tactical air defense threats. These were not only the SA-6s, SA-7s, and ZSU-23s of October War fame, but also the new SA-8 short-range, radar-guided missile. As for the TASVAL aircraft, there were eight Nellis AFB A-10s, an Army helicopter regiment, a small group of aggressor fighters, and some FAC planes. For its size, TASVAL was the most heavily monitored test yet; laser tracking recorded and computers assessed all of the players' shots.[36]

A-10s flew nearly five hundred TASVAL test missions, and ratified the previous exercises' outcomes. Kill ratios for separate attacks were better than expected. The Hog drivers worked very hard to prove their plane, and the defenses they faced ensured maximum effort. Randomly varied low-level approaches to the target were a tactical necessity, and actual wings-level firing passes had to be five seconds or less. To minimize target area exposure while looking for targets, the pilots strove for good target orientation and often made a quick reconnaissance pass before commencing attacks. Indeed, if they could see nothing distinct, they exited quickly. Additionally, their formations featured one "shooter" pilot who primarily attacked the armored vehicles and a "cover" pilot who watched for and shot at threats. At times they even designated "decoy" pilots to distract defenses. Any in-cockpit radar warning indication generated instant evasive maneuvers and use of chaff and flares. The intense air defense environment led the A-10s to rely on the Maverick guided missile more than the gun, because its longer range enabled greater standoff. If the defenses slackened, all A-10 formation members switched to shooter roles.[37]

The combined tactics—now called JAAT, for Joint Air Attack Team (or Tactics, depending on the source)—now yielded even better results. This was good for all concerned, because the participants later observed that at least initially, TASVAL had the feel of a competitive flyoff. Hog drivers found that the roving Cobra battle captain provided even better target information than their own FACs. The battle captain also positioned other Cobras as visual references to cue Hog drivers on the best attack axis to follow or the best spot to climb for the attack run. The Cobras and Hogs covered each other in their respective attacks, and their continuous, random axis attacks confused the defenses and also directed maximum sustained firepower upon enemy armor. JAAT beat TASVAL's surrogate Soviet forces, and helped end any congressional action against the A-10 and attack helicopters due to an alleged inability to fly CAS against European defenses. It was also a remarkable Air Force–Army effort to prove their respective weapons instead of the traditional zero-sum interservice squabble.

Further Challenges for the A-10

Congress still wanted an N/AW CAS plane, even though testimony by Undersecretary for Defense, Research, and Engineering William Perry and TAC Commander Dixon reduced some of that pressure. Perry observed that a night requirement did not exist when the OSD and the Air Force conceived the A-10. He also reminded the legislators that since they had wanted to cut this plane's costs, they cheated themselves later when extra requirements surfaced. Dixon told Congress that a rugged, simple attack plane like the A-10 met CAS requirements well enough, but added that the INS the pilots wanted could enhance its capabilities.[38]

Fairchild acted upon congressional interest by proposing in 1977 a N/AW A-10. It would carry the extra equipment required for such operations, as well as the two-member crew required to operate it. The company needed the business. Congress still balked at funding the full purchase of approximately seven hundred planes, and the Air Force believed that this amount amply fulfilled CAS commitment within its force structure. Further, the company could not sell A-10s overseas, though there were fleeting sale possibilities with Ecuador, South Korea, and Thailand. No country was willing or able to buy a plane dedicated to one mission—especially one that most air forces considered low in war-fighting priority. The U.S. State Department would have fought such a sale anyway. One of the Hog's main functions was to resist armored assaults, but its designation of *A*, for "attack," made it an offensive weapon in the eyes of government officials responsible for keeping allies from starting trouble.[39]

By 1979, Fairchild had produced a N/AW prototype model for the Air Force to test. Though it flew night, low-altitude, high-threat CAS better than any other plane, such a mission remained impractical. The crew had to match their radar image and INS information with a small infrared image of the outside world, and then gingerly maneuver toward the target. The extra precautions were necessary because the infrared scope's narrow field of view made the outside view akin to that seen through a soda straw. It could look only so far ahead into a turn, and sometimes blanked if the turn was too sharp. These limitations hobbled the Hog's vaunted maneuverability, and in spite of the extra crew member, the workload remained high. Further, these were only the problems associated with low-level night visual CAS, and not the more daunting low-altitude, all-weather, nonvisual version. Nonetheless, Congress funded the project through 1982, when the Air Force confirmed that no further A-10s were needed and that N/AW air support indeed required further development. Even if N/AW CAS remained impractical, the A-10 was still the best fixed-wing airplane choice for CAS in marginal visual conditions, such as those seen in Europe. By 1977 the OSD was so concerned about the Soviets' tank superiority over NATO forces that it ordered studies on permanently basing A-10s somewhere on the Continent. Exercises like TASVAL answered tactical questions about the A-10 and European warfare; but a more basic issue involved how to base A-10s there. The Hog could operate from austere locations, but how far did the service want to go with this concept? A diverse basing scheme allowed closer proximity for A-10s and Army units, and it also answered NATO planners' concerns about air base availability after a Soviet air assault. Conversely, austere basing's major challenge involved consistent logistical support of isolated units. This issue arose in the 1971 Senate CAS Hearings concerning the Harrier, and again in the later Enforcer hearings.[40]

After several studies, the Air Force compromised. A large A-10 wing at a Main Operating Base (MOB) would routinely deploy flights of roughly seven aircraft and occasionally squadrons of approximately twenty planes for training at various Forward Operating Locations (FOLs). Thus, the A-10s would have a central location for training and logistical support. Routine FOL deployments allowed practice of combat logistical resupply, and guaranteed the pilots' familiarity with the terrain and friendly ground units of their particular sector. Such operations, combined with the A-10's slow speed and maneuverability, enabled pilots to fly confidently in weather that would ground other air support planes.[41]

The service validated the FOL concept in April 1978, when it deployed Hogs from its first operational A-10 unit, the 354th Tactical Fighter Wing at Myrtle Beach AFB, South Carolina, to simulate such operations at Shaw AFB across the state. That January, the 354th had passed its Operational Readiness Inspection to ratify the combat-ready status that it had attained ahead of schedule the previous October. Most impressive during this build-up was the plane's high sortie rates, which its simplicity and easy maintainability guaranteed. (A sortie is one aircraft mission.) The Shaw evaluation demonstrated this as well—available pilots and not flyable aircraft determined sortie totals.[42]

Although the A-10 and its pilots had shown that they could handle such European challenges as weather and logistics, the gun issue arose once again for the skeptics, who included members of Congress intent upon limiting the total A-10 buy. The A-X concept planners had devised gun specifications that apparently accounted for the more low-threat arena of Vietnam-style CAS. The gun's shells worked best in medium-altitude, high-angle dive attacks against a tank's more vulnerable upper sections. (Tank designers put the thickest armor on the front of the turret and upper front chassis, which normally face the enemy's direct ground fire; they reduce weight and improve mobility by reducing armor everywhere else.) The 1974–1975 live-fire attacks used higher angle deliveries, and at least one observer wondered how A-10s could survive conducting such attacks in a high-threat environment. Therefore, the question arose whether the gun could destroy tanks in the shallow-angle attacks required to avoid antiaircraft coverage in Europe. A second gun issue concerned holding down the cost of the large ammunition buys needed for proficiency training and combat reserve. Since this was still the time of the Enforcer hearings and the TASVAL test, it appeared that congressional and other enemies might stop further A-10 purchases if the gun failed this requirement.[43]

As chief of the A-10 SPO's Armaments Directorate, Col. Bob Dilger responded by orchestrating a "save the program test." His Lot Acceptance Verification Program (LAVP) not only validated the rounds' effects but also proved that the gun could destroy Soviet tanks in these shallow-dive-angle attacks. LAVP used Soviet T-62 tanks captured by the Israelis, as well as the services of Dr. Russel Stolfi, a Marine Corps Reserve officer and Ph.D. historian who had surveyed captured Arab tanks after both the 1967 and 1973 Arab-Israeli Wars. Stolfi provided to LAVP the same rigorous damage assessment that he had done in the Mideast. He did not want a "technical event," such as a computer study. He wanted "an effectively constructed simulation of an historical event," and insisted that the tanks, the planes, and the tactics match combat conditions as much as possible. Dilger obliged him, as the A-10

Armaments Directorate overcame the now-familiar resistance from Eglin Armaments Lab people. LAVP tests ran intermittently between 1978 and 1982, and proved that the GAU-8/A worked as advertised. Shots hitting the front of a T-62's turret were not very effective, but given the GAU-8/A's shell size and velocity, no one expected them to be. However, unlike U.S. ground troops, an A-10 could maneuver quickly for a better shot. During LAVP, A-10 shots on the tanks' sides and rear were very successful, as shells tore through the treads, engine, fuel tanks, and less-armored turret sections.[44]

Dilger reduced ammunition costs further, and used the LAVP testing to guarantee good products. The service used civilian competitive contracting for shell purchases, but the Army still produced the rounds' percussion caps. When testing revealed cap defects, Dilger "fired" the Army from producing even these and used another civilian company to make them. LAVP discovered that the shell tips of some rounds broke away during firing; instead of performing a long and costly study, Dilger and the SPO worked the fix directly through the contractors. This not only saved money but also eliminated potential serious firing problems. Bad caps caused barrel explosions, while shell tip breakage destroyed accuracy and created particles that damaged jet engines. Concerning the ammunition contractors, Dilger modified the competitive bidding procedure—the contract was renegotiated yearly during the late 1970s—to give the winner an extra 10 percent of the total production amount. Since the total was over 1 million rounds, the incentive was significant. Dilger also oversaw a reduction in the cost of shell packaging to one-fourth of the original estimate, and even worked around military specifications if he could maintain quality while reducing cost. As a result, the GAU-8/A round cost of less than fifteen dollars remained at one-fifth the original cost estimate. He used the resulting budget surplus to sustain large ammunition lot buys, and then shifted to multiyear contracts in 1980. These actions created an economy of scale and prevented the common feast-famine procurement cycle that induced high costs. By all accounts, Dilger ran around, over, and through rules and officials, and used sympathetic Pentagon contacts to help; but his aggressive, driven style antagonized superiors and earned him an early retirement.[45]

The gun beat the skeptics, but the more high-technology part of the A-10's antitank arsenal, Hughes Aircraft's Maverick missile, had an uneven reception. Its concept was attractive: use a television camera in a missile's nose to guide that missile to a target that visually contrasted with its background. Thus, the pilot could lock onto a target via a cockpit TV screen displaying the seeker's view, fire the missile, and escape while the missile did its business. The IAF had some success using Mavericks during the October War, and the U.S. Air Force, seeing an answer to the tactical air defense problem, pressed for large-scale Maverick production. The problem concerned acquiring and locking targets. At longer ranges, one could not recognize the objects on the TV screen. Additionally, the target might not contrast visually with its surroundings. The Israelis shot vehicles that stood out in the desert, but hitting camouflaged tanks on an overcast day in Europe was another matter. In 1977 Hughes introduced the Imaging Infrared (IIR) Maverick, which used an infrared seeker to track objects that thermally contrasted with their surroundings. The same difficulties remained. Many factors determined a vehicle's thermal appearance, and these often combined to create an unrecognizable image. Further, the vehicle might not contrast thermally with its surroundings any more than it would visually. These issues generated a series of congressional hearings and critical news reports—at forty thousand dollars apiece, Mavericks were high-interest weapons. However, the service and the Hog drivers adamantly supported a weapon that they considered the best available antidote for tough air defenses.[46]

The A-10 Wing in Europe

The European A-10 MOB was actually two closely located U.S. bases in eastern England: Royal Air Force base (RAF) Bentwaters and RAF Woodbridge (the "RAF" designation was due to basing agreements with the British). The setup required two bases because the command in charge of A-10 operations, the 81st Tactical Fighter Wing (TFW), would be the largest tactical fighter wing in the service. The 81st already existed as a three-squadron F-4 wing, but it expanded to six squadrons as it switched to A-10s. The transition began in January 1979, as the 92d Tactical Fighter Squadron began operations that included an immediate deployment to support an Army exercise in Germany. As it received its full A-10 complement, the 81st deployed A-10s to four German FOLs: from north to south, Ahlhorn, near Bremen; Nörvenich, near Cologne; Sembach, near Wiesbaden; and Leipheim, near Ulm. Specific responsibility for one of the bases was assigned to one squadron, while the two remaining squadrons assumed a roving commitment to two of the bases. This achieved the MOB-FOL aim of having A-10 pilots who were intimately familiar with both the terrain and the NATO ground units of their squadron's sector.[47]

The Air Force and the pilots of the 81st understood the challenges facing good CAS plane operations in Europe and rose to meet them. In an ironic twist, the A-10's excellent maintainability and fuel endurance became potential problems, as pilots found that repeated long, low-level flights in poor weather induced high fatigue. The Hog surpassed other jets in such operations, but this type of flying required skill and concentration. The 81st thus took care that the A-10 transition included a judicious balance of experienced and junior pilots. Another A-10 strength that created a small problem was its gun, which quickly destroyed training targets that had withstood many hits from other aircraft guns. Tight European defense budgets did not allow for frequent target replacement, and weapons ranges restricted A-10 strafing until they could obtain suitable targets.[48]

As the 81st accepted the Hog, other active-duty and Air National Guard units also accepted the new plane. The 23d TFW at England AFB, Louisiana, switched from A-7s to A-10s in 1978, and both it and the already formed 354th TFW at Myrtle Beach trained for combat contingencies in Europe and elsewhere in the world. Guard and Reserve A-10 units formed in various states, to include Connecticut, Louisiana, Maryland, Massachusetts, and New York. In the Pacific, the Air Force established one A-10 squadron apiece in Alaska and Korea. Since the North Koreans lacked the sophisticated air defenses and mobile warfare opportunities of the Soviets in Europe, A-10 CAS tactics recalled those used in Vietnam, in utilizing higher altitude dive attacks. Further, the rugged Korean terrain channeled any communist assault into well-known avenues of approach, thus easing overall the U.S. and South Korean defense effort.[49]

Still, Europe dominated American thinking. Thus, the Air Force always ensured that the 81st was the first A-10 unit to receive all of the best and newest tactical equipment: chaff and flare dispensers, radar jamming pods, the IIR Maverick, and the INS. For their part, the pilots of the 81st focused hard upon success in European combat. Adopting TASVAL's tactics, including JAAT, they almost exclusively practiced low-level operations. They understood that a European war might entail heavy losses, but they believed that their training and tactics and their plane's unique construction would prevent loss rates greater than those of other NATO aircraft. After all, if the defenses were as fearsome as some estimates indicated, one wondered what attack plane could survive. Further, Hog drivers believed that they would inflict as much damage as they received. They knew that their plane was rugged enough to withstand hits, and they planned to use their gun to exploit the weaknesses of Soviet tanks—even the ZSU-23/4, which earned such infamy in the 1973 October War. Still, they liked having the potential for standing off from the enemy and striking an accurate blow; the 81st TFW's preferred combat weapons configuration included both gun and Mavericks.[50]

The Air Force not only tried to make the A-10 a success in Europe, but also did its best to ensure that others accepted the idea as well. The Air Force Association published several articles in its periodical, *Air Force Magazine*, touting the A-10's chances in that theater. These efforts during the 1970s and early 1980s represented the zenith of Air Force interest in traditional CAS: supporting troops at or near the battlefield. Observers also noted that the time was a high point for the traditional CAS mission's foundation, Air Force–Army relations.[51]

Ironically, the condition that motivated these good feelings, the Soviet threat in Europe, would also generate new interservice war-fighting ideas that threatened the Hog just as it achieved fully operational status. The two services wanted a more flexible battle doctrine that emphasized deep attacks against Soviet rear-echelon forces, as well as a new airplane to accomplish this. The A-10 would thus find itself in yet another fight for its life.

8 FIRE IN THE MINDS OF MEN

Even as Air Force leaders pressed the Hog's case in the late 1970s, commentators used European considerations to criticize the plane, its mission, and the style of warfare they believed both represented. Writing in the U.S. Air Force professional journal, *Air University Review*, two RAF officers disputed the efficiency of CAS while praising their own service's preference for interdiction. They both scoffed at the Americans' cumbersome CAS control procedures, and one directly impugned the A-10's ability to fly CAS in a European war. In other journals, American military experts decried a perceived national disposition for attrition warfare, and cited CAS reliance as an example. All these writers believed that interdiction yielded greater returns for the risk of facing Soviet air defenses, and praised the RAF's approach. Perhaps foreshadowing the CAS debate of the 1980s, these criticisms inspired rebuttal. U.S. military officers answered that their country preferred firepower to losing lives. CAS was the most important mission in certain situations, and the United States had other theater concerns besides Europe.[1]

However, changes occurring in U.S. war-fighting attitudes favored the critics. From its inception in 1976, the U.S. Army's Active Defense doctrine was criticized by officers within that service who believed that it promoted a defensive mind-set and depended too much upon attrition to stop a Soviet attack. The doctrine's emphasis upon committing all necessary forces to blunt the initial Soviet attack also created dissension. Some officers worried about the massed Soviet follow-on units—one might stop the first assault only to be overrun or outmaneuvered in subsequent attacks. The leader of the doctrinal change movement was General DePuy's successor as TRADOC commander, Gen. Donn Starry. Starry had supported Active Defense and had participated in its development, but a 1976–1977 tour commanding the U.S. V Corps in Germany brought home the problem of handling Soviet follow-on assaults. After assuming TRADOC command in 1977, Starry began to develop a replacement doctrine, later called AirLand Battle. His aim was to use both firepower and maneuver in the air and on land to disrupt the Soviets' attack tempo and eliminate their follow-on forces threat. The resulting fight would be a pell-mell affair with no traditional battle line.[2]

Though AirLand Battle was an Army doctrine, that service invited Air Force leaders to comment upon its development as early as 1979. Through their ALFA joint war-fighting directorate, the two services explored how to attack Soviet reserves in an effort known as Joint Attack on the Second Echelon (J-SAK). As this occurred, NATO altered mission definitions within its air support doctrine. The British preferred interdiction not only because they considered it a more efficient use of their relatively small air force, but also because they still let the ground commander assign air support units to preempt reinforcing attacks. U.S. Air Force commanders opposed the British concept because it interjected a ground officer into the air leader's decision-making turf. The difference caused haggling over air support definitions, but a compromise was reached in 1979. CAS continued to be defined as the support of engaged troops that required detailed control and coordination, while battlefield air interdiction (BAI) was defined as the attack of enemy forces "which are in a position to directly affect friendly forces." NATO's definition of BAI kept the required coordination intentionally fuzzy to satisfy both the Americans and the British, but the big development was that air commanders recognized BAI as a separate air support activity. The change contributed to the ongoing U.S. Army–U.S. Air Force air support talks.[3]

ASSOCIATED DEVELOPMENTS

Before its publication in 1982, Starry and TRADOC generated widespread support for AirLand Battle, and this included the military reform movement of the time. Upset with rising weapons costs and poor military performance in fiascoes like the 1980 Iranian hostage rescue attempt, the reformers demanded a more efficient and competent military. Lacking sharply defined membership and aims, they consisted of several groups: (1) former "Fighter Mafia" people, such as Tom Christie, Pierre Sprey, and John Boyd (considered by many to be the movement's intellectual leader); (2) OSD staff members, such as Chuck Spinney and, again, Tom Christie; (3) journalists, such as Dina Rasor and James Fallows; (4) a bipartisan group of members of Congress, such as Sen. Gary Hart and Rep. Newt Gingrich; and (5) sympathizers, such as OSD staff member Don Fredericksen.[4]

The reformers believed that President Ronald Reagan's administration misspent its large military budget on poor weapons; to them, a prime example was the Cheyenne's long-awaited successor. Development work had begun as soon as the Army canceled Cheyenne in the early 1970s, and in 1976 Hughes Aircraft's AH-64 Apache won a competitive flyoff over Bell Helicopter's AH-63. The winning design was not as ambitious as the Cheyenne, but it was grand enough that inflation and development problems drove its prototype cost above $6 million per copy. For this, Congress and President Jimmy Carter's administration cut its budget in half. It then staggered through a negative GAO report, a test mishap, and attacks from political leaders before it received full production approval in 1982.[5]

In spite of its development problems, the Apache represented another reason for the Army's doctrinal change: growing confidence in the ability of its own troops and weapons to halt attacking Soviet units at the point of contact. The Army now had its super attack helicopter, along with other new, highly lethal weapons, such as M-1 Abrams tanks and Multiple Launch Rocket System artillery. Indeed, a prominent author of the published Air-Land Battle doctrine, Huba Wass de Czege, observed that "the final impetus for change in doctrine comes from the near term introduction of new systems which increase our mobility and firepower."[6]

If the Apache did not inspire confidence among reformers as it did with the Army, the reformers applauded an Air Force plane that made its operational debut in the early 1980s and became the Hog's nemesis. Pentagon staff members, such as John Boyd, Pierre Sprey, and Tom Christie, had spurred creation of the petite F-16, which embodied their reaction against the desire for large, costly, and complex fighters, such as the F-14 and F-15. These officials, nicknamed by some the Fighter Mafia, wanted a smaller, cheaper plane dedicated to dogfighting superiority, and they ramrodded their project, entitled the Lightweight Fighter, through the defense procurement maze. One milestone was the 1974 Schlesinger-Brown agreement, which not only helped the A-10 achieve operational status but also made the Air Force accept the Lightweight Fighter. General Dynamics's F-16 beat Northrop's YF-17 in a 1975 flyoff competition.[7]

The Air Force may have had some initial doubts, but in the 1970s several NATO countries, as well as Israel, bought the F-16, whose initial cost per plane was a relatively cheap $10 million. General Dynamics's bantam fighter performed well beyond expectations. In 1981 Israeli F-16s successfully bombed the Iraqis' Osirak reactor in Baghdad. During Israel's 1982 "Peace for Galilee" Lebanon incursion, F-16s were the top MiG killers. The U.S. Air Force's first operational F-16 wing, the 388th from Hill AFB, Utah, scored a spectacular victory over dedicated interdiction planes in the RAF's 1981 tactical bombing competition. The initial F-16 units enjoyed good maintenance reliability as well. Though the fighter lacked the loiter and weapons carriage abilities of dedicated attack planes, its maneuverability, high thrust-to-aircraft-weight ratio, and fire control system enabled it to dogfight better and bomb more accurately than nearly any other fighter.

With successes in other activities besides dogfighting, the F-16 seemed an ideal plane to fill any air force's order of battle. Though other aircraft purchases continued on a yearly basis, F-16s were purchased via a multiyear production plan. Sources indicate that the OSD pushed hardest for this action because, like Bob Dilger's GAU-8/A ammunition policies, multiyear buys achieved economies of scale—in this case, for both domestic and foreign purchases. An increasingly large number of tactical air leaders embraced the F-16 because it seemed to be the multi-role fighter that they had so long sought. "We are back to an airplane that can do it all," gushed one. Thus, Air Force officers planned staged modifications to incorporate ever more sophisticated fire control and navigation avionics— and ever more diverse missions.[8]

MOVING AWAY FROM THE TRADITIONAL CAS PLANE CONCEPT

All these developments had ominous implications for the A-10 or any follow-on, dedicated CAS plane. In a congressional budget hearing early in 1982, Air Force chief of staff Lew Allen told the legislators that the A-10 was an excellent weapon, but that the force requirements for this specialized CAS plane had been met. As such, the Air Force's fighter procurement plan, labeled the Fighter Roadmap, did not include more A-10s. To tactical air leaders such as Robert Russ and Larry Welch, who helped write the plan, the N/AW, two-seat A-10 was not worthwhile. Both officers envisioned making the Hog a FAC plane, where it could switch roles between coordinating CAS missions and doing them itself. However, Welch later said that the idea for an "A-16" arose during this time, since consultations with Army leaders about F-16 CAS missions provoked no objection as long as the Air Force provided CAS upon demand.[9]

The beginning drift away from the A-10 was not some unsanctioned staff scheme. The J-SAK studies showed that both the Air Force and Army wanted more emphasis upon interdiction, a mission that better suited the F-16. A later congressional hearing on the topic revealed that the "Air Force began analysis and documentation of A-10 deficiencies to support the new Army doctrine shortly after AirLand Battle was published." In 1984 the two services ratified their new war-fighting view in the "31 Initiatives," which were agreements on implementing AirLand Battle. These accords pledged to develop procedures for "close air support (CAS) in the rear area" and to ensure that Air Force fixed-wing and Army rotary-wing air support capabilities did not overlap.[10]

Through late 1984 and early 1985 the Air Force conducted an informal study, which suggested that a modified F-16 would be a good follow-on CAS plane candidate. According to General Russ, the budget picture did not allow for a completely new CAS plane and all the other Fighter

Roadmap tactical aircraft purchases. Air Force leaders also observed that the long development time for a new plane would not allow their service to support the Army's AirLand Battle doctrine soon enough. Finally, the service rejected the A-10 or a similar plane as unsuitable for the interdiction mission, given European-style defenses. "We need a faster, more maneuverable aircraft," one Air Force staff member asserted. In 1985 the Air Force and Army chiefs gave interservice blessing to the new CAS plane design philosophy when they published an agreement requesting that any replacement machine fly air support missions "on the non-linear battlefield across a broad spectrum of combat scenarios and threats ranging from the friendly rear area to the traditional battle area and the deep maneuver area." The agreement specified that the design "also provide for the inherent capabilities needed for air interdiction." In keeping with budget and development considerations, it further stated that the "timely fielding of a follow-on CAS aircraft dictates that . . . [the program] focus on existing airframes available for procurement in the late 1980s."[11]

Immediately after the agreement, the Air Force issued a request for industry to determine again if there were any other designs, but the service seemed sold on the F-16. Industry leaders expressed skepticism about Air Force sincerity, and only companies with already available fighter and attack planes responded. In July 1985 the service submitted a project justification for the new plane, but the OSD rejected it for restricting CAS plane options. The Air Force also omitted new CAS plane funding from its FY 1987 force posture statement, and it allocated no money for one in its FY 1988 budget plan. Meanwhile, Air Force officers published the occasional article claiming that AirLand Battle meant a more interdiction-oriented air support. Finally, in early 1987, TAC commander Russ announced that the Air Force's CAS plane would be a modified F-16, and that the A-10 force would be reduced and transferred to FAC units.[12]

The Air Force's actions had not gone unopposed, and at least while the plane and the company that built it seemed viable, congressional A-10 supporters mounted some of the first challenges. Addressing an Air Force witness, Maj. Gen. Robert Russ, in March 1982 hearings, Sen. Barry Goldwater agreed that the service had enough A-10s, and then added, "I know what you are up against. You have the parochial problem of Massachusetts, New York, Pennsylvania, and Maryland, all wanting to keep that A-10 going just like they bought A-7s to keep Texas happy." Russ had earlier faced some of these states' legislators. Rep. Sam Stratton of New York wondered why, if Russ said the A-10 was still important, the service did not want any more of them. Rep. Beverly Byron of Maryland asked Russ what plane would replace the A-10 as it aged and its numbers dwindled. Russ replied, "We will use the other planes, the F-16's, that are dually capable," and insisted that the F-16 could carry an A-10's weapons complement. "And still gets off the ground?" quipped Byron, who apparently had seen the small fighter. Other lawmakers did more than ask questions. A pork-barrel deal between Congress and President Reagan bought twenty extra A-10s—a deal that, given the infrastructure costs for this special purchase, raised their price to $18 million per copy.[13]

Some reformers and former Fighter Mafia members also resisted the Air Force's tack from the start. Key players were Donald Fredericksen, who became deputy undersecretary of defense for tactical air warfare programs in 1985; Tom Christie, director of program integration; and Chuck Spinney, who worked in the Program Analysis and Evaluation Office. These officials did not occupy the government's highest positions, but they were all well placed to affect Air Force procurement efforts, especially Fredericksen. They also had friends and connections. Pierre Sprey no longer worked in the OSD, but he gave advice. Military reformers in the press and in Congress also helped.

Some of these reformers helped develop the F-16, but they would not part easily with the A-10 that they believed was another of their inspirations. Indeed, one asserted that the Hog was a "symbol of the Reform Movement." Sprey later opined that the attempt to build a CAS F-16 to replace the Hog was "one of the most monumentally fraudulent ideas that the Air Force has ever perpetrated." If anything, these people wanted a similar plane to replace the A-10, and Tom Christie offered the practical air support reasons for that approach: "We were absolutely convinced that you had to do a new airplane. The F-16 was not going to do close air support. It was not the right airframe for close air support; and [with] what weapons? It didn't have the weapons [carriage capability], and with high speed, and tracking targets and all that, it just didn't go together, in our book." In the late 1970s they had answered Air Force preliminary plans for an expensive fighter-bomber with their "Blitzfighter," an attack plane smaller, simpler, and even cheaper than the A-10. They also proposed such a plane as an A-10 replacement in the early 1980s.[14]

As the Air Force moved forward with its A-16 plans in 1986, the friction increased between the OSD—particularly Fredericksen's office—and the Air Force. Both sides stuck to their positions amid a flurry of staff studies (these were the advance gust of a coming blizzard). Fredericksen later said that during this time, industry gave him affordable design alternatives that convinced him the Air Force was not looking hard enough for a suitable A-10 replacement. In summer the service proposed a mixed force of F-16s and modified A-7s, and though the OSD allowed studies to continue on the modified A-7, it rejected the F-16 outright. That December an angry meeting between Air Force and OSD representatives in the office of Deputy Secretary of Defense William Taft IV led to an OSD directive to create an oversight committee to ensure that the Air Force studied other CAS plane candidates. Don Fredericksen was to chair the assemblage, called the Close Air Support Mission Area Review Group (CASMARG). Up to this point, the friction between the two sides had simmered, but CASMARG's formation and General Russ's public endorsement of the F-16 CAS plane shortly thereafter brought it to a full boil.[15]

Bureaucratic infighting underlay a public debate that lasted through the end of the decade. The focal point during much of the organizational scrapping was Fredericksen's CASMARG, whose other members included OSD staff, a general from the Joint Chiefs of Staff office, and two Air Force Headquarters representatives who knew the issues of the debate: Lt. Gen. John Loh and Lt. Gen. Jimmie V. Adams. Loh had worked for the legendary John Boyd during the F-16's genesis, and Adams had been TAC's action officer during the A-7/A-10 flyoff as well as an A-10 wing commander. Both officers aggressively pursued their service's point of view. Apparently intending to re-create the same grounds that justified the Air Force F-16 choice, Loh demanded studies of current threats and operational concepts to help produce yet another CAS plane design request. The two generals also told Fredericksen that they wanted their own group to study alternative designs and produce the request, and the CASMARG chairman let them do so. This group called itself the Close Air Support

Aircraft Design Alternatives (CASADA) study. CASADA submitted a draft design request in autumn 1987, but the CASMARG rejected it as a "requirement for an F-16."[16]

CLASHING BACKGROUNDS, CLASHING HISTORY, AND REAL HISTORY

The Air Force leaders' historical perspective on war-fighting, technology, and budget imperatives guided their choice. Just as the bomber generals emphasized nuclear strategic bombing, the fighter pilots had their own war-fighting priorities, such as high-speed performance. As the debate developed, General Russ and other Air Force officers asserted that speed would keep the F-16 from suffering hits on the battlefield, though it certainly did not do so for fast planes in Korea, Vietnam, and the Golan Heights. Like past Air Force leaders, however, they often embraced modernity and the idea that new technology could render null and void previously learned war-fighting lessons. AirLand Battle assisted their return to the 1950s opinion that European war with the Russians would be so different as to annul traditional CAS. Finally, they preferred accomplishing the air superiority, interdiction, and CAS missions with the same fighter; and in the F-16, they saw the multi-role plane desired by earlier generations. Of course, CAS would be the third priority in training and interest for these multi-mission aviators, and that did not escape critics' notice. One commented, "A glance at history suggests that the US Air Force needed no prodding from the Army or its doctrine to pursue the counter-air and air interdiction roles with enthusiasm."[17]

The air leaders also understood that from Billy Mitchell's days, the budget had dictated Air Force structure and compelled attention to doctrinal priorities. The CAS plane debate erupted and Welch was chief of staff when the Reagan administration's military buildup ended and forced hard procurement choices. Welch told an Air Force Association (AFA) gathering that he would sacrifice none of his higher priority acquisitions—a fighter, a bomber, a transport, and an air-to-air missile—to buy a dedicated CAS plane. As TAC commander, General Russ still wanted to make the A-10 a dual FAC/CAS plane, but this plan and other parts of the Fighter Roadmap fell victim to the budget cuts of the late 1980s. He later complained about the budgetary and political conundrum: "As you know, you've got support [in Congress] to buy any airplane. . . . [but] it's not a matter of them adding the money." To other Air Force officers, it appeared that the leaders made careful decisions based upon realistic assessments. One criticized the F-16 opponents as operational neophytes who disrupted a well-considered procurement decision.[18]

To these opponents, it seemed that the Air Force F-16 decision on CAS defied the historical lessons of combat CAS experience, which dictated a design such as the A-10. Christie remembered, "Suddenly, we who had pushed the F-16 were viewed as being anti–F-16. . . . We weren't anti–F-16. We were just saying that's not the airplane that was going to do close air support well." The critics of the F-16 CAS idea maintained that in spite of Air-Land Battle's emphasis upon individual units prevailing in free-style combat, Army troops still wanted maximum available firepower applied against the enemy facing them. Fredericksen's Army officer contacts seemed uncertain about the Air Force plan. Thus, the opponents saw the F-16 CAS plane gambit as at best a shortcut purchase—and at worst, yet another example of air leaders' traditional neglect of CAS.[19]

Ironically, the two sides and the defense press focused so intently upon European war and the F-16 that nearly all failed to examine or cite wars occurring throughout the 1980s. This was perhaps understandable for the Air Force, because these events demonstrated that other war theaters, as well as the wars actually being fought, still required traditional CAS and its planes. AirLand Battle advocates believed they could apply their high-technology, maneuver-oriented doctrine elsewhere besides Europe. However, the U.S. invasion of Grenada in 1983 featured infantry fighting on a small, hilly, tropical island. Swirling tank battles and an extensive interdiction campaign against "follow-on forces" could not happen in this case. Instead, the troops appreciated the CAS provided by Air Force AC-130 gunships and Navy A-7s. (A-10s were deployed to Puerto Rico, which, at a distance of over five hundred miles, was the nearest U.S. land base. Hostilities ended before they could participate.)[20]

Korean war-fighting conditions also highlighted CAS and AirLand Battle's relative applicability around the world. Here, both sides were on a constant war footing that required as much of a constant, vigilant U.S. presence as Europe, but at least through 1988 the commander in chief of the UN Command/U.S. Forces Korea, Gen. Louis Menetrey, still considered the A-10 his "primary ground support aircraft" because it constituted a "vital part of our ability" to thwart a North Korean invasion. The rugged Korean terrain and in-place defenses on both sides made AirLand Battle's war of wide-ranging maneuver a tough proposition. At least one military scholar believed that Air-Land Battle did not fully apply to defeating a North Korean attack.[21]

Foreign air forces that slighted CAS encountered war-fighting situations that highlighted its necessity. British forces had to reestablish ground-air coordination procedures while wresting the Falkland Islands back from the Argentines in their 1982 war. Still, RAF Harriers contributed to winning the Battle of Goose Green and helped knock out Argentine artillery positions. The Israelis had better CAS success during their 1982 "Peace for Galilee" Lebanon invasion, but lack of practice gave rise to coordination problems, including inadvertent attacks upon their own troops. In the conclusion of their history of late-twentieth-century wars, military scholars Anthony Cordesman and Abraham Wagner noted that the world's air forces had trouble with CAS because they were "almost congenitally incapable of honestly assessing and improving" upon it.[22]

Perhaps the "congenital" disregard meant that air leaders wanted attack helicopters to assume the mission. Israeli helicopters flew air support missions in Lebanon and later assumed sole responsibility for Israeli CAS. U.S. attack helicopters flew in Grenada, and throughout the 1980s and beyond both the Army and Air Force viewed the Apache as an air support asset. It joined Army units in early 1986, and A-10 pilots appreciated how it enhanced JAAT operations. Its Hellfire antitank missiles provided stand-off ability equivalent to the Maverick, and its laser designator could mark targets for the Hog's laser spot tracker. Apache pilots also praised the A-10's contributions, but some increasingly saw the Apache operating in its own right. At least one Army officer opined that his service, "through its armed helicopter, now has the CAS system it had searched for since 1942." Another regarded the aircraft as a more efficient use of firepower than Air Force fighters. Military journal articles described how Apaches would singlehandedly conduct interdiction assaults in support of the AirLand Battle concept—that is, a mission for which the Air Force disqualified the A-10 because it was too slow.[23]

Others stuck with the team concept. One Army War College student praised the Apache's capabilities, but concluded that combined fixed-wing and helicopter operations provided the best air support. Former Army general and helicopter advocate John Bahnsen wondered if the Apache crews' cocksure, go-it-alone attitude would ultimately hinder the ready air support they were supposed to provide. In a 1986 defense journal article, he wrote:

> Aviation service is slowly being divorced from the combined arms team. Take a look at the National Training Center [the Army's combat training range at Fort Irwin, California]. . . . On a typical morning, as a battalion task force rolls into battle, an aviation voice comes on the command net and announces that he is God's gift to the upcoming fight. No . . . fully synchronized concept of operation, just a rather disinterested taxi driver come to fly around the edge of a battle he knows nothing about.

The Army generals who promoted Apache interdiction missions qualified their position by saying that they wanted Apache pilots to exploit air defense creases for operations similar to old horse cavalry raids. Indeed, when pressed, they conceded that they would use their Apaches cautiously in order to avoid heavy losses.[24]

It was well that they should do so. Helicopters provided valuable service on the battlefield, whether through reconnaissance, logistics, or fire support, but they required careful handling. Of the 107 helicopters used in the Grenada invasion, 9 were either shot down or damaged severely by defenses consisting of only light, non-radar-guided AAA. Most were cargo aircraft, but the rest were Marine AH-1 Cobras that attempted unsupported attacks upon defended positions. The losses were significant enough that nearly all accounts cited helicopter vulnerability as a cautionary lesson. Foreign military arms also encountered these pitfalls in varying degrees. The British used helicopter fire support sparingly in the Falklands War, and while the Argentines downed only a couple of helicopters, the presence of enemy aircraft over the barren islands made British helicopter pilots nervous. One recalled, "There was nowhere to hide, we stuck out like dog's balls. . . . Had I been the Argentine squadron commander I would have said to my guys: 'Their advanced troops are here, their base is there; in between there will be helicopters, go and shoot them down!'" Overall Israeli aircraft losses in the 1982 Lebanon invasion were perhaps too small to be significant, but the Israelis lost six helicopters versus two fixed-wing planes.[25]

A graphic example of the pitfalls involved in recklessly using helicopters or any aircraft in the CAS environment came with the Soviet experience in

Afghanistan. From the late 1960s through the 1970s the Soviet Union, like the United States, oriented its forces away from overreliance on nuclear warfare technology and toward conventional weapons. This move also reflected the Soviet Union's growing confidence in projecting military muscle into world trouble spots—as demonstrated in Africa in the 1970s and 1980s. This behavior reached its climax with Russia's occupation of its troubled client state, Afghanistan, in 1979. The ensuing Afghan insurgency was ready-made for CAS. Only CAS could provide the quick firepower superiority for the small-scale battles that erupted in the remote Afghan mountain fastness.

The Soviets seemed to have the equipment, though their air support doctrine and aircraft setup differed from those of the Americans. For fixed-wing planes, they emphasized what NATO called BAI. In 1982, they produced a plane for that mission, the Sukhoi (Su)-25 (NATO nickname, "Frogfoot"). Some Western observers initially thought the Frogfoot was the A-10's CAS arena adversary, and pilots called it the "A-9ski," thanks to its visible similarities to Northrop's losing entry in the A-X flyoff. However, the Su-25 was smaller, did not carry a powerful gun, and flew faster than the A-X candidates. (It was a design that the U.S. Air Force might have accepted as a follow-on CAS plane.) The Frogfoot could and did fly CAS in Afghanistan; its straight wing gave it an ability to maneuver at slower speeds and deliver more weapons more accurately than the MiG-21 and MiG-27 fighters, which also tried to fly the mission. However, for direct aerial fire support of troops in contact, the Soviets relied upon an attack helicopter, the Mil (Mi)-24 (NATO nickname, "Hind"). Ugly, heavily armed, and durable for a helicopter, the Hind came into its own in the Afghan War's early days. The Frogfoot's appearance garnered some notice since it was a new Soviet air support machine, but the Hind's performance held the rapt attention of Western military leaders.[26]

That was the case until late summer 1986, when the Americans supplied the Afghan rebels with a U.S. Army shoulder-launched, heat-seeking SAM called "Stinger." Until then, Soviet aircraft had occasionally used some self-protection tactics in situations where they expected Afghan resistance, which largely consisted of old guns and the occasional stolen Soviet SA-7. But the Stinger was a far more capable man-portable SAM than the SA-7, and it downed one hundred Soviet helicopters in the next eighteen months. The Stinger did not drive the Soviet Air Force from the skies, as some asserted, but it definitely affected the Soviets' ability to carry the war to the Afghan rebels. Fixed-wing pilots had to adjust their attack altitudes high enough to avoid the SAMs—so high that disgruntled ground troops started calling them "cosmonauts." Though Su-25s made some adjustments to reduce losses, the effect upon both Hinds and transport helicopters was nearly catastrophic. They had already suffered the lion's share of Soviet air losses before Stinger—the final tally was 80 percent at war's end—but the American SAM's arrival denied entire areas to helicopter action, since a helicopter could neither outclimb, outmaneuver, nor outrun the missiles.[27]

Finally, in 1988, aviation writer Jeff Ethell used a war of that decade to support the CAS F-16 opponents. Commenting in *Armed Forces Journal International* on the long-running conflict between South Africa and the Angolan/Cuban alliance, he noted that the South African Air Force had used highly trained pilots, a smooth ground-air coordination system, and slower but rugged, well-armed, and highly maneuverable jets to conduct effective CAS. In conclusion, Ethell asked, "Are we listening or will we have to learn some very hard lessons all over again when the next conventional shooting war starts?" A reader attacked Ethell in a letter response, pointing out that the South African airmen had recently taken more serious losses. They also had been unable to influence the outcome of a recent long battle.[28]

Some resolution of the mini-debate came via Helmoed-Römer Heitman, a South African who used personal interviews and archival research to recount the war's wind-down stage in the late 1980s. Heitman declared that all air forces needed to reevaluate the CAS mission's usefulness given modern defenses—but he painted a picture of airmen who did not fight a hardnosed air war. He reported that during this period, AAA and SAMs made "traditional close air support impossible"; thus South African troops increasingly saw little of their air force, whose air support consisted of occasional BAI missions by fast jet fighter-bombers. However, Heitman admitted that the "remarkable reluctance of the South Africans to accept casualties" was the reason that country's air crews were so cautious. South African fighter pilots had even conceded command of the air to Cuban/Angolan MiG pilots, who in turn would not attack South African troops due to their own fear of air defenses. Heitman concluded that air forces could no longer operate unmolested in the skies, but the Angolan War that he described did not apply to the air war-fighting approach contemplated by either side of the American CAS plane debate.[29]

The Conditions of the Debate

Meanwhile, the dispute continued within bureaucratic, political, and defense media circles from 1986 through 1990. The defense media debate was the most colorful, and disparate voices made it alternately the most confusing and the most informative. As David Packard and the senators had discovered in 1971, CAS was a complex mission, and the debate threatened to spin out of control even though it primarily involved one war theater and one plane. TAC commander Russ assigned one of his top staff members to join Army TRADOC officers in determining a joint Air Force–Army CAS definition from which to argue their case publicly. After a few months the officers gave up, overwhelmed by the myriad opinions. A former Army general wrote a defense journal article criticizing the airplane debate for emphasizing form over function. Like some others, he prescribed more attention to practicing the mission so the Air Force could provide CAS when needed. Ensuring smooth coordination was essential to CAS, but the type of plane could determine the amount of effort made. At times, someone asked why the Air Force could not do CAS like the Marines—just as uninformed legislators had asked in past hearings. Others wanted the Army to get CAS planes. As the dispute intensified, each side overstated its case, generating criticism from both inside and outside its own ranks. Indeed, dissension within the Air Force/Army side made its case hard to present, while the opposition was itself a loose coalition of groups and individuals.[30]

Yet the defense press debate showed that the CAS mission and any plane assigned to fly it faced turbulence. It also reflected the more serious bureaucratic and political fight under way at the same time. One could not really understand some of the official moves without first knowing the battle for opinion. A good starting point is the supported service in the CAS equation.

The Army leadership officially supported the Air Force. Army Chief of Staff Carl Vuono stressed that the 1985 Army–Air Force CAS plane memorandum was the Army's position, as it followed AirLand Battle doctrine and the ensuing 31 Initiatives. Weighing in with a 1989 public communiqué, the

U.S. Armed Forces in Europe Commander in Chief John Galvin stated, "It seems to me that the F-16 provides the flexibility and capability that I need." As mentioned, others asked that the Air Force skip the messy CAS plane details and simply support the Army when needed. As one officer put it, "The Air Force doesn't tell the Army how to fight the land battle, and the Army doesn't tell the Air Force how to fight the air battle."[31]

A factor that the Army leadership did not mention openly was fear of interservice rancor. An argument with the Air Force entailed the risk not only of losing air support but also of attracting unwanted congressional attention, which threatened both services' wishes. Brigadier General Bahnsen later said that the Army could not fight the A-16 because it might trigger Air Force moves against the Apache. At least one other debate player asserted that the Army accepted the Air Force's CAS F-16 in return for support of its own purchase of guided missile artillery and even more advanced helicopters.

However, not everyone in the Army agreed with the Air Force's CAS plane choice. There were public examples, such as the attack helicopter zealots who wanted CAS for themselves, and General Menetrey in Korea who thought the A-10 did well in his theater. Also, a former Army secretary, James Ambrose, claimed that the Air Force did not support the Army enough if it was unwilling to fund a new dedicated CAS plane. Most opposition seemed to come from the lesser ranks. An *Aviation Week & Space Technology* article noted that junior officers still thought the F-16 less capable than the A-10. Even the Air Force Association's *Air Force Magazine* noted that "a substantial faction in the Army bitterly opposes the Air Force's plan to employ a modified fighter, the A-16, for close air support rather than designing a new airplane from scratch for that role." But true to military discipline, no one directly challenged the leaders' course. Instead, the opposition remained nameless, but nonetheless potent, as it bubbled forth in research papers, informal conversations, and the occasional article. It was not much, but it contributed to the more public opponents' conviction that they were right.[32]

Air Force leaders understood that winning the public debate could secure the congressional dollars that in turn powered this most technology-oriented of all the services. Past budget, doctrinal, and procurement policy battles had honed this awareness: status struggles within the Army, the bomber-versus-aircraft-carrier controversy of the late 1940s, the Korean CAS squabble, later bomber funding, and even the selling of the A-10 in the 1970s. Pursuing its argument through the defense press, formal studies, a flight test, and even one professional historian, the service still had difficulty making its case.[33]

Though there were exaggerated claims, the Air Force pursued a fairly consistent line of argument based upon AirLand Battle doctrine, European warfare, and the ensuing joint service BAI emphasis. Air Force sources predicted a lethal, "nonlinear" battlefield—a term that appeared along with "high-tech battlefield" or "the battlefield of the 90s." This warfare scenario required fast, multipurpose planes like the F-16. Besides, interdiction was always a more lucrative Air Force mission, especially when shrinking budgets forced hard procurement choices. And even though they now described CAS as something resembling interdiction, Air Force partisans also asserted that fire-control improvements made the first-attack destruction of either BAI or traditional CAS targets a cinch, regardless of aircraft speed.[34]

AIR FORCE MEDIA EFFORTS

As if to test the defense media waters, the Air Force held a CAS-oriented Aerospace Power Symposium just after Russ's CAS F-16 announcement. The meeting's conduct and reception foretold the difficulties to come. Although some speakers expressed support for interdiction or focused upon how to incorporate the F-16 into its new mission, others still preferred traditional CAS and the A-10 that flew it. To muddy the CAS plane issue a little more, an RAF Harrier pilot extolled his jet's virtues.[35]

Air Force officers took up their service's public case. One of the symposium speakers, Lt. Col. Price Bingham, repeated his assertions in a September 1987 article entitled "Dedicated Fixed-Wing Close Air Support—A Bad Idea." By 1988 former A-10 pilot and then Air Force Headquarters staff member Maj. Pat Pentland wrote in *Armed Forces Journal International* that the Air Force amply supported the Army and that the CAS concept was too wedded to the Vietnam War. Air Force FAC Mark Barrett's *Infantry* magazine piece concluded that A-10s could not meet Europe's anticipated combat challenges. Extolling to *Air Power Journal* readers the march toward better technology, Air Force Maj. Robert Chapman wrote that the "handwriting was on the wall" for

CAS as it had been for horse cavalry earlier in the century. Pressing the F-16's case in a 1989 article in *Military Review*, Lt. Col. Bruce Carlson averred that "the A-10 never was the optimum solution for CAS in a high threat environment." Air Force Col. Robert Foglesong and Army Lt. Col. Edward Shirron emphasized the Air Force's budget angle.[36]

TAC Commander Russ made perhaps the most overt assertions. He assured readers of two defense journals that the F-16 was the best CAS plane choice and that the Air Force would always meet its CAS commitments. In a July 1988 *Air Force Magazine* article entitled "No Sitting Ducks," Russ seemed to abandon any previous intention to retain A-10s. Instead, he extolled the F-16 and condemned the Hog, which led others to call his piece a sales job. (*Armed Forces Journal International* opined that Russ gave General Dynamics, the F-16's maker, several pages of free advertising.) Even Air Force officers wrote *Air Force Magazine* contesting the general's logic.[37]

The service also got help from sympathetic journalists. Steven Daskal, a "widely published defense systems analyst in Northern Virginia," told readers that "attrition warfare is finally being pushed out of the limelight," and endorsed the interdiction reemphasis. Former marine aviator and then LTV Products Group employee Allan Bloom praised the Air Force proposal to modify A-7s as F-16 supplements and thus still avoid buying a dedicated CAS plane. In 1989 the *Air Force Times* carried a "Defense Trends" special section with articles touting the F-16 and the modified A-7 as the service's new CAS planes. In one piece, defense experts stated that in the "battlefield of the future," Army firepower would do CAS and the jets would do interdiction. Another observed that many Air Force leaders detested the A-10; yet another described how the F-16's avionics would help it do CAS.[38]

Air Force Magazine made an about-face to support its service's tack. It had published a favorable segment on A-10 European CAS in July 1983, and its November 1986 issue contained an air support article discussing the plane in favorable terms. It changed its tune in March 1987, just after Russ's F-16 decision. In "New Roadmap for the Airland Battle," Senior Editor Edgar Ulsamer discussed the doctrinal and budgetary imperatives for the service's choice, as well as how the A-10's thick wing condemned it to slow speed and replacement by the F-16 and modified A-7. As the debate escalated in 1988, AFA writers presented the Air Force position four times. These articles discussed not only the familiar imperatives behind the service's decision, but also asserted that the A-10 was too slow to accompany the massed fighter-bomber formations that NATO airmen claimed were necessary to beat Soviet air defenses.[39]

The threat vulnerability issue appeared often, recalling some defense observers' accusation that the Air Force sometimes inflated threat capabilities in order to convince Congress that it needed a new plane. SAC had cited increased Soviet threats to justify new bombers, and the defense reformers asserted that the service recently did the same thing to buy the most expensive and sophisticated fighters. Similar justification for a CAS F-16 was not entirely convincing. The only Soviet ground-to-air defense improvements mentioned during the debate were a new radar-guided AAA gun and a man-portable, heat-seeking SAM. The Air Force had tested the Hog against similar defenses during the A-7 flyoff and TASVAL. There were new Soviet fighters, but the Soviets probably would not use their top interceptors to shoot fleeting attack planes at the front when greater threats existed from other NATO air missions. If NATO lost frontline air superiority, air support would be very difficult for any plane. The CAS F-16 proponents claimed that their plane would use countermeasures and judicious attack routing, but one wondered why this did not apply to other planes. Rebuttals to Russ's "No Sitting Ducks" article enumerated other qualifications. SAMs were so fast and maneuverable that the F-16's speed made no difference (further, its external weapons load limited it to subsonic speeds). Ability to travel with a strike force applied to interdiction raids, which was relevant only if one accepted the new interdiction-oriented definition of CAS. Finally, assertions about the unprecedented nature of nonlinear battlefields made one wonder if this was another case of asserting modernity's unique circumstances. Most other battles from the past defied anyone's capacity to describe an orderly, linear front line. Avoiding catastrophic surprises in such circumstances was why CAS required positive coordination.[40]

Conclusions from academic works also supported the Air Force position. In 1986 RAND Corporation tactical air warfare analyst Benjamin Lambeth announced the end of single-purpose planes due to rising costs of tactical jets. Officers at the Air University wrote papers declaring the end of traditional CAS and hailing the F-16, though writers who were former A-10 pilots tended to qualify their support for the F-16. Noted aviation historian Richard Hallion supplied a remarkable academic effort on the Air Force's behalf. In the late 1980s he was a visiting professor at the Army's Military History Institute; while studying how air power interacted with the Army, he conducted extensive research both at the institute and in Air University archives. Work at the archives led him to Lieutenant Colonel Bingham and other fighter interdiction proponents. Their influence showed in his 1989 book, *Strike from the Sky*, as well as in a 1990 *Airpower Journal* article. Though Hallion covered air support through World War II, he included a set of maxims that he claimed also applied to air support as it was accomplished through the mid-1980s.[41]

Some of Hallion's maxims had merit. Hallion observed that N/AW CAS was difficult, that CAS and BAI had tremendous synergistic value, and that good communications were essential to CAS success. He also overstated his case, especially when he used the word *always* in some of his assertions. BAI "always" yielded more results than CAS, and multi-role fighters "always" performed better in the CAS/BAI role than special-purpose attack planes. He cited only a few historical instances where CAS was decisive, and then proclaimed that it was a mission "which typically reflects more desperate or peculiar circumstances." He followed an Air Force tendency to deemphasize non-European war scenarios, and also mixed up facts. Recalling many Air Force leaders' impression of the A-10's genesis, he claimed the Air Force purchased the A-10 due to congressional pressure. The Su-25 was not a "mirror-image" of the A-10, as Hallion stated.[42]

Hallion focused heavily upon the U.S. Army Air Force in World War II Europe, which helps explain his multi-role fighter bent, given the P-47's and P-51's air support successes there. However, his cursory coverage of post–World War II air support rendered his universal conclusions about CAS planes suspect, since the low-intensity fights that he discounted characterized this later period. He singled out the F-16 as the preferred air support plane, while ignoring attack planes that performed yeoman service for many air arms. A former Vietnam War FAC reproved his fighter bent in a letter to *Air Power Journal*: "The ubiquitous F-4 was the last airplane I really wanted to see for a CAS mission. Any aircraft in the attack category was better."[43]

AIR FORCE OFFICIAL STUDIES

Both to rebut the CASMARG and to support its public campaign, the Air

Force conducted various studies and tests, with results as mixed as much of the rest of its efforts. It had commissioned an inquiry by its Scientific Advisory Board (SAB) back in 1985, when it first announced its CAS plane plans. In this case, electronics expert and retired Air Force Maj. Gen. John Toomay led a seven-member committee that included both scientists and aviation business leaders. Using familiar verbiage, the service's task order for Toomay's group stated that doctrinal changes and the European combat environment necessitated a review of CAS.[44]

Toomay and company did not take the hint. They pointed out that in any CAS scenario, one had to assume air superiority and acceptable air defense suppression. The latter meant that ground defenses would not be completely eliminated because one attacked armed soldiers. The panel observed that the CAS and BAI missions required performance characteristics different enough to demand two different planes. The SAB felt that multi-role planes guaranteed neither a more efficient air effort nor better CAS. A European war would place critical demands on all air missions simultaneously. Also, the Air Force would not want to cripple its air campaign by losing expensive multi-role planes in rough-and-tumble CAS combat. The SAB recommended purchasing a new dedicated CAS plane for the distant future and modifying the A-10 with new avionics and better engines in the short term (though the board felt that new engines would introduce problems, such as greater fuel consumption and bigger heat signature for heat-seeking SAMs).[45]

Some sources later claimed that the Air Force suppressed the SAB report, for when some of its findings were leaked to the press in 1987, it embarrassed the service. When later asked about Toomay's SAB, General Welch said that anyone who thought Toomay would not support the A-10 had not "read his resume." This was not fair to the SAB because as Toomay later said, it included a mix of people with diametrically opposed views, and the report reflected a consensus. The report also emphasized that the plane needed modifications to improve its performance. During this time, the Air National Guard examined the engine modification because budget reductions meant that it might have to keep its A-10s for a long while.[46]

The Air Force Headquarters staff, as well as two subordinate major commands, TAC and Systems Command, also undertook studies. The CAS-MARG rejected their findings as too biased toward the F-16. The Pentagon study's combat scenario disregarded small-arms fire, denied A-10s very low altitudes, and gave F-16s an unproven heat-seeker countermeasures capability. TAC wanted a plane that could fly both CAS and BAI. The Systems Command inquiry assumed rugged F-16s.[47]

These and other studies mentioned modifications that the Air Force thought would prove the plane's CAS viability. Its CAS F-16 demonstrator planes carried a digital terrain monitor that guaranteed precise navigational system accuracy, as well as a data link system, called the Automatic Target Hand-off System (ATHS), which allowed ground parties to transmit CAS target information instantaneously to the F-16's weapons computer and cockpit head-up display. The OSD scouted the 1987 CASADA study because it not only specified conditions favoring the F-16, but also assumed the ATHS's perfect operation. Before the debate ended, the Air Force also equipped its demonstrator F-16s with laser spot trackers, a new radar, a cockpit map display, and infrared gear. Planned modifications for CAS F-16s included hardening certain vulnerable areas and carrying a detachable 30-mm gun pod. The Air Force felt compelled to include the pod since it enhanced the fighter's "tank buster" image relative to the A-10.[48]

Remembering how TASVAL quelled skepticism about A-10 and helicopter survivability during high-threat CAS, General Russ ordered an F-16 CAS demonstrator plane evaluation in 1988 and 1989. To help secure credibility, the Air Force assigned former Hog drivers who had moved to the F-16 as demonstration pilots. A report on the portion flown at the Nellis ranges stated that it went well. The other part of the test was flown at Fort Hood, Texas, and according to defense press reports, F-16 CAS earned praise from previously skeptical Army helicopter pilots. Shortly afterward, Air Force Chief Welch and Army Chief Vuono jointly stated that "we should therefore proceed to deliver some part of the F-16 buy in the A-16 configuration." However, a couple of sources later commented upon its staged aspect; as one F-16 pilot noted, these systems required a lot of effort to work. The implication was that without them, the F-16 would not do CAS well at all. Indeed, one had to wonder how these features might further improve the A-10's effectiveness, but the A-10 was then considered a nonplayer for future air support. By 1990 the service had evaluated only a couple of these items on the Hog in a tentative and much less publicized effort—and even this was accomplished with funds remaining from another project.[49]

Securing a good operational venue for proving the F-16 over the Hog, the service upgraded the New York Air National Guard's A-10 squadron at Syracuse to F-16s in late 1988. As part of the 174th TFW, a dedicated CAS unit flying A-10s, this squadron had shown pride in the Hog and its mission. The unit had instigated weapons modifications that the regular Air Force then incorporated into its Hogs. According to a June 1990 *Aviation Week & Space Technology* feature article, the F-16 did well, though some of the former Warthog pilots' comments sounded unusual. They were not concerned about the F-16's lack of armor, because its speed allowed them to avoid hits. They liked the plane's versatility, and added that it was maneuverable enough to allow one to hit targets that suddenly appeared on an attack run. Though the 174th had not received the equipment used by the demonstrator planes, the F-16's other features seemed to impress its pilots. The 174th's commander, Brig. Gen. Mike Hall, declared the F-16's air-to-air radar a necessity for CAS missions (though he also emphasized that he would use it only to maintain tactical situational awareness). And though Hall boasted about the amount of ordnance the F-16 could carry, his planes carried no bombs when flying the *Aviation Week* reporter on a demonstration mission. The pilots also claimed that they would soon carry the 30-mm gun pod. It was much smaller and carried far less ammunition than the GAU-8/A, but they asserted that its tank-killing power was twelve thousand feet slant range. No one claimed such a capability for the Hog's more powerful gun—indeed, A-10 critics were already proclaiming its obsolescence against newer Soviet tanks.[50]

The F-16 Opponents

The opponents never mustered anything like the Air Force's public relations effort. Further, their own efforts diverged and at times were unrealistic or simply ludicrous. Other than a few articles, the odd editorial, and a couple of CAS-MARG-sanctioned studies that offered alternatives to the Air Force position, their voices remained disparate shouts that never reached a crescendo.

The most visible figure was CASMARG chairman Fredericksen, who opposed the F-16 mostly on the plane's specific weaknesses in traditional CAS. He gamely expressed his views at an AFA symposium in late 1987. "I love the F-16 as a fighter. . . . [but] I just think it's too soft an airplane," he said, pointing out that the diminutive fighter's vulnerability to the Soviets'

23-mm AAA was nine times that of the A-10. He continued, "If you can do CAS without getting in close, that's one thing. But I don't think you can. The good guys won't call you unless they're getting overrun, and you've got to know exactly where they are. You've got to worry about fratricide. So you've got to get down in there, and you're going to take an awful lot of fire."[51]

Fredericksen could have told the symposium even more. He could have criticized the plane's lack of weapons carriage and loiter capabilities. He could also have pointed out that both the F-16 and its pilots labored under multi-role responsibilities. He had visited a U.S. F-16 base in Europe and found that the pilots did not consider CAS an F-16 mission. Even an Air Force liaison officer to the Army noticed the same thing; the F-16 CAS he saw indicated that the pilots' "hearts simply aren't in it." In a 1990 *Fighter Weapons Review* article, an F-16 weapons officer told his fellow pilots that CAS required more practice or there would be the usual problem of unprepared F-16 pilots fumbling through the mission. A 1988 Air Force safety article attributed a high percentage of F-16 crashes to loss of situational awareness induced by such things as cockpit task saturation, distractions, and the plane's lack of aircraft attitude cues.[52]

Though Fredericksen believed that SAMs like the Stinger made traditional CAS a tough proposition, he still wanted a new dedicated CAS plane. One of the few takers was Fighter Mafia veteran Chuck Myers, who wrote an article pushing what he called a "Mud Fighter." Myers revived the Blitz-fighter concept by proposing a small, simple, high-subsonic plane that could maneuver well at all speeds and possess STOL capability. Several months later, innovative aircraft designer Elbert Rutan proposed such a machine in response to Fredericksen's call for dedicated CAS plane designs.[53]

Myers and others wanted the Mud Fighter to carry only a gun and rockets—and almost no avionics or defense countermeasures gear. Unlike the A-10, which carried a variety of weapons as well as countermeasures equipment, and was big enough to accept equipment modifications, the design was too wedded to anti-tank warfare and not to other contingencies. However, Myers asserted that the Mud Fighter's simplicity and low cost would yield a large CAS force that could swarm over the battlefield and endure losses better than other planes.[54]

Fredericksen had already sanctioned a study of the follow-on CAS plane issue by the Institute for Defense Analysis (IDA). At about the same time that the CASMARG rejected the CASADA findings, the IDA offered three design options, all of them favoring a lightweight dedicated CAS plane similar to Myers's Mud Fighter. An *Aviation Week* article announcing the IDA results featured a design drawing very similar to Myers's design, though the optimum IDA design was somewhat more substantial.[55]

However, the Mud Fighter concept allowed some CAS F-16 supporters to contend that any dedicated CAS plane compared poorly to the F-16; even reformers opposed it as too simple. Chuck Spinney observed that had Myers expanded the design, it might have had a better chance of acceptance. Critics asked how the Air Force would support the operation and maintenance costs for such large numbers of planes and pilots. General Russ especially disliked the disposable plane aspect, perceiving Myers's argument to be that "in World War II, a lot of guys bailed out. . . . And I kind of said, 'Why should we do things like we did in World War II? We used to have horses too. Should we have horses?'" For his part, Myers questioned why pilots should think that they could not suffer loss rates similar to those of the troops they supported. His query conjured an issue that had been around since World War I: unacceptable losses of hard-to-train pilots.[56]

Perhaps this attitude would change if the pilots were themselves troops, and Myers said that he wanted the Army to get the plane. In this, he was like other F-16 opponents who believed that since the Air Force wanted to shirk its CAS responsibility, another service should assume sole custody for the mission. An *Aviation Week* editorial expressed this view just after General Russ's 1987 F-16 announcement, and reader letters addressing the piece wanted the Marines to be the CAS service. Air Force defenders quickly pointed out that the Army could not fund extra fixed-wing units, and that its leaders had made their preference clear via the 31 Initiatives. As for the Marines, their charter differed sharply from those of the other services. One had to wonder how Marine aircraft would divert from supporting their own units to help some other service.[57]

Such positions as these reveal that the anti–F-16 side had its share of overly confident assertions. One military reformer claimed that "by 1989 the A-10s had been moved to the reserves." The Air Force may have planned such an action, but in 1989 all active A-10 units were intact. In 1990 Jack

Anderson and Dale van Atta asserted in the *Washington Post* that the Army opposed the CAS F-16 decision. As previously mentioned, some Army people opposed it, but others, particularly the leadership, supported it. The worst example of anti–Air Force hyperbolic claims was a *Washington Monthly* piece by Robert Coram, an *Atlanta* magazine editor who had "written extensively on military affairs."[58]

Other than Anderson and van Atta's quarter-page editorial, Coram's piece was the nearest thing to an in-depth study of the debate by the mainstream press. Just as it had ignored previous CAS plane and mission controversies, it passed on this one as well. The CAS plane candidates' procurement money paled in comparison to more publicly recognizable defense controversies of the time, such as the B-2 Stealth Bomber or the "Star Wars" Strategic Defense Initiative. (The total price for modifying F-16s to fly CAS was $1 billion—less than the cost of just one B-2.) Nor did the debate have the obvious strategic import of these other defense programs. Indeed, the CAS argument was hard for outsiders to understand. Coram apparently decided to present it to his readers by following the popular press's usual simplistic assessment of military controversy.[59]

Coram asserted that Air Force neglect of CAS still "forced" the Army to buy attack helicopters. The airmen's 1950s apathy toward the mission had caused an initial Army turn to these aircraft, but their tactical value and ardent service constituency cemented the Army's rotor-wing commitment. Coram told readers that subsonic design was the primary reason for the low cost of dedicated CAS planes, but other factors such as avionics also determined a tactical plane's expense. Coram implied that the airmen did not want to do CAS because they would have to take orders from Army enlisted FACs. FACs were nearly always Air Force or Army pilots because the most important FAC quality was conveying target descriptions in a manner that the person in the air could comprehend—poor communications was often the reason for CAS targeting snafus.

Coram made some valid points. Like Hallion, he observed that N/AW CAS was a very difficult and expensive task. He also pointed out that an F-16 could not carry an A-10's weapons load, and that the external stores it did carry hobbled its otherwise nimble performance. He showed that the service was historically infatuated with a high-technology vision of future war while ignoring the prosaic but tactically important CAS mission—and that this fixation found it wanting in real conflicts. Unfortunately, his delivery ruined the message.

THE A-10 STANDS ALONE

For all the opposition to the F-16, the F-16 opponents thought little of keeping the A-10. In spite of some improvements for the Hog during this time, one observer had commented as early as 1986 that "upgrading the A-10 is dismissed by almost all observers." Later, there were strong rumors that all A-10s would be retired. Fredericksen ignored the SAB's endorsement of the Hog, as well as support expressed in another study that he sanctioned by a Beltway think tank, BDM. Even Robert Coram opted for the Mud Fighter. One could say that General Russ showed more respect for the Hog when he singled it out for unfavorable comparison with the F-16 in his "No Sitting Ducks" article. The Hog's only support during the debate appeared in occasional articles describing the A-10's potential for certain war scenarios. General Menetrey's Korea comments were one example, and a couple of articles promoting A-10 operations in Norway were another. Air Force Reserve Maj. Darrel Whitcomb used AirLand Battle's chaotic battle predictions to suggest the Hog's value for CAS in the friendly force rear area. As the Cold War subsided, A-10 Fighter Weapons School instructor Robert "Muck" Brown pointed out in *Airpower Journal* that the A-10 would excel in low-intensity fights. However, Menetrey was one theater commander, Norway was one country, and the debate participants were too focused elsewhere to listen to Brown or Whitcomb.[60]

As for Hog drivers themselves, operations proceeded normally during the debate, and no units were disestablished until the post–Cold War drawdowns that affected all of the Air Force. However, several general officer briefings to A-10 units implied or even emphasized that the Hog and most of its pilots had little future in tactical aviation (though the service already had a program called "Constant Carrot" that allowed a small number of "deserving" Hog drivers to make the transition to fast fighters). Unit briefings by TAC staffers tried to sell the Air Force position, and pilots scheduled to meet with congressional staff members investigating the debate received strong hints to support the service. The heavy-handed and disparaging approach only stimulated many Hog drivers' bloody-minded pride.[61]

Thus, the Hog clung to an uncertain life, as the views expressed in the debate highlighted its fragile and shifting support. Budget constraints and European combat demands seemingly validated Air Force leaders' choices. Certainly, they could not be blamed for selecting the F-16 if Army leaders agreed that it was the best option. One could say that like CAS itself, the F-16's CAS avionics modifications represented yet another technological promise for troop support by fixed-wing planes.

Yet there was no reason why these devices could not also apply to a plane dedicated to this mission's traditional concept—which evidence from most 1980s wars indicated was not passé. There was likewise no reason why improved tactics or countermeasures could not beat new defenses. The one thing that the dispute did for the A-10 or any such follow-on plane was to delay the Air Force's action until other circumstances broke the deadlock—if not among the advocates, then at least in the real world of weapons procurement and actual combat.

9   THE BUREAUCRATIC WAR, THE POLITICAL WAR, AND THE REAL WAR

Approximately thirty formal studies of the CAS plane issue were completed up through 1990. People familiar with them observed that most backed their sponsors' opinions. The OSD used the Institute for Defense Analysis two more times for research; RAND produced a report; and there were independent studies by Lockheed and Northrop, both of which endorsed using Stealth technology for CAS planes. This was not surprising, since Lockheed built the F-117 Stealth plane and Northrop the B-2. (The SAB CAS study rejected this option due to that technology's limitations and inapplicability to CAS.)

CASADA released its formal report in autumn 1988, and its ensuing project requirements package climaxed the debate's study phase. By this time, however, other forces arose to control the course of the debate. CASADA solidified others' impression of the Air Force's commitment to the CAS F-16, since the study's conditions supported that plane. CASADA assumed perfect operation of automatic target hand-off system (ATHS) technology, even though the OSD questioned that premise as well as how well the F-16 would fly CAS if ATHS failed. The Air Force had already caused bad feeling when it raised CASADA's security classification to a level where some CASMARG members could not attend deliberations. One OSD official remarked, "The Air Force wants us to get out of the ballgame."[1]

Indeed, by late 1987 the Air Force was gaining the upper hand in the bureaucratic fight. Only the chairman of the Joint Chiefs of Staff could advise the secretary of defense on policy, and the airmen charged the CASMARG with usurping that function by making recommendations on CAS matters. One source asserted that this constrained Fredericksen's CAS plane efforts, but the bureaucratic bottom line was that the service could always execute its plans and try to withstand the resulting criticism. Other conditions also contributed to Fredericksen's weakening position. Force reductions made the service's desire for an existing multi-role aircraft more compelling. The Air Force's adamant insistence upon the F-16 discouraged contractors from making serious bids. Therefore, Fredericksen compromised his stand. Where before he had accepted only design proposals featuring simple and inexpensive planes, he now expanded consideration to all planes.[2]

The Air Force's ensuing actions showed that Fredericksen's move did not change its position very much. After all, 1988 was when CASADA released its conclusions and General Russ made his brazen and criticized pitches for the F-16. Yet, having given the service some leeway in the CAS plane search, Fredericksen's CASMARG still pushed the service to seek alternative design studies from six companies. The service pressed forward unfazed, however, and its tack was so obvious that an *Aviation Week* editorial called it a "Close Air Charade." In fact, the aviation journal reported in June that Brig. Gen. Joseph Ralston of TAC Headquarters had told a Lockheed senior official that the company "should not waste money on a new CAS aircraft for the next twenty years" because the Air Force's CAS plane would be the F-16.[3]

Congress, Its Interests, and the Fortunes of Constituent Planes

The service's behavior spurred congressional intervention, though the legislators had varied motives. The 1986 Goldwater-Nichols Act required, among other military reform measures, a periodic review of service roles and missions by Congress and the Department of Defense. The CAS issue was a good topic for the upcoming first assessment. Also, TAC commander Russ's fighter force structure plan lacked credibility with both his Air Force superiors and Congress, since it asked for so much. The plan made tactical sense, in that Russ wanted a large force of multi-role fighters to respond flexibly to any contingency. Smaller numbers of specialty planes, whether F-15s, F-111s, or A-10s, would enhance Air Force capability in situations demanding their strengths. In spite of his debate-related hyperbole, he still wanted to hide the A-10s in the FAC role so Congress would allow him to maintain a large air support force, but given the budget reductions under way, Congress rejected these proposals as unrealistic. Further, the service's efforts on behalf of the F-16 upset various key groups: military reformer legislators, ex-Marine congressional staff, and A-10 pilots who discreetly registered their opinions with Congress. Some lawmakers and staff had their own, individual motives. Rep. Les Aspin always searched for military budget savings and saw opportunities here. Sen. Alan Dixon had the bedroom communities for McDonnell-Douglas among his constituency, and thus pushed that company's AV-8 Harrier as a follow-on CAS candidate. (McDonnell-Douglas had assumed the contractor role from the British.)[4]

Congressional attention manifested itself in other ways as well, including GAO investigations. A 1988 GAO inquiry gave up in the face of the complex issue, and only recounted the history of the debate, venturing no recommendations. As congressional hearings focused more upon the debate, the GAO investigated a familiar question: since the Marines had such a great CAS reputation, why did the Air Force not buy their Harrier attack jet? Things had not changed since the 1970s investigations, as the GAO found that the services had specific needs and made no recommendation. The office was more decisive concerning General Russ's FAC A-10 plan. With the Air Force touting the value of ATHS, Armed Services Committee Chairman Aspin had the GAO investigate whether airborne FACs were necessary. The GAO cited the Air Force's own assertions about the lethal modern battlefield and ATHS's worth to conclude that they probably were not needed, and it recommended against creating FAC A-10s until the Air Force resolved the contradiction.[5]

One other GAO report revealed the distinctly different constituent sponsors of two 1970s CAS plane competitors, the A-7 and the A-10. The Texas congressional delegation still applied pressure on the A-7's behalf; in the early 1980s a *Washington Post* article complained about how Congress pushed A-7 acquisitions far in excess of those required. During the CAS plane dispute, however, A-7 manufacturer LTV submitted a modification proposal attractive to the Air Force. It featured an afterburning engine, low supersonic speed, and the most advanced avionics to answer the service's call for an off-the-shelf, high-speed BAI attack plane. General Russ and other Air Force leaders saw the new A-7 as an interim acquisition until A-16s came on line, and then as an air support supplement. From the mid-1980s onward, the defense press published articles hailing its anticipated comeback, thus fueling an already complex debate. In 1987 Congress had the GAO study the plane's potential, and the office's 1988 response was cautious. The GAO did not like the uncertain costs and returns involved in upgrading the Vietnam-era jet. Further, it wondered if the modifications would rectify the current model's lack of survivability features. Still, the GAO wanted the modified version tested, and the CASMARG had already allowed it to continue development. The Air Force awarded LTV a contract to build two prototypes for further testing, and by 1989 they started flying, to some acclaim from the Air Force and other supporters.[6]

The A-10 no longer had high-powered sponsorship. Fairchild's president in the mid-1980s, Robert Sanator, later said that a closed production line was nearly impossible to reopen—and the A-10's line had shut down in 1984. The CASMARG asked the company to propose either a new design or a modified A-10, but the company's ability to participate ended with two developments. One was the Air Force's blatant desire for the F-16, and the other was the company's ignominious departure from military aircraft production in 1987. The Air Force selected Fairchild to build a new trainer jet, the T-46, but the company failed to deliver the plane on time and within budget. The service refused to fund the T-46 in 1985 and, after some controversy, terminated the program in 1987. With this, Fairchild-Republic shut down its Long Island production facility and ceased military aircraft production operations. Grumman, a Long Island producer of Navy combat planes such as the F-14, assumed responsibility for the A-10, but an out-of-production Air Force plane was a low priority for the company, to say the least.[7]

With the Hog virtually on its own, its equipment upgrades were mostly small scale. A more extensive avionics modification, called the Low Altitude Awareness and Targeting Enhancement (LASTE) program, only occurred due to an unusual combination of circumstances. The Air National Guard—especially the Syracuse outfit before its transition from A-10s—sponsored a GE fire control modification for the A-10. Incorporating a ground collision avoidance warning system into the modification helped gain acceptance by the Air Force for flight safety reasons; otherwise, LASTE probably would not have survived initial review. The late 1980s found it undergoing test and development. The Air National Guard's other initiative for the plane, incorporating more powerful engines, yielded disturbing conclusions. (TAC also studied this issue.) Installing GE F404 engines—the same ones powering the Navy's F/A-18 strike fighter and the Air Force's F-117 Stealth fighter—introduced the kinds of problems anticipated in earlier studies. Fuel consumption at the high power settings required to increase the Hog's speed was high enough to threaten loiter capability. One proposed solution was to avoid using the high power settings except in serious combat situations, but as one ANG flight test officer observed, the temptation to overuse it would probably be too strong.[8]

Congressional Hearings and Congressional Action

The year 1988 witnessed more congressional probing of the debate via budget hearings. In March a House Armed Services subcommittee called in various military and OSD leaders to explain their positions. There was a brief flash of congressional A-10 advocacy when New York's Sam Stratton asked Don Fredericksen if the CASMARG had given Fairchild-Republic a chance to propose a modified A-10. Fredericksen answered that Fairchild's difficulties rendered it unable to respond, and there the matter rested. (Fredericksen mentioned that the Air National Guard was the only agency still interested in A-10 modifications.) Marine generals praised their Harrier so well that the lawmakers pressed OSD and Air Force representatives to study it more closely for possible purchase. Fredericksen answered that he did not like its combat damage vulnerability, and Air Force witnesses explained that they had twice rejected it for lack of desired payload and range. Air Force and Army representatives reiterated their agreement, based upon AirLand Battle, that the "battlefield of the 1990s" required an interdiction plane like the F-16. Even Fredericksen conceded that the Army's confidence in defeating enemy ground forces made it more interested in interdiction. The same witnesses appeared before the Senate Armed Services Committee that same month, with the same result. The congressional testimony and other factors, such as the service's brazen F-16 sales attempts, an *Aviation Week* article commented that opinion "had hardened on Capitol Hill that the Air Force has willfully obstructed an on-the-merits evaluation" of follow-on CAS planes.[9]

By mid-1988, the opinion became action. In its report for the National Defense Authorization Act for Fiscal Year 1989, the House Committee on Armed Services indicated that it wanted the OSD and the Air Force to consider the AV-8 and other existing aircraft in their CAS plane studies. The National Defense Authorization Act for Fiscal Years 1988 and 1989 demanded that the secretary of defense provide a master plan for CAS/BAI by the end of 1989. Further congressional action that fall augmented this directive. The Defense Authorization Amendments and Base Closure and Realignment Act ordered the secretary of defense to examine transferral of the A-10 and the CAS mission to the Army. It also called for the OSD's director of defense operational test and evaluation (OT&E) to plan a competitive flyoff between CAS plane candidates.[10]

Though one publication stated that Congress aimed to resolve perennial CAS problems, other sources later claimed that the legislators wanted to scare the Air Force away from choosing the F-16. Although most lawmakers seemed to be against the fighter, opinions differed in Congress about the best alternative, which might be no plane at all. In response to concerns about the national budget deficit in the late 1980s, as well as to a reduction in U.S.–Soviet tensions, all government agencies worked to reduce the military budget far more significantly than anticipated. In late 1988 Fredericksen discerned this potential threat to CAS plane procurement, and proposed another CAS plane compromise: reject the modified A-7 proposal and accept a mix of currently operational A-10s and F-16s, which would both be modified with FLIR and ATHS technology. OSD's OT&E director, John Krings, also understood the budgetary exigencies and tried to make the flyoff test a relatively quick and inexpensive undertaking—at least for its ambitious aims. He proposed a two-phase test to include only the A-10, A-7 (as a modified A-10 "surrogate"), and F-16. Phase one would last several months and address three different European battle scenarios. Krings hoped that it would answer many of the questions concerning each competitor's suitability and thus eliminate much of the rest of the test. If not, then the much longer second phase would address more diverse scenarios. Krings planned to stage the test at Army maneuver ranges, using large ground units to simulate modern combat as much as possible. Further, he wanted the most sophisticated tracking and recording devices to track the proceedings.[11]

If the OSD or Congress thought that they could change the Air Force leadership's mind through compromise, a mandatory test, or intimidation, they were wrong. As the public debate revealed, the service did not stop plugging the F-16 through 1989. Regarding Fredericksen's latest compromise offer, *Aviation Week* reported in late February that "senior officers are not happy with the prospect of retiring A-7 aircraft and retaining A-10s." Service leaders carped at Krings's test plan, saying that it did not account enough for the "ability of high-speed aircraft to hit targets in a fast-pass [*sic*] attack and the Army's ability to provide . . . accurate target information." They also protested Krings's exclusion of BAI missions and his choice of the Army to conduct the test using its own facilities.[12]

Air Force opponents in Congress did not help their case in House hearings on close air support, held in April 1989—unless their intent was to highlight the Air Force's stance. Various staff members and legislators wanted to formalize the recent pressure upon the service, but the panel members' opening statements indicated that they wanted to think about the issue before acting. Perhaps they had heard enough of the anti–F-16 critique, and wanted to hear the other side. Whatever the reason, they called three witnesses who mostly pushed the service's viewpoint: Air Force historian and retired general Perry Smith, former Air Force secretary Hans Mark, and the OSD's OT&E director, John Krings.[13]

Smith claimed that he would be objective and use the "lessons of history," but he supported his service. His historical examples cited only air support efforts in which air superiority was contested or where interdiction played a more prominent role than CAS. He cautioned the House panel not to rely too much upon the Korean and Vietnam conflicts for air support lessons. (However, in advocating fast fighter air support, he cited his experience flying F-4s in Vietnam.) In his conclusion, he recommended that the representatives read such authors as Richard Hallion.[14]

Opening his testimony, Hans Mark confessed his lack of CAS expertise, saying, "My experience is more in the laboratory than on the battlefield." He claimed that new technologies, such as more effective SAMs and precision-guided munitions (PGMs), would change air support. To Mark, the Air Force's F-16 choice correctly followed its World War II practice of using fighter planes for CAS. In this, he ventured from discussing technology to citing combat results, and thus made dubious claims. He stated that Soviet jet fighters did not suffer losses to Stinger missiles like slower aircraft, but they did not fly much CAS either. In discussing the Falklands War, he cited the Argentines' cruise missile successes, but these weapons attacked ships in the open sea, not troops in contact. To Mark, the Argentine A-4's performance showed the worth of high-performance aircraft, but the U.S. Air Force had rejected the A-4 as too slow. Mark claimed that the Argentines' Pucará twin-turboprop counter-insurgency plane suffered heavy losses to British fighters and flak, but it did not.[15]

Krings thought the test might reveal that with new weapons, such as the Apache, the Army would probably need less CAS. Asked to define the CAS to be flown in the test, Krings said it occurred from the line of troop contact to approximately twelve kilometers behind that line. Smith added a definition oriented more toward interdiction and told the legislators that they should consult the Europeans. Although the witnesses had supported the Air Force position, the hearings report included a cartoon that perhaps reflected legislators' exasperation with either the competing CAS plane concepts or the Air Force's attempt to make the F-16 do too much. It portrayed an aircraft that looked like a flying armored tugboat carrying every avionics feature and weapon in the U.S. defense inventory.[16]

As part of a DoD budget hearing, the Senate Armed Services Committee brought in various witnesses who could address the debate. Former U.S. Army in Europe commander Glenn Otis supported the Air Force, since he wanted planes that could shift from one mission to another. TAC's director of requirements, Brig. Gen. Joseph Ralston, conceded that his service would retain a mixed CAS plane force of A-10s and F-16s through century's end, which was an interesting departure from his 1988 assertion that the Air Force's future CAS plane would only be the F-16. His testimony signaled a change in Air Force fortunes.[17]

The 1989 House and Senate hearings marked the zenith of the Air Force effort to procure CAS F-16s. Its congressional witnesses delivered virtually uncontested testimony, and the Army supported it. The joint statement that Welch and Vuono made after the F-16 air support demonstration reaffirmed their stance as well as their desire to test the modified A-7. However, they accepted some of the Fredericksen compromise, in that they considered keeping A-10s not only for the FAC role but also for CAS in low-intensity conflicts. The OSD and Fredericksen tried to push the compromise further by including it in a budget proposal to Congress. Fredericksen hoped that it would persuade Congress to cancel the test, since the flyoff would impose a long delay upon CAS plane selection and on needed modifications to existing CAS planes. As one legislator observed, the planned full test would last longer than World War II. The compromise proposal also aimed to retain a mixed force of planes that could complement each other in the air support spectrum. Finally, it would preserve the F-16 production line, which, with the budget cuts, faced termination if the CAS plane buy fell through. Still, in early 1989 the future augured well for the service's preferred CAS plane, as an *Armed Forces Journal International* headline proclaimed: "A-16 Stretches Lead in Race for Close Air Support Role."[18]

However, the service had overplayed its hand. It did not moderate its message in the on-going CAS debate, and in spite of apparent compromise concessions, rumors surfaced about the Air Force retiring all Hogs in the near future. OSD staff who believed that Fredericksen should not have compromised chafed anonymously to the defense press and to Congress about the Air Force's attitude. One told *Aviation Week* reporters that the service's credibility suffered due to its F-16 obsession. The Air Force's design stipulations and ensuing high cost estimates for accepting the Harrier or a re-engined A-10 (separate from Fredericksen's compromise upgrades) irked the lawmakers. The representatives may not have asked Air Force witnesses hard questions during the earlier hearings, but they disliked the service's disregard of their wishes. In September the House denied funding for the A-10 and F-16 modifications requested to support Fredericksen's compromise effort, and by this time Congress had added the AV-8 to the planes studied in the big test. "Nobody really believes the Air Force," said a congressional staff member. "The blinders they have on the F-16 are unbelievable."[19]

THE POLITICAL WAR

In November 1989 the CAS debate reached its climax in a series of open clashes between the Air Force, the chairman of the Joint Chiefs of Staff, and

Congress. That same month DoD completed a report on the roles-and-missions review required by the 1986 Goldwater-Nichols Act. In it the chairman, Adm. William Crowe, tried to resolve the CAS plane debate by declaring that all services performed CAS in their own way. One source stated that Crowe's Navy background perhaps limited his awareness of the Air Force–Army CAS legacy. He might have been directing a parting shot at the perennial issue, since his office published the report one day before he retired. In any case, his report suited neither Air Force Chief Welch nor Army Chief Vuono. Such an indefinite statement of responsibilities risked upsetting the delicate situation with Congress, and threatened yet another interservice squabble. Congress stirred the pot by demanding an answer from DoD by the end of 1989 about transferring the CAS mission to the Army—something both services vehemently opposed. Vuono and Welch rebutted Crowe in attachments to his report. Informing the chairman that their services had a long air support relationship and that they precisely defined CAS because of it, they reminded him that they saw CAS as a function performed by Air Force fixed-wing aircraft and not by Army helicopters. The new chairman, Gen. Colin Powell, backed the service chiefs. He was more aware of the issue's background and did not want to upset matters.[20]

Although the Air Force showed its mettle in dealing with OSD critics and Admiral Crowe, its resolution failed against Congress, which controlled the federal budget. November saw the Senate endorse the House's earlier denial of most of the money requested for the A-10 and F 16 modifications, but that was not all. By this time the legislators had also directed that the test include yet one more plane, the Navy's F/A-18 Hornet strike fighter, as well as the Army's attack helicopters and guided missile artillery. They wanted everything involved with CAS analyzed to solve its riddles—recalling Congress's and the OSD's desires for a similar test in the 1970s. Meanwhile, the estimated cost of the test ballooned from Krings's initial $125 million estimate to over $400 million. Its length stretched to six years, and its start date slid back to 1995.

When the lawmakers encountered what they called Air Force "lobbying" against expanding the test and letting the Army conduct it, they brought the service to heel by denying funds for any F-16s until it produced a test plan including the additions. An anonymous, retired, and apparently senior ranking Air Force officer told *Defense News* that a "few creeps in Congress" and "Washington-based, think-tank officials" were to blame. General Russ told reporters, "We need to . . . write down what we *think* Congress wants us to do, and then take it to Congress and have them sign up to it" (italics in the original). The congressional side delivered blunter rejoinders. "This is an Ayatollah Khomeini strategy," said one staff member, pointing out that, as the notorious Iranian leader had done with U.S. diplomats, Congress had taken the Air Force's prize plane hostage. Another warned, "What Congress is saying to the Air Force is, 'Don't jerk us around any more. That's not the way to win the hearts and minds of the people on the Hill.'" Yet another specified that the funding ban would remain until the service "shapes up and takes close air support seriously." As an Air Force Pentagon staff officer later recalled, "Congress controls the force structure." It did so with a vengeance in this case.[21]

Interestingly, Air Force leaders and others later dismissed congressional actions such as the CAS mission transfer to the Army. Welch claimed that he gave it little thought. And though the transfer issue lingered through 1991, then Air Force Chief of Staff Gen. Merrill McPeak later said, "I spent a grand total of zero time worrying about the problem."[22]

However, the Air Force did pay attention to Congress, as it began cooperating with that body on some actions. Responding to the demand for a completed, modified test plan, the service finished it by means of an around-the-clock session at TAC Headquarters in spring 1990. The lawmakers reciprocated by releasing contractually committed F-16 funding. Air Force leaders also showed their awareness of congressional pressure by continuing public complaints about the test plan. They were probably concerned about the ever shrinking defense budget and the effects of any further congressional action. The European affairs backdrop to the late 1989 CAS confrontations contributed to this precipitous economic slide. Throughout the autumn, the people of the Warsaw Pact nations overthrew their communist masters and thus undermined the vision of the big European war that justified the CAS F-16, AirLand Battle, and much of the U.S. military force structure.[23]

The shrinking budget also affected the on-going OSD attempts to resolve the CAS plane question. When Congress ordered the secretary of defense to deliver a master plan for a CAS/BAI plane, the secretary in turn ordered the Joint Chiefs of Staff to prepare a Mission Needs Statement (MNS) supporting it. Their April 1989 MNS matched Air Force studies, in that it dismissed the A-10 as unsuitable for BAI and opted for the F-16. Perhaps reflecting the recent Fredericksen compromise proposal that aimed to guarantee upgraded CAS planes in the mid-1990s, it also stipulated a 1995 delivery date, which ruled out any plane not already in production. However, the service chiefs knew they had an MNS that too much resembled the Air Force plan that aroused congressional ire, so they called for a Defense Acquisition Board (DAB) to review it. Similar to the DSARC that held developmental milestone meetings for new weapons systems, the DAB would present a DoD judgment to the lawmakers instead of one that appeared to be one service's partisan stance. (The DAB consisted of leaders from the Joint Chiefs of Staff office, the OSD, and the services.) In the meantime, the Air Force continued a Cost and Operational Effectiveness Analysis (COEA) of the CAS plane issue, which still pushed for the F-16.[24]

The OSD announced that the DAB would make an acquisition decision even though planning for the big flyoff continued, but the DAB's first milestone meeting leaned toward compromise. It resolved to make decisions for both short- and long-term CAS plane procurement, and aimed to do this by April 1990. It might have met this goal, but the Air Force delayed submitting all required inputs, apparently until it could see how the budget situation developed. One of the inputs it did submit was the COEA, which provoked another face-off with OSD officials. Thus, the DAB pushed its decision milestone back to autumn 1990, which allowed the shrinking military budget to dictate the choice even more. The recent fall of communism sparked discussion over whether the world really was about to enter a new millennium of peace, liberal democracy, and consumerism. The implications for U.S. defense policy were obvious, as force reductions increased. In 1984 there had been plans for an Air Force of forty tactical fighter wings by 1990, but that year started with the service holding at a peak of thirty-six and anticipating a reduction to twenty-five by 1997. These circumstances induced a game of musical chairs for planes and missions within the service, not to mention any proposed acquisitions.[25]

Things started looking up for the A-10 in this shuffle. In spite of the cuts, Congress voted funds for upgrading the plane's engines. The A-7F faced an ominous future, for though Congress included it as one of the test candidates, the Air Force had by this time backed away from it. Two A-7F prototypes began developmental flight tests in late 1989, but shrinking force projections left room only for existing planes, and most of those would be F-16s. Also, the existing A-7 airframes were old, and suffered fatigue problems that would negate any performance advantages conferred by a more powerful engine. Interestingly, the Texas congressional delegation did not push the A-7 as hard as previously. A possible reason was that the F-16's manufacturer, General Dynamics, was also based in Texas. And if the ill fortunes of such Texas lawmakers as ex-senator John Tower and House speaker Jim Wright were any indication, the Texas congressional delegation did not have its usual clout.[26]

The DAB's fall 1990 final decision further demonstrated how the options had decreased. The board made no long-term choice, apparently due to the budget situation and Congress's apparent desire for the big flyoff. Instead, it decided to modify approximately four F-16 wings and retain approximately two A-10 wings, whose planes would also be modified with LASTE and FLIR equipment. This was a sharp drop from the nearly seven hundred Hogs available, but at least the dedicated CAS plane survived the debate and the sharp force cuts then under way. Overcome by the exigencies of force reduction, which included a greater reduction of tactical wings to twenty, the Air Force accepted the DAB decision without complaint. "It's an affordability issue," said Maj. Gen. Joseph Ralston, now the director of tactical weapons procurement in the Air Force Headquarters acquisition office.[27]

Still, given its outspoken promotion of the F-16 at the A-10's expense in previous years, service leaders could have insisted that multi-role F-16s were even more necessary in the smaller Air Force and could have rejected A-10s entirely. That they did not do so was probably because of Congress's and the OSD's scarcely veiled desire for another CAS plane besides the F-16—as well as the views of people like General Russ who still wanted a mix of planes. Russ later observed that DABs normally made compromise decisions, and in this case the DAB apparently opted for already operational planes, which satisfied all sides to some extent. Although it wanted a flyoff, Congress did not object to the DAB decision. This may have been due to one other factor, which no one at the time or later would admit drove the A-10's reprieve. That was a potential real world assessment of air support aircraft—the Operation Desert Shield deployment to face the Iraqis.

Desert Shield, Congress, and the Hog Drivers

Iraqi dictator Saddam Hussein sent his army into Kuwait on 2 August 1990. Three days later President George Bush proclaimed Operation Desert

Shield and ordered U.S. troops to the Middle East to protect Saudi Arabia and drive the Iraqis out of Kuwait if they did not leave voluntarily. Although the United States organized a large coalition of nations that augmented the U.S. force and isolated the Iraqis, Saddam did not back down. Weeks dragged into months as diplomatic initiatives and economic sanctions failed to resolve the impasse. President Bush ordered more forces to the region throughout autumn 1990.

A-10 units were deployed, though some sources claimed that the Air Force did not want to use them. A story popular among Hog drivers and other A-10 supporters related how the overall commander for coalition air forces, U.S. Air Force Gen. Charles Horner, briefed the coalition commander, U.S. Army general H. Norman Schwarzkopf, on the air unit deployment. Schwarzkopf asked why no A-10s had arrived, and Horner replied, "Sir, we believe that our F-16s can do air support," or words to that effect. Fearing further advance by Iraqi troops, Schwarzkopf shot back, "Skip the sales job; I want my A-10s!" Another account had Defense Secretary Richard Cheney ordering a reluctant Air Force to send Hogs. Research reveals no solid basis for either story, but they are both understandable given the service's lobbying campaign for F-16 close air support.[28]

The Air Force deployed units to Desert Shield in a sequence that followed its doctrine. This made sense, since the Iraqis had an air force with the latest Soviet- and French-made tactical jets. Thus, some air superiority fighters and early warning planes were deployed first, followed by some interdiction planes, and then some A-10s. Horner was concerned about Iraqi ground units poised at the border between Saudi Arabia and Kuwait, and the time between the first air deployment and the first A-10 arrival was ten days, not weeks or months. In Desert Shield's initial stages, there were only small ground units between the U.S. bases and Iraqi forces; Horner demonstrated his anxiety when Air Force Headquarters planner, Col. John Warden, briefed him on a proposed air war plan. Upset with what he perceived as Warden's flippant disregard of the Iraqi ground threat, Horner kept the plan but sent Warden back home.[29]

The Air Force could not deploy everything at once. The initial deployment represented a fraction of the entire air component that later fought the war, but all the early air units to arrive encountered problems with air base availability and suitability. Even so, Horner's irritation with Warden recalled the Hog driver maxim quoted earlier: "You can kill all the MiGs you want, but if you return to base and the enemy tank commander is eating lunch in your snack bar . . ." The incident also underscored flaws in the Air Force and Army's stance in the CAS plane debate. America's diverse worldwide defense commitments meant that it could encounter a war-fighting scenario favoring other weapons and tactics besides those foreseen in AirLand Battle. Interestingly, some post–Gulf War Air Force histories criticized Horner and others for being too blinkered by their air support commitment to grasp the genius of Warden's master plan.[30]

The mobilization had its effect at home as well. There was yet another congressional hearing on close air support, this one conducted by the Investigations Subcommittee of the House Committee on Armed Services. Like the 1989 hearing, it lasted only one day and featured only three witnesses, but they were sympathetic to the mission and the dedicated planes that flew it. Though Chairman Nicholas Mavroules declared that he wanted to discuss the mission itself, both the panel members and the witnesses moved toward addressing CAS plane technology because the witnesses asserted that the mission required certain aircraft characteristics. Also, with Desert Shield under way, the legislators asked what airplane could best stop Iraqi armored divisions, and the ensuing discussion indicated that this was the A-10. Finally, panel member Andy Ireland and some committee staff publicly opposed the Air Force's F-16 tack, and the lawmakers apparently wanted a hearing that highlighted their concerns.[31]

Two of the witnesses supported the CAS mission and dedicated planes outright. David Isby, a consultant with BDM, stated that the mission remained important because of America's many commitments around the world and possible war-fighting situations. Pointing out that the fixed-wing CAS plane was the one heavy firepower weapon that could deploy itself, he mentioned that the A-10 appropriately backed up the first Army units to be deployed in Desert Shield. Recounting various historical instances where CAS decisively affected battles, he stated that "no general or admiral can permit a battle to be lost or avoidable casualties to be sustained through a reluctance to use fixed-wing close air support." Relying upon multipurpose planes for CAS denied the ground commander the potential for decisive firepower application on the battlefield. But Isby did not want only Mud Fighters or Hogs. He appreciated the "broad spectrum of firepower" provided by attack helicopters, A-10s, and other attack jets, and mentioned that this air attack mix would vex Iraqi defenders. Concerning the modern air defenses that the Air Force used as a justification for F-16 CAS, he believed that they were not as strong as anticipated. Isby cited his visits to Afghanistan as evidence that even the Stinger SAM could be defeated if air forces used good tactics and countermeasures.[32]

Tom Christie agreed with Isby that a multi-role plane for CAS risked loss of expertise. Without mentioning the A-10 specifically, he asserted that the mission required a rugged plane that could loiter and carry large weapons payloads. Later he said that he was uncertain about how much longer the Hog could usefully serve, but added that the end of the Cold War had certainly changed the criteria for determining its fate. The third witness was retired Air Force general Robert Herres, who presented his service's party line. He believed that the Army's attack helicopters fulfilled the traditional CAS plane's role, and saw Air Force planes providing complementary firepower only when needed. As for the A-10, he condemned it as unsuitable in the modern air combat arena due to its lack of speed and technological sophistication. Citing the Goldwater-Nichols Act's augmentation of a wartime theater commander's control, he praised the F-16 for giving that commander the flexibility to accomplish CAS or BAI with one air asset. To him, Air Force critics were lost "trying to solve a problem of the 'fifties'" because Army firepower and the theater commander's control over all units precluded firepower deficiencies.[33]

Christie rebutted Herres, saying that the F-16 lacked the characteristics necessary for effective CAS. Ireland agreed, asserting that multi-role fighters would accomplish no mission well. To him, Herres's arguments were a "kindergarten discussion." The representative also accused the Air Force of ignoring a request by Don Fredericksen that CASADA include traditional CAS plane requirements. Mavroules questioned Herres's claims of A-10 obsolescence by citing the varied war-fighting conditions around the world.[34]

The hearing witnesses were not the only ones to believe in the Hog's chances against the Iraqis. Hog drivers told the *Air Force Times* that they hoped the deployment would justify keeping the plane. They believed that the Iraqi air defenses were no worse than any simulated air defense setup that they had faced in training exercises. *US News & World Report* noted that, though the Air Force had previously "dumped on the A-10," the service now praised it "as the close air support weapon of choice against Iraqi tanks."[35]

By early September 1990 the 354th TFW at Myrtle Beach AFB, and the 23d TFW at England AFB, Louisiana, had each deployed two squadrons to King Fahd Air Base in Dhahran, Saudi Arabia. At first, the Hog drivers thought they would employ the standard low-level tactics to account for the Iraqi air defenses, which included the latest Soviet-made missiles. However, special conditions demanded an initial BAI campaign using high-altitude dive-attack tactics. The Iraqis seemed content with holding Kuwait, which would require an offensive to dislodge them. Until that ground advance began, air attacks would favor interdiction to soften the Iraqi Army infrastructure. Next, air defense suppression units were deployed into the theater, easing concerns about air defenses. Finally, Kuwait and southern Iraq's flat and featureless terrain precluding using it to achieve surprise. Pilots also complained that poor visibility due to blowing dust eliminated any horizon reference, which degraded their ability to judge height above the terrain anyway. Horner and his chief planner, Brig. Gen. Buster Glosson, confirmed the new tack when they told all U.S. Air Force air crews to trust the defense suppression effort and avoid low-altitude attacks, which encountered the small arms fire and AAA that historically caused most ground attack losses.[36]

There was some debate about whether the air planners wanted to hold the A-10s in reserve for potential CAS tasking instead of assigning them to fly BAI missions with other jets. Some A-10 pilots claim that this was so, though Glosson later said that his staff always planned to use A-10s at the start of hostilities. Either way, Hog drivers welcomed the opportunity to demonstrate their plane's abilities, though many initially questioned BAI missions, higher altitude attacks, and another initiative—night flying.[37]

Night action was not an easy sell to the pilots, who were conditioned by many years of day air support operations. Indeed, this was one of the primary arguments against retaining A-10s made by advocates of F-16 CAS, since the service had so far made no serious effort to upgrade Hogs with night vision equipment. But the Hog drivers overcame their own initial opposition and improvised, for when they examined the issue a little closer, they saw certain advantages. Flying at altitude removed the concerns about terrain avoidance in darkness, and reduced defenses meant that the pilots could handle what remained without achieving nighttime flying task saturation. Indeed, they fashioned night formation procedures and tactics that closely resembled those used by A-1s in Vietnam. The Hog drivers solved the night target acquisition problem by using their Maverick missiles' infrared sensors to locate targets. Though this was a primitive night vision source, the BAI mission tasking meant that they did not have to worry about hitting their own troops. When they found an enemy position, they illuminated it with flares and attacked it.[38]

Once the war commenced, the Hog drivers found that night operations provided another important benefit. The Iraqis could neither see them nor, with their quiet engines, hear them before they attacked. In contrast, the Iraqis' gunfire was very visible, especially if they used tracers. This allowed the A-10s to locate and destroy these trouble spots during their missions. The day mission A-10s had no such luck, and there were also later complaints that the leadership did not work hard enough to repaint Hogs in a lighter color. The paint scheme had nearly always been dark green— apparently to show "solidarity" with the Army and to provide camouflage against Soviet fighters as Hogs flew low in a European war. Higher altitude tactics required a color that would help hide A-10s from gunners on the ground; a proposal for a light mottled gray scheme came with one of the three other A-10 units deploying to King Fahd in late 1990 and early 1991— the Air Force reserve's 917th TFW from New Orleans. (The other units were the 511th TFS from RAF Alconbury, England, and the 23d Tactical Air Support Squadron from Davis-Monthan AFB, Arizona, which was an A-10 FAC unit.) The leadership of the two A-10 wings that formed at King Fahd addressed the issue, but it apparently got lost amid other war preparations.[39]

The Real War

Called Operation Desert Storm, the war began on the night of 17 January 1991, and A-10s were among the first attack waves. From that date to war's end on 28 February, they flew eight thousand combat sorties, while losing only 6 planes. To prevent false estimates of Iraqi Army losses, they conservatively estimated their "bomb damage assessment" (BDA) kill claims, and due to their lack of sophisticated targeting gear, the coalition staff devalued these even further. Still, they received credit for destroying 987 tanks, 926 artillery pieces, 501 armored personnel carriers, and 1,106 trucks. Hogs also destroyed other targets, such as Scud missile sites, SAM sites, and two helicopters.[40]

The varied targets reflected the diverse missions that Hogs flew in this war. Not only did this recall technology historians' comments about unintended uses and consequences, but it also reflected this war's—any war's—special conditions. The Air Force cowed most of the Iraqi air defense radars throughout much of Kuwait, allowing A-10s to roam more freely. The coalition needed many air tasks accomplished, and the A-10 with its excellent loiter and weapons carriage capacity answered the call. Hog drivers often flew their aircraft on three missions of roughly three hours' duration daily. And since the coalition air leaders took advantage of the Hog's easy maintainability and positioned A-10s at forward bases, their loiter time over target during these missions was even greater. It was certainly better than that of the F-16s used for BAI missions—and the weapons load difference was even more glaring. An average A-10 load during the war was six bombs of 500 pounds each, either two or four Maverick missiles, and one thousand rounds of 30-mm ammunition. This compared quite favorably with the F-16's load of four bombs of 500 pounds each and five hundred rounds of 20-mm ammunition.[41]

In the BAI missions, A-10s flew as directed to "kill boxes" in southern Iraq and Kuwait and attacked targets therein. When the Iraqis launched Scud long-range artillery missiles at Israel and Saudi Arabia, Hogs became the chosen airplane for the long hours spent searching for Scud launcher and stowage sites. As the historian of A-10s in the Gulf War, William Smallwood, observed, "They [the coalition air leaders] didn't want F-16s. . . . The F-16s would get out there and have a two-minute TOT [time over target]—and they still wouldn't know where the Scuds were." Hogs flew armed escort for combat search-and-rescue (SAR) helicopters—though, as with the A-10, SAR was supposed to be obsolete thanks to advanced air defenses. Desert Storm demonstrated that rumors of the Hog's and SAR's demise were greatly exaggerated. On 21 January, A-10s provided defense suppression in the successful rescue of downed F-14 aviators inside Iraq.[42]

There were other, diverse missions. Detachments from the first A-10 FAC outfits also served in the war, assisting not only the few available CAS missions but BAI Kill Box missions as well. As the ground offensive date neared, coalition commanders sent A-10s on low-level reconnaissance forays deep into Kuwait to reconnoiter the Iraqi Army's status. In yet another repudiation of Air Force claims in the CAS plane debate, one account stated that they were chosen for these hazardous flights because "the A-10's slow speed, armor, and survivability qualified it best to perform the mission." Although coalition air forces intimidated Iraqi radar SAM operators in Kuwait, a few installations fired the occasional pot-shot missile at overhead planes. The leadership directed the Hogs to help destroy these sites. They normally accomplished this mission using their Maverick missiles, which provided some stand-off capability and good accuracy. Indeed, Hog drivers were the champion Maverick missile shooters in Desert Storm; their five thousand shots accounted for 90 percent of the Air Force's total Maverick usage. After the Hogs downed two helicopters, A-10 historian Smallwood reported that Hog drivers jokingly called their ugly aircraft the "RFOA-10G," to denote all of its newly assumed missions. These initial successes inspired General Horner—who allegedly once wisecracked about his Hog driver son, "I think he died from brain damage"—to tell his staff, "I take back all the bad things I have ever said about the A-10. I love them! They're saving our asses!"[43]

However, the Hogs' spectacular success and ever more daring assignments bred an aggressiveness not only among A-10 pilots but also within the coalition air command. Pilots were encountering difficulty finding undamaged targets, in part due to a problem with confirming their destruction from high altitude. Thus, coalition air leaders allowed lower A-10 attack altitudes on 31 January to resolve the problem. However, this change meant that enemy gunners now had a better chance to find and hit A-10s. One A-10 had fallen on 25 January, and since Desert Storm's first day, a few pilots had brought back A-10s with repairable battle damage. The Hogs were proving the worth of their survivability features. The Hog driver who led the successful SAR effort to save the F-14 pilots, Capt. Paul Johnson, later brought a plane home with much of its right wing shot away. On a couple of other occasions, the Hog's redundant flight control system allowed crippled planes to fly home. Aircraft Battle Damage Repair (ABDR) crews repaired in-theater all but one of the estimated seventy A-10s damaged during the war—and of those, twenty suffered significant damage. These repairs were usually quick, and used cheap, accessible materials. In one case, an ABDR team used a broomstick to replace a flight control rod! Through mid-February, the A-10's loss rate compared most favorably with that of the Marines' AV-8 Harrier, which paid the price of survivability design neglect predicted by Don Fredericksen.[44]

Horner and Glosson were confident enough to sanction Hog missions ever deeper into Kuwait to attack the heart of the Iraqi military machine, the Republican Guards divisions. Initial strikes against the Tawakalna Division went well, but on 15 February, a flight of two A-10s fell during their attacks on the Medina Division. These losses upset the A-10 commanders, one of whom observed in a message to Horner that F-16s were attacking targets south of the A-10s' target sets. Horner responded by restricting A-10 operations to southern Kuwait, and the A-10 leaders seconded his action by restricting gun usage to SAR and CAS missions. These were wise moves, if one accepts that the leadership wanted to quell overaggressive behavior and eliminate as far as possible the chance for air casualties. (Horner said it was a "gut call.") The reception of these moves by Hog drivers themselves was mixed, but in any case the A-10 commander's letter and the ensuing restrictions formed the foundation of later criticisms against the A-10.[45]

By this time the coalition ground offensive was about to commence, giving the A-10 pilots a chance to fly traditional CAS. So far, CAS opportunities had come through occasional calls to terminate Iraqi shelling along the front lines and, more prominently, during the 29 January Iraqi incursion at the Saudi Arabian border town of Khafji. A-10s acquitted themselves well during this fight, but one A-10's Maverick missile malfunctioned and accidentally hit friendly troops. When the coalition ground advance began on 24 February, General Horner lifted the altitude restrictions and told A-10 commanders, "That's the time to hang it out if you have to, but just don't do anything foolish. . . . I'm in awe of what your guys are doing with that airplane. So I know you'll do the job that has to be done when the friendlies need you."[46]

As the A-10 unit chronology and subsequent *Gulf War Air Power Survey* put it, the Hog was now "in its element." The lower altitudes meant that Hog drivers could use the GAU-8/A gun at ranges where it was most effective. During this time two Hog drivers, Capt. Eric Salmonson and First Lt. John Marks, used both Maverick and gun to destroy twenty-three tanks and set a single-flight tank-kill record. The ground advance reduced (but certainly did not eliminate) the defenses the Hogs faced. Warthog historian Smallwood recounted several cases where the plane's ability to operate in marginal weather helped coalition ground units, though the weather and burning oil well smoke got so bad that it affected even A-10 operations.

A-10s sometimes removed potential Iraqi Army obstacles, such as tank and artillery units, before the ground forces arrived. In one case, Lt. Col. Jack Shafer and Capt. Tom Atkins used their GAU-8/As to maul a defensive line of T-72 tanks. (They not only killed enemy armor but also disproved assertions made in the CAS plane debate that the GAU-8A could not kill the latest model Soviet tanks.)[47]

The Hogs' success was so overwhelming that the A-10 command's 25 February situation report ended with the sentence, "Having a wonderful day." The debrief of a captured Iraqi regimental executive commander described the Hog's success even better: "The single most recognizable and feared aircraft was the A-10. This black colored jet was deadly accurate, rarely missing its target. Seen conducting raids three and four times a day, the A-10 was a seemingly ubiquitous threat. Although the actual bomb run was terrifying, the aircraft's loitering around the target area prior to target acquisition caused as much, if not more, anxiety since the . . . soldiers were unsure of the chosen target." Though CAS opportunities remained fleeting even during the ground war due to the coalition armies' overwhelming success, the Hogs' performance and capabilities impressed General Schwarzkopf enough that he used them for top cover during the truce talks. The Warthog had left this war with a good reputation —or so it seemed.[48]

ECHOES OF THE CAS DEBATE: ASSESSING THE REAL WAR

True to the shifting, often controversial nature of the CAS mission and its dedicated planes, the Hog received mixed combat reviews in the years following Desert Storm. The immediate press reaction was favorable. In the *Air Force Times*, Hog drivers expressed pride in their plane and negative opinions of their service's treatment of it. "The A-10 has been the red-headed stepchild of the Air Force," Lt. Col. Danny Clifton told reporters. "Others see us as second-class citizens. We don't see it that way." Another pilot summarized the Hog's success in these words: "It's the A-10's day in the sun." The mainstream press paid attention to CAS planes this time. The *Washington Post's* Jack Anderson and Dale van Atta wrote a short piece on the A-10 entitled "The Hero That Almost Missed the War." The team quoted Horner's "They're saving our asses!" and described how the plane's design helped it succeed. John Fialka of the *Wall Street Journal* went further, using Pierre Sprey to comment about the plane's development. Fred Kaplan of the *Boston Globe* did the same, and also cited the influence of Avery Kay, Richard Yudkin, Bob Dilger, and James Schlesinger. *Aviation Week & Space Technology* carried a couple of articles praising the Hog's durability and announcing that its war performance guaranteed its operational survival and equipment upgrades.[49]

For all his wartime enthusiasm, Horner did not include A-10 pilots in the group of aviators he brought to his postwar congressional appearance. Sen. John McCain expressed concern that people might conclude that the A-10's success meant that cheaper weapons could have won the war. Horner agreed that this was wrong, and dwelt upon the 15 February losses. Horner at least stated that he had "great admiration for the A-10, [and] the job it did," but he added that it must be retired as it became old.[50]

One of the first major official studies of the war also followed the Air Force's line in the CAS debate. DoD's *Conduct of the Persian Gulf War* lauded the plane's durability and the pilots' skill in handling diverse missions such as SAR, but it also criticized the A-10's slow speed and recommended that any future CAS plane fly faster. It rated the Hog's performance using two criteria seen in later studies: precision guided ("smart") weapon usage and night fighting capability.[51]

Any Desert Storm air war study was correct to focus upon smart weapons such as laser-guided bombs (LGBs), for they did well and rated serious development. The DoD report discussed the Hog drivers' Maverick missile exploits, as well as their praise of that weapon's accuracy and lethality. However, it also incorrectly asserted that the gun was virtually useless. Altitude restrictions limited the gun's effectiveness, and conservative BDA policies minimized its apparent impact. To declare an armored target truly destroyed, one had to observe it burning or exploding. The Maverick had a warhead large enough to achieve the desired pyrotechnics instantaneously, but the GAU-8/A's armor-piercing rounds often could not do so from long ranges. Indeed, at any range, they were designed to penetrate armor and wreak havoc inside. Thus, Hog drivers reported hit and smoldering tanks, which they did not claim destroyed because they did not see large fires or major explosions. Desert Storm Hog driver Mark Koechle told the author that his long-range gun shots still made some tanks explode. Other shots would make the tanks "start to smolder and smoke and finally there would be secondaries from . . . [the] ammo." When the Hog drivers fired their guns at closer ranges, they got better results. Also, determining which coalition weapon destroyed what tank became so muddled that the Army's BRL quit trying to assign credit during its postwar study of Iraqi tank relics. Advancing ground forces shot any enemy tank that did not appear completely destroyed, thus confusing the issue of whether air or ground forces hit it.[52]

The DoD report criticized the A-10's lack of night capability—but A-10s delivered promising service in this arena. As anticipated, their quiet engines and dark color made them invisible to any Iraqi gunner lacking radar, and the Air Force lost no A-10s during night operations. True, Hog drivers had to use their IIR Mavericks to obtain a primitive night vision capability. And sometimes they returned to base without using all of their weapons because they could not locate targets as they could during daytime. Still, the Maverick and illumination flares guaranteed Hog drivers some success. The report did not add that these deficiencies were due to the Air Force's neglect of A-10 night-fighting development in favor of the F-16.[53]

An even more critical assessment of the Hog's Desert Storm record came from its old academic nemesis, Richard Hallion. Apparently wanting to justify his negative prewar judgment, Hallion criticized the Hog's performance in his 1992 work, *Storm over Iraq*. He dwelled upon the gun's developmental problems and followed the DoD report in emphasizing the Maverick's importance. As for the Hog itself, he acknowledged that it received a lot of attention as the Cinderella plane that overcame its own service's contempt. However, he quickly added that the Air Force justly denigrated it because of "legitimate questions about its survivability." Hallion's own conclusion was similar to his previously published maxims: "Historically, the fighter-bomber had always proven a more survivable airplane than the special purpose battlefield attacker." Citing a Hog driver's letter expressing concern about missions in north Kuwait, he emphasized the plane's operational cost in battle damage and aircraft losses.[54]

Hallion's approach raised the issue of what he said and did not say about air support and the planes that flew it. Praising the productive "tank-plinking" missions flown by F-111s using laser guided bombs, he observed that they achieved more armor kills per sortie than the A-10s. However, given the BDA policies, a video of a 500-pound smart bomb destroying a tank guaranteed confirmation. Hogs were diverted to other missions besides armor attacks—including a couple that F-111s normally flew—but Horner was so concerned about scoring confirmed tank kills to support the upcoming ground offensive that he diverted every F-111 with LGB capability from the strategic interdiction campaign to tank plinking.[55]

In *Strike*, Hallion favored the F-16 over the A-10 for air support, but in *Storm* he conceded that the war revealed certain F-16 flaws. The fighter had limited range without tanker support and carried a limited payload regardless. He also observed that too much faith was placed in its ability to drop bombs accurately. It did well with deliveries at lower altitudes during peacetime, but the high release altitudes of the combat restrictions exceeded even its fire control computer's abilities. (Also, its gun pod was useless, due to vibrations encountered during the few occasions it was fired.) What went unsaid was the Air Force's tacit contradiction of its own claim that the F-16 was fast enough to beat heavy AAA and small arms fire. Hallion also failed to mention that two F-16s fell in one raid on Baghdad. Afterward Horner pulled F-16s from Baghdad raids just as he pulled Hogs from north Kuwait, thus proving that the air defense "level of pain" could become unbearable for even the vaunted F-16. Although their maintenance availability was impressive, the F-16s also encountered problems with airborne reliability. During the war, five went down due to "other" causes, one of which was an engine failure over southern Iraq.[56]

As if to accentuate the Army's desire for Air Force CAS in spite of Air-Land Battle and Apache helicopters, Brig. Gen. Robert Scales's official Army history of the war, *Certain Victory*, expressed Army frustration over occasional air support shortfalls. In spite of the thirty-eight-day aerial maelstrom that pulverized the Iraqis before the ground assault, Scales cited some ground commanders' complaints that the air leaders did not assign enough airplanes to air support before the advance began. He also wrote that air task system rigidity, fratricide incidents, and bad weather made ground commanders reluctant to use CAS. As one scholar noted, Scales's and other Army leaders' complaints were at least in part based upon Air Force–Army differences over airspace control. Further, Scales's assessment was not completely negative. He praised the "tenacity and skill" of fixed-wing CAS pilots, and he mentioned only the A-10 when discussing fixed-wing air support, describing it as a "devastating" weapon.[57]

Scales's perspective inadvertently questioned claims made in the CAS debate concerning the Apache and AirLand Battle doctrine. Except for a dramatic first-night raid on Iraqi radar facilities, the Army did not let its Apaches conduct any serious attacks until the ground war began. After that, they flew attacks out front of friendly forces and destroyed many Iraqi vehicles (there were also a few productive JAAT operations with A-10s). However, AirLand Battle's Air Force and Army advocates spoke of a wild battle in which the most aggressive and technologically advanced would win. The most aggressive and advanced side did win, but the war began with no hell-for-leather ground melee. Reflecting the traditional American emphasis upon firepower, it started with a nearly six-week aerial bombardment; then when the ground advance commenced, one source observed that the Army reined in its Apaches due to uncertain intelligence on defenses. The ground assault featured a brilliant turning thrust toward the enemy's rear, but later accounts noted a seeming wariness that allowed Iraq's best forces to escape. One can conclude that the heavy firepower preparation and caution were justified, as they often are in American wars, by the leaders' genuine concern for their citizen-soldiers.[58]

Other government studies also delivered guarded, if generally fair, A-10 judgments. The Air Force's *Gulf War Air Power Survey* praised the A-10's versatility and excellent work during the ground war. It also offered nuanced explanations of Horner's and Glosson's decisions to send the A-10s against the Republican Guards and to later withdraw them. However, it seemed more concerned with the A-10's combat losses than with those suffered by planes such as the Harrier and AC-130. Fred Frostic's RAND study, "Air Campaign against the Iraqi Army in the Kuwaiti Theater of Operations," described the A-10 as a workhorse. Though he too dwelled upon the A-10's losses more than on those of other planes, he also made some sensible observations: the Hog's survivability features helped it in combat, slower planes flying at low altitudes experienced more problems, and small sample size required careful handling of any loss analysis. Concerning the last point, Frostic observed that though fratricide statistics showed that A-10 units were involved in more such incidents, they worked more frequently with friendly troops than other planes did. However, he dismissed GAU-8/A tank claims, saying that the altitude restrictions made the gun ineffective.[59]

Other studies offered more generous praise. Another RAND study by James Winnefield, Preston Niblack, and Dana Johnson called "A League of

Airmen" described the A-10 as a plane maligned by its own service both before and after the war. The authors mentioned that Air Force officers told them the GAU-8/A was overrated, and in an observation ironically supporting the tale about the Air Force being forced to deploy Hogs during Desert Shield, they wrote that "some Air Force planners have . . . [said] that A-10s were either not needed in the Gulf War or were moved into the theater too soon." The authors noted that the Hog did well; they believed that the DoD report's negative assessment was based "more in Air Force and hardware acquisition preferences" than in how well the Hog flew in the war.[60]

Whether they criticized or praised the A-10, most studies saved their highest praise—in part to ensure good budget support—for the tour de force delivered by sophisticated U.S. airpower weapons such as smart bombs and Stealth Fighters. A 1997 GAO report, *Operation Desert Storm: Evaluation of the Air Campaign*, debunked some of the technological enthusiasm. Though it occasionally overstated its case as much as the pro–high technology accounts, it gave the Hog sympathetic treatment. It pointed out that many sophisticated targeting systems required good weather to operate, belying claims that the A-10 was one of a few planes lacking an all-weather attack capability. When discussing aircraft losses, the GAO was the only source that mentioned the problematic A-10 paint scheme. Finally, it observed that changing combat environments made sweeping conclusions about specific aircraft unreliable. It cited the variety of airplane types as a factor in the air campaign's success.[61]

Because of the hope that tactical jets might someday enjoy more flexible basing, many studies gave sympathetic treatment to another air combat technology, the Marines' Harrier. While criticizing the Hog's lack of sophisticated weaponry, they glossed over the Harrier's lack of same. The separate standard also applied to combat loss discussions. As the ground advance date neared, Harrier units flew at lower altitudes. They lost five planes partly as a result of their lack of survivability features. Although Hallion dwelled upon Hog losses in similar circumstances, he focused only upon the Harrier's potential in *Storm over Iraq*. John Guilmartin in the *Gulf War Air Power Survey* was one of only a couple of sources who mentioned the plane's survivability difficulties. The other was the Marines' own Battlefield Assessment Team members, who also gave the A-10 perhaps its greatest praise. They enumerated Harrier weaknesses, and then ranked the Hog second to the attack helicopter as the number-one tank killer. To them, it was ideal for CAS work, and they noted that Marine Desert Storm FACs, "acknowledging some degree of heresy for their comment, stated that the A-10 was worth as many as 10 sections [two-plane formations] of AV8Bs."[62]

A SURVIVOR AND A VICTOR

The A-10 Warthog passed flyoff tests and an effort by its own service to retire it. The late 1980s found it facing yet another major flyoff against several planes and weapons. That exercise was preempted by a real-life test called Desert Storm—a trial in which the Hog gave its best performance. It should have. Its creators had incorporated every CAS combat lesson in designing it from the ground up to fly air support.

Due to the bitter fight over CAS planes that preceded the war, the Hog's own service mostly dismissed its wartime success. The Air Force had cited Army doctrine, air defense threats, and budgetary constraints to claim that only multipurpose fighters like the F-16 could best fly air support. The Hog remained because people in the OSD and Congress thought otherwise. Although some preferred a replacement dedicated CAS plane, their efforts delayed Air Force action until the vagaries of the budget, politics, and war itself guaranteed that the best dedicated CAS plane yet built would live on.

10 CONCLUSION

The Hog and its pilots capped 1991 by winning the Air Force's biennial autumn air-to-ground gunnery meet, called "Gunsmoke," for the first time. The Maryland Air National Guard's 175th Tactical Fighter Group used the brand-new LASTE fire control modification to beat F-16, F-15E, F-111, and A-7 regular Air Force, Reserve, and Air National Guard outfits. Lt. Col. Roger Disrud of the reserve's 442d TFW also flew a LASTE-equipped Hog to win individual honors. Unavailable to Desert Storm units, LASTE installed in operational units led to victory in an event that F-16s routinely dominated. *Air Force Magazine* reminded its readers of the Hog drivers' advantage in using their plane's slower speed to give themselves a steadier aim and to get closer to the target. Of course, in CAS, that was the intent.[1]

One might wonder why Hogs had not always won if they could fly closer to the target. Gunsmoke was a sport competition that until some time after the 1991 contest emphasized bomb scores over tactical considerations. The fast jets used what were for them unrealistically close ranges to allow their more sophisticated fire control systems to beat the A-10s' advantages of speed and firing distance. When the Hog received advanced fire control equipment, these performance advantages reemerged.[2]

In the 1990s the Air Force reduced the number of operational A-10s, but at least the Hog survived the post–Cold War drawdown—unlike the A-7, F-4, and F-111. The service's long-range plans included keeping Hogs until 2028! Its operational life seemed secure, and following the DAB's wishes, service leaders bestowed upon it some of the technological largesse given to other jets. Night-vision goggles (NVGs) were an initial post–Desert Storm modification. Accounting for the new higher altitude attack preference and tacitly admitting that the green paint scheme had been a factor in Desert Storm losses, the service repainted Hogs light gray. ATHS and the Global Positioning System (GPS) satellite navigation system were slowly added. Studies continued on such modifications as making an engine change, reducing heat sources to defeat heat-seeking missiles, and covering surfaces with radar absorbent material. In 2000 the service introduced an impressive modification plan called Hog Up. It reinforced the airframe and landing gear to prolong airplane life, and increased electrical power generation to handle several avionics improvements. These included equipment for various precision weapons, and cockpit video screens that, among other things, provided tactical situation updates using data link information. Hog Up was scheduled to occur between 2000 and 2010; if funding allowed for further modifications, it would include a targeting pod for functions such as FLIR and laser designator usage.[3]

As one Air Force senior procurement staff member observed, the A-10 also required tactical upgrades because the Air Force planned for no direct successor to it. The transfer of the CAS role to the Army and the planned big flyoff did not happen either. By 2000 the service's tactical force had shrunk to just over half of what it had been in the late 1980s. This factor combined with the Air Force's pursuit of higher priority weapons—precision munitions, the F-22 air superiority fighter, and the F-35 Joint Strike Fighter—to make any ambitious air support action seem too extravagant.[4]

The F-35 was primarily an attack plane, but its interdiction bent and $40 million price tag meant that the Air Force would be unlikely to commit it to the traditional, rough-and-tumble CAS arena. Further, the F-35 was supposed to replace the A-10 as it was phased out. A limited budget partially drove these mixed signals about Air Force CAS, but a big reason was the issues of the CAS plane debate, which still resonated.[5]

If CAS was a technological solution to ground battlefield problems, then many observers believed that technological advances solved the CAS "problem" itself. Interdiction advocate Price Bingham touted the Desert Storm services and future promise of the Joint Surveillance Target Attack Radar System (JSTARS), a Boeing 707 modified to provide radar surveillance of vehicular traffic behind enemy lines (conditions permitting). JSTARS detected the Iraqis' Khafji incursion and later Kuwait City mass retreat, allowing coalition air power to smash both actions. Former Syracuse Air National Guard commander Mike Hall's 1998 *Aviation Week* editorial claimed that JSTARS, advanced avionics, data link transfer, GPS, and precision weapons made traditional CAS obsolete. For example, programmable coordinates and GPS guidance on the new Joint Direct Attack Munition (JDAM) bomb gave it amazing accuracy regardless of weather conditions. (JDAM was part of the Hog Up modification.) JDAM was so accurate that the Air Force planned to build smaller bombs that could be carried by fighters and still have high destructive effect. Some people believed that the Army's attack helicopters and precision-guided missile artillery threatened to make fixed-wing CAS irrelevant or interservice arguments over supporting firepower too hard to resolve. Others asserted that air defense weapons such as the Stinger rendered CAS impractical.[6]

These developments and attitudes underlay some Air Force leaders' continuing claims that CAS was an emergency procedure that should rarely be flown if a campaign's military players all did their jobs. Air Combat Command (ACC; in 1992, TAC and SAC merged) commander Mike Loh answered lingering criticisms of the Air Force CAS policy by saying that technological advances in high-tech battlefield surveillance and interdiction would "relieve to a great extent the Army's direct contact." Academic works by various Air Force scholars asserted the same thing. In proposing which functions each service should embrace in future years, Air Force chief Merrill McPeak saw the Air Force performing a campaign's "deep" and "high" missions, such as interdiction and air superiority, while the Army and Marines would take various remaining "close" missions, including CAS. Thus, although Congress's plan to transfer CAS responsibility to the Army had been stillborn and no more than half-serious to boot, an Air Force chief now embraced it. McPeak's position might have pleased some partisans in the recent CAS debate, but it resurrected several questions of force structure and campaign coordination. Like the congressional threat, it remained on the shelf and was dismissed by later chiefs.[7]

Yet through the 1990s others did not care for such confident assertions about future air support. Even General Welch and other Air Force and Army generals interviewed for this book saw a need for the mission. In the same military journals where the air power visionaries pressed their case, other observers questioned what they regarded as the air power advocates' utopian vision. They saw the Air Force once again seizing upon the budget and technological determinism to abandon the mission, and they wondered if the service would be too busy with its preferred missions to answer desperate calls for CAS when things went wrong.[8]

The CAS debate continued, though at a subdued level, and again actual combat generally supported the case for the mission and its dedicated planes. CAS might have averted the tragic casualties and even the disastrous outcome of a battle on 3 October 1993 between U.S. Army infantry and clan gunmen in Mogadishu, Somalia. Here was an urban firefight where JSTARS, interdiction, and the Army's artillery and tanks were not factors. A-10s, AC-130s, or even attack helicopters could have served nicely, since heavy firepower was needed.[9]

A-10s were often used in both potential and actual flash points around the world during the 1990s. They patrolled Iraq's no-fly zone, remained a part of the U.S. order of battle in Korea, and were heavily used in war-torn Bosnia. In Bosnia, political circumstances prevented a large Army ground presence and an all-out interdiction air campaign. Also, the enemy was too dispersed in mountainous terrain for JSTARS and interdiction to work as well as they had in Desert Storm. However, the Americans did not want losses, and there was enough of an air defense threat to require more maneuverable planes than AC-130s (which were used less frequently). Using NVGs, Hog drivers flew night SAR missions for downed pilots (which, by the way, involved the loss of other planes, not A-10s). In 1995 they figured prominently in attacks against Bosnian-Serb army positions. This work was not precisely CAS, for there were no troops in contact, but it did involve pinpoint military targets amid civilian settlements and among the chaotic patchwork of various factions' positions. With LASTE, the A-10's bomb and gun accuracy was excellent, as the GAU-8/A destroyed trucks at a range of 2.5 miles.[10]

The Warthog repeated the performance in the spring 1999 war with Serbia over Kosovo, even though controversy arose over actual damage that NATO planes inflicted upon the Serb Army, as well as over the issue of whether an early strategic bombing campaign could have ended the war sooner. If there was a problem with destroying Serb Army forces, then two outside factors played a large role. An extreme concern about preventing NATO casualties created a high (fifteen thousand feet AGL) minimum operating altitude. Acquiring targets and authenticating their destruction thus became difficult—especially since Serb units intermixed with the Kosovar population, and strict rules existed about identifying targets before attacking them. (JSTARS encountered problems with force detection.) However, observers still cited A-10 successes, and they were also allowed to break the altitude restriction in certain cases. At least one Hog took hits, but its durability allowed it to return to base. The Hogs' fortunes compared favorably with an Apache force deployed to the war. That unit encountered logistical problems and was held in reserve throughout the conflict due to concerns about air defenses and ensuing casualties. Indeed, national command authorities considered the A-10 as a better air support option.[11]

As this book was being completed, Hogs flew in yet another conflict: the war against the Taliban and al Qaeda in Afghanistan. Initially, their participation did not receive more than a brief comment on the Air Force Web site and passing notice in some *Washington Post* articles. Weak to nonexistent air defenses and a relatively isolated enemy ground force in defensive positions allowed other planes to provide adequate CAS. F-15Es, F/A-18s, B-1s, and B-52s all exploited excellent air-ground communications and precision weapons like JDAM to deliver accurate strikes. The light air defenses allowed the AC-130 to showcase its considerable CAS strengths. Helicopters still occasionally struggled with light arms fire, but their battlefield utility was undeniable. Unmanned drones delivered even more noteworthy air support performance. These craft had served ably as observers and decoys from the Vietnam War onward, and in Afghanistan they were equipped to fire Hellfire missile at military targets.[12]

Of course, these conditions generated familiar comments about how modern technology would change the nature of CAS. Observing that "the bomber also brings lots of muscle to the . . . [battlefield] because of its long loiter time and high payload," the Air Force's on-line news service noted that electronic target data transfer allowed these planes to provide accurate air support with programmable JDAMs. A *Washington Post* article pointed out that bombers could do this as well as the A-10. *Aviation Week* mentioned that the war spurred more modifications for the AC-130. Some articles speculated that drones had a bright combat future if electronic data transmission was as good as seen in this conflict. Indeed, by 2000 Congress was urging the services to build drones for traditional combat missions. (One contractor even proposed turning A-10s and F-16s into drones!)[13]

As the Afghan ground campaign progressed, the A-10 came to the fore. The anticipated CAS strengths built into the Hog made it the primary choice for an Afghan-based tactical air support plane. It had already delivered good support in several firefights, and its easy maintainability in austere conditions was required to operate from local bases. Also, the war's other CAS stars possessed many characteristics desired for good air support. AC-130s, bombers, and drones all possessed loiter or weapons carriage capability. They were not fast fighters on a short fuel leash. The Afghan ground war's deliberate pace in rooting out terrorist cadres allowed these aircraft to fly equally deliberate missions.[14]

The evolving air support needs within this war underscored the varied conditions that one could encounter in any conflict. Although JDAMs and heavy bomber CAS worked well, this occurred in specific circumstances. Given America's wide and unpredictable foreign policy scope, air defenses might not always be nonexistent, bombers might have other commitments, or the combat situation might be too fluid to allow long, time-consuming bomb runs. War's friction could produce occasions when either the air control party lacked the data link equipment or the enemy somehow jammed electronic transmissions. Indeed, one finding of a 1990s fratricide prevention study was that new communications and targeting gear faced even more potential problems with reliability and interservice compatibility than traditional equipment, such as radios. Drone technology had already proven itself in the interdiction role in the form of cruise missiles. But its applicability to the more fluid CAS and air combat arena would be highly dependent upon how well one can guarantee discernment of friend and foe. Further, any drone would require other CAS strengths—simplicity and survivability—

since their cost and loss rates were so far rather high. Finally, as the Hog deployment into Afghanistan revealed, a conflict might require austere basing for sustained air support operations.[15]

Other observers were more cautious about the war's impact. Apparently they understood that combat's fluid nature and dire exigencies might force reliance upon more traditional means. The best such assessment actually was made years earlier, when reporter Anatol Lieven covered Russia's brutal war with Chechnya and observed that the United States might encounter similar scenarios:

> A senior US official told me recently that if I went around saying this kind of thing in Washington, I'd be ridden out of town on a rail. He meant it as a joke—I hope!—but there was an element of truth in his words. A belief in the all-powerful nature of high-tech weapons and long-range bombardments is wonderfully appealing to contemporary Americans. It both flatters their . . . pride in American technology and suggests the possibility of repeated victories without the risk of serious casualties. However, the US will not always have the ability to pick and choose its wars, and the key lesson of Chechnya is that there will always be military actions in which a determined infantryman will remain the greatest asset.[16]

Technological Constituencies

As Anatol Lieven pointed out, reducing technological enthusiasm would be a tall order not only for the Air Force but also for America in general. Some "postmodernist" intellectuals have proclaimed Americans' loss of faith in technology, but I think mostly otherwise. One need only ponder the unquestioning enthusiasm for computer technology, such as the Internet. One could say the A-10 was a "postmodern" plane in that it defied normal notions of progress. As such, its existence offends many people, as evidenced by the resistance it has encountered.[17]

Notions of technology constituencies certainly can be applied in studying the history of the dedicated CAS plane. People become attached to a technology due to investment of resources, professional concerns, or extensive usage. Political scientists call this "cybernetic behavior." But constituents for any technology are not mindless robots. All see some kind of advantage in it; one might say that they are a loose, conditional mix of enthusiasts, salesmen, and opportunists. There are times when one might accuse either CAS opponents or proponents of cybernetic behavior—even myself, whose A-10 background certainly colored my choice of book topic and overall CAS outlook. However, all of these people had concerns that they believed experience, economic wisdom, studies, and so on proved were legitimate.[18]

Air Force leaders are an obvious air technology constituency. Having the best equipment prevents wartime disaster, and they cite historical examples of how air power stumbled or prevailed because of the machinery's relative quality. Money is the engine of air power, because the airmen learned early on that first-rate airplanes were expensive to build, maintain, and fly. Training and proficiency for pilots, technicians, and other supporting players were likewise expensive. Congress controls the budget for all of these things, and in order to sell their programs to legislators, Air Force officers learned equally quickly that they needed to conduct ironclad, compelling sales campaigns. This was often a source of the sweeping claims for air power or aerospace weapons. The Air Force's proposed new technology of the moment would pass congressional muster if the service's leaders create a sense of urgency by making the new device seem the superior answer to the imminent dangerous obsolescence of the current device. Air Force leaders' statements during the 1980s CAS plane debate offer a prime example of this behavior.[19]

The CAS plane story shows that constituencies within military services have existed and developed in various ways. For example, Air Force bomber advocates believed that devoting attention to their particular planes at the expense of others was necessary. For these officers as for Army helicopter supporters, loyalty hardened through initial bureaucratic insurgency, followed by increased resource commitment, and then actual combat use. The last factor was important in fighter leaders' insistence upon high-speed performance. It was axiomatic that this quality equaled survival and victory in combat. CAS defenders in the Army and Air Force believed that historical lessons indicated that the F-16 would be a relatively poor CAS plane, and their opposition spurred more vocal resistance by others. This last point shows that military constituent groups can have outside backers. For example, Army helicopter leaders cultivated support from senior officers and key politicians. Outside supporters become involved because of their enthusiasm for a particular military technology. A prime example involves those military observers who touted the Harrier as a means of keeping the VTOL concept alive. Military reformers in the 1980s threw their support behind various weapons, sometimes with unintended results.

As the Air Force's concern for congressional opinion makes clear, Congress is the most significant outside constituent. As former member of Congress Robert Giaimo told me, it is the representatives' duty to give their constituents' concerns—whatever they may be—a public hearing. If the hearing decides against them, then so be it. (Giaimo added that the last time Americans gave up on this process, the Civil War occurred.) One reason why the senators could not resolve the interservice CAS plane impasse in the 1971 hearings was because some of them favored a particular service or aircraft company. The Texas congressional delegation continually pushed the A-7 and even helped secure for it a flyoff against the A-10. Sen. Strom Thurmond's advocacy of the Enforcer forced the Air Force to ensure that the A-10 met its cost and performance goals. Congressional reformers and ex-Marine staffers fought the CAS F-16 in the 1980s. Indeed, the 1980s debate opened the gates for congressional representatives who wanted constituent planes like the Harrier examined for the role of follow-on CAS plane. The A-10 had congressional support early in its life, but lost it during the CAS debate when Fairchild ceased military aircraft construction. Fortunately for the Hog, its constituency then came from elsewhere.

The CAS mission and airplane constituency was a coalition of individuals and groups. In the A-X's formative days, it featured a mix of former A-1 pilots, Air Force leaders intent upon preempting the Cheyenne, and activists like Pierre Sprey, Tom Christie, and Bob Dilger. Christie and Sprey formed part of the Fighter Mafia, skilled staff insurgents who pushed development of the A-10 and F-16. They later formed the backbone of the 1980s military reform movement, a loose alliance of reporters, members of Congress, and government staff who also supported CAS planes. Although many reformers favored a machine like Chuck Myers's Mud Fighter during the 1980s CAS plane debate, they still played a role in delaying Air Force action against the A-10. During this time, the A-10 received discreet support from the Hog driver community itself. Some observers pointed out that the Hog's existence ensured a CAS advocacy within the Air Force. There were also anonymous Army officers who contradicted their leaders' support of the Air Force's CAS plane tack.

TECHNOLOGICAL IRONIES

The A-10's eclectic, shifting support brings up the issue of technological irony. For instance, the Warthog's very existence is accidental. Another service's aircraft project and a war led Air Force leaders to abandon their disdain for such a machine. Other factors intruded to ensure that the service remained supportive and that the plane worked as planned. Later, fortuitous circumstances defended the Hog against a concerted attempt to retire it.

Many other ironies exist in this story—too many for all of them to be discussed here. First, there is the obvious paradox in Air Force leaders' desire to substitute their own firepower solution—first, strategic bombing, and later, high-tech interdiction—for other firepower missions, such as CAS. Second, the reformers who supported military developments that matched their notions of war-fighting and their vision of military technology, such as AirLand Battle and the F-16, found themselves in a quandary during the CAS plane debate. Both items formed the basis of a threat to another reformer favorite, the dedicated CAS plane.

The Army's attitude toward the CAS plane is yet another irony. Over the years, the impression arose that the Army unanimously and consistently supported the Hog because, after all, here was an airplane designed to support the troops. As shown repeatedly in the preceding chapters, this was not true. After the A-X program got under way, Army leaders saw it as the preemptive move against the Cheyenne that it was. Their support remained at best lukewarm until the Cheyenne failed and they needed firepower to halt the Soviets in Europe. After they promulgated the AirLand Battle doctrine, they backed away from CAS and the A-10 yet again. Many Army leaders believed that their new weapons, such as the Apache advanced attack helicopter, rendered CAS unnecessary. Also, evidence exists that Army leaders also accepted the CAS F-16 to preempt Air Force resistance to their follow-on attack helicopter plans.

Congress was not always friendly to the CAS plane, either, though it is easy to fall into the trap—as both A-10 opponents and defenders have done—of thinking that the legislators always favored the Hog. Congress is a highly diverse body, and the lawmakers' support for the CAS plane waxed and waned depending on circumstances and who could make their influence stick. After Rep.Otis Pike's 1965 hearings criticized the Air Force's neglect of CAS, the 1970s brought legislators who opposed the A-X program because of their support for a constituent's plane, alarm over military spending, and concern about recent combat results. (This created another irony in that it inspired Air Force leaders to deliver a superb CAS plane!) Then, congressional military reformers fought for the CAS plane's survival in the 1980s.

The A-10's existence represented irony in the evolution of tactical plane design. As designer Hans Multhopp observed in 1966, ever faster speeds were not necessary for all tactical planes. Although advanced fire control avionics helped the Hog, its designers preferred simplicity, which lowered costs and improved maintainability. Some legislators' search for an air support plane bargain yielded the Enforcer—which was cheaper, slower, simpler, and more underpowered than the A-10.

There is also the irony of helicopters, which the Army initially procured to have a cheap means of aerial fire support. That service encountered the same pitfall that sometimes afflicted the Air Force: the pursuit of ever greater technological capability at the penalty of exploding cost. General Momyer predicted something like this in 1971, when he observed that Army aviators, like any others, would always want greater performance. Also, as helicopter advocate John Bahnsen observed with alarm in 1986, some of the Army people assigned to the attack helicopters increasingly viewed themselves as separate from the other Army combat arms.

The Army's expanding plans for its helicopters brings up a more basic technological irony: the difference between what a device is designed for and how it is actually used. The P-51 began its World War II operational life as an air support plane, became one of the war's greatest fighters, and then ended its days as a Korean War CAS plane. Conversely, the P-47 was the Army Air Force's choice for air superiority missions, but was an even better air support plane. The A-1's designers probably never imagined at its 1945 rollout that it would fly in a war twenty years later and even shoot down jet fighters. Even the A-10's supporters did not claim for it anything other than a close air support capability, but in Desert Storm it flew other missions that Air Force leaders during the CAS plane debate had said it could never do.

TECHNOLOGICAL MEDIA

The planned and actual uses of technology recall how people argue their case for technology's proper utilization. Although the mainstream news media ignored the CAS plane debates of the early 1970s and late 1980s, the defense media became a free-style forum as the various sides jockeyed for advantage. Some of these publications, such as the AFA's *Air Force Magazine*, featured pronouncements by the principal actors or their allies. The Air Force had advertised the A-10's strengths in the 1970s, but during the CAS plane debate of the 1980s the service's public relations effort faltered, as even people within the Air Force questioned some of its more hyperbolic claims. Congressional hearings provided opportunities for people on both sides of the issue to make their case, whether their facts were correct or not.

The CAS plane story featured many studies, tests, flyoffs, and investigations that addressed topics whose complexity almost forced one to take sides ahead of time. In the early 1960s the OSD's systems analysis reports asserted that slower attack planes were more operationally efficient than the Air Force's fast jet fighters, but the airmen's own systems analysis studies favored the F-4. Computer models were popular at the time, and these also contained traps. In 1965, as General McConnell wrestled over deciding for attack planes or another jet fighter, his staff's massive computer studies produced conclusions supporting both choices, depending upon which inputs they supplied. Differing opinions over how to predict weapons effectiveness generated serious friction between service weapons labs and GAU-8/A proponents. The GAU-8/A side suspected the validity of the labs' computer models and introduced the concept of realistic testing to prove the gun's worth. However, real tests could sometimes have suspicious underpinnings too. The A-10's flyoff against the A-7 featured conditions that obviously favored the A-10's performance characteristics. One may counter that the test aimed to prove the A-10 the better choice for antitank CAS in Europe's often poor flying weather. However, one wonders what the setup would have been had the Air Force and Army needed a plane for CAS elsewhere. The TASVAL exercise was supposed to be the most heavily instrumented test in

U.S. military history. As such, it could have been the flyoff between A-10s and helicopters that Congress and others wanted, but instead the services used it to demonstrate that JAAT enhanced the survivability and effectiveness of both aircraft. The finding was true, and ultimately benefited the services and the nation. However, it also implied a rejection of congressional scrutiny.

The studies conducted during the CAS plane debate of the 1980s were one more example, for even those people involved later observed that many of the inquiries backed their sponsors' intentions. This time, the Air Force used flight testing against the A-10 when it conducted CAS F-16 technology demonstrations. Perhaps the most unbiased and scientific study during the debate, the Toomay SAB study, did not receive much attention from either side since its recommendations did not fit their preconceptions.

War itself was an obvious source for both pro- and anti-CAS arguments, but individual wars and conditions—especially European ones—could be both compelling and tricky. Given America's concern for Europe during the Cold War, Air Force leaders often cited that region's World War II results to justify their coolness toward the CAS mission and its planes. However, Army discontent about Korean War air support helped fuel that service's helicopter development; and Air Force leaders could not dodge Representative Pike's pointed questions about early CAS problems in Vietnam. Both neutral observers and CAS critics concluded that the Israeli Air Force's heavy losses in the October War showed that tactical air support and the A-10 were obsolete. However, they disregarded the IAF's unreadiness for densely packed modern defenses and the great lengths to which Western air arms would later go to counter them. Further, CAS opponents endorsed the IAF's dismissal of the mission, even though Israel's military situation was not the same as America's.

The debate in the 1980s featured many appeals to combat history, but both sides focused upon specific cases, sometimes made errors when citing them, and rarely cited contemporary wars. To support its case against traditional CAS and the A-10, the Air Force insisted that Korea and Vietnam were anomalies, while European war was the only valid scenario. As a single example, Desert Storm did not completely resolve the debate. It quelled some of the anti-Hog feeling, but opponents played up the plane's losses and cited the limited CAS accomplished as proof that the mission was obsolete. The recent Kosovo and Afghanistan conflicts have spurred even greater claims for high-tech air power. This book presents an extensive examination of combat results, but unlike others, it does not stick to one war or scenario and then make sweeping "always" assertions about them.

This work has a wide scope in order to show that on balance, the CAS mission and its dedicated plane are important to Americans and worth air service attention. U.S. foreign policy has exposed the nation to a wide variety of opponents in diverse environments and scenarios. The situation continues. The United States does not resemble a country such as Israel, which fights specific enemies in a desert environment. U.S. war-fighting scenarios remain virtually unlimited, as a result of the nation's global role. Further, Americans have come to expect victorious wars with few if any casualties. In spite of recent wars that have featured remarkably low losses, a future conflict may entail serious sacrifices, just as some situations have demanded in the past. Even so, this country has expected and will expect that every measure should be taken to ensure the success and survival of its citizen-soldiers. The numerous controversies, complaints, and debates about CAS involved this concern, and the use of heavy CAS firepower in nearly all air-age American wars confirms its significance.

Likewise, expanding air technologies seen after World War II entailed greater mission performance specialization, and the fighter was no longer as good a CAS plane choice as it had been previously. Instead, the dedicated attack plane designs and crews were preferable for CAS and other air-to-ground missions. New technologies, such as advanced weapons guidance systems and rapid data link transfer, may solve basic CAS problems, such as target acquisition and poor weather conditions, and thus allow fighters to fly the mission well. However, they place a great premium upon availability, reliability, operator proficiency, and absence of enemy countermeasures. If any of these factors intervene, then weapons carriage, loiter, slow-speed maneuverability to attain visual confirmation, and the ability to withstand hits while doing it will remain as necessary design features. Even with new developments, the aircraft that have received air support acclaim in recent wars possess many such CAS strengths. Bombers, gunships, and drones all possess good loiter and weapons carriage abilities. These planes and helicopters all contribute to air support.

However, a dedicated CAS plane like the A-10 has the best all-around abilities to handle a wide variety of CAS situations. Its maneuverability and durability allow it to handle more serious air defense environments than helicopters or gunships. Its weapons carriage and loiter capabilities are better than those of fighters. Its simplicity and maintainability make it more easily deployable to austere locations than most other aircraft. The pilots are already proficient in handling a difficult mission. Addition of the latest avionics and smart weapons features increases its versatility in the air support arena.

Air Force leaders should not act like some of their early predecessors: blinded by technological promise, anxious about their service's status, paranoid about the budget, and yearning to win wars singlehanded in climactic air battles, as earlier sea power advocates wanted navies to do. They should realize that the country relies upon the Air Force as its premier air power tool in the varied situations where air power can be applied. If there is a sea power analogy for air power, it is that the U.S. Air Force has assumed a station roughly similar to the British Empire's Royal Navy. Its leaders should embrace the chance to exercise air power in whatever manner necessary, instead of using the first available technological, status, budget, or doctrinal excuse to flee one of its missions—close air support.[20]

I do not challenge the Air Force's traditional tactical priorities. Indeed, I disagree with those who say that the current purchases of the F-22 air superiority fighter should stop because all future conflicts will not need its services. Such a comment is as wrongheaded as the demands to retire the Hog, and for the same reason—the country's warfare liabilities prevent such cocksure foresight. However, if the service has worked so hard to achieve air superiority, it should not refrain from exercising it. I agree that a central tactical force of multipurpose fighters augmented by specialist planes is the best setup, for it allows the U.S. Air Force the best opportunity to meet the wide spectrum of tactical air war possibilities created by aviation technology advances and U.S. foreign policy contingencies. Attack helicopters, the AC-130, and the A-10 currently fill the CAS niche nicely, but the A-10 is the best "multipurpose" plane in this specific arena. I hope that the U.S. Air Force will eventually recognize this continuing need before even the rugged A-10 becomes too fatigued to answer the call.

ABBREVIATIONS

A	attack
AAA	anti-aircraft artillery
AAD	Air Assault Division
AAF	Army Air Force
AAFSS	Advanced Aerial Fire Support System. Initial project name for AH-56 Cheyenne.
AAH	advanced attack helicopter. Initial project name for AH-64 Apache.
AATC	Air National Guard and Air Force Reserve Test Center
AC	attack/cargo
ACC	Air Combat Command
ACSC	U.S. Air Force Air Command and Staff College
ACTS	Air Corps Tactical School
AFA	Air Force Association
AFHMP	Air Force History and Museums Program
AFHRC	Air Force Historical Research Center
AFJ	*Armed Forces Journal*
AFJI	*Armed Forces Journal International*
AFM	*Air Force Magazine*
AFMC	Air Force Material Command
AFRDQ	U.S. Air Force Headquarters Directorate of Research & Development, Operational Requirements, and Development Plans Office

AFT	*Air Force Times*
AFTEC	Air Force Technology and Evaluation Center
AGL	above ground-level altitude
AGM	air-to-ground missile
AH	attack helicopter
AJ	*Airpower Journal*
ALFA	Air-land forces application
AMHI	Army Military History Institute
ANG	Air National Guard
ASCIET	All Services Combat Identification Evaluation Team
ASD	Armaments System Division
ATHS	automatic target hand-off system
ATP	allied tactical publication
AUP	Air University Press
AUR	*Air University Review*
AV	VSTOL attack
AVM	Air Vice Marshal (RAF)
AWACS	airborne warning and control system
AW&ST	*Aviation Week & Space Technology*
A-X	Designation for dedicated CAS plane project
B	bomber
BAI	battlefield air interdiction
BDA	bomb damage assessment
BDM	Braddock, Dunn, & McDonough Corp.
BRL	Ballistic Research Laboratory
C	cargo
CAS	close air support
CASADA	close air support aircraft design alternatives study
CASMARG	Close Air Support Mission Area Review Group
CDC	Combat Developments Command
CFP	Concept formulation package. Formal expression of design concept.
C&GSC	U.S. Army Command & General Staff College
CHECO	Contemporary Historical Evaluation of Combat Operations
COEA	cost and operational effectiveness analysis
COIN	counter-intelligence
CONARC	Continental Army Command
CPGW	*Conduct of the Persian Gulf War*
DAB	Defense Acquisition Board
DCP	Development concept paper. Defines the project prior to actual aircraft development process.
DDR&E	Director of Defense Research and Engineering
DN	*Defense News*
DoD	Department of Defense

DOL	dispersed operating location
DoT	Directorate of Training
DSARC	Defense System Acquisition Review Council. Project oversight group that reviews development progress milestones.
DTG	date-time-group
DVIC	Defense Visual Information Center
ECM	electronic countermeasures
F	fighter
FAC	forward air controller
F-Kill	firepower kill
FLIR	forward-looking infrared
FM	field manual
FOD	foreign object damage
FOL	forward operating location
FOLTA	forward operating location training area
FWIC	Fighter Weapons Instructor Course
FWR	*Fighter Weapons Review*
FWS	Fighter Weapons School
FWW	Fighter Weapons Wing
F-X	Designation for F-15 project
GAU	gun, automatic
GC	Group Commander (RAF)
GPS	global positioning system
GWAPS	*Gulf War Air Power Survey*
HAAC	History of Army Aviation Collection, Army Military History Institute, Carlisle Barracks, Pa.
HASC	House Armed Services Committee
IAF	Israeli Air Force
IDA	Institute for Defense Analysis
IDR	*International Defense Review*
IIR	imaging infrared
INS	inertial navigation system
IOC	initial operational capability
JAAT	joint air attack team (or tactics, depending on context)
JAWS	joint attack weapons systems
JMSNS	justification for major system, new start
JSTARS	joint surveillance target attack system
Ju	Junkers
K-Kill	catastrophic kill
KKMC	King Khalid Military City
LACAS	low-altitude close air support
LARA	light armed reconnaissance aircraft
LASTE	low-altitude safety and targeting enhancement
LAVP	Lot Acceptance Verification Program

LTV	Ling-Temco-Vought Corp.
MACV	Military Assistance Command, Vietnam
MAFB	Maxwell Air Force Base, Ala.
MCPL	Members of Congress for Peace through Law
Me	Messerschmitt
Mi	Mil (Russian aircraft builder)
MiG	Mikoyan Gurevich
M-Kill	mobility kill
MLRS	multiple launched rocket system
MNS	mission needs statement
MOB	main operating base
MR	*Military Review*
MRP	mission requirements package
N/AW	night/all-weather
NTC	National Training Center
NVG	night vision goggles
OA	observation/attack
OAFH	Office of Air Force History
OSD	Office of the Secretary of Defense
OT&E	operational test and evaluation
OV	fixed-wing observation
PACAF	Pacific Air Forces
PA&E	program analysis and evaluation
PSAC	President's Science Advisory Committee
RAD	Requirements action directive. An aircraft project's initial concept document.
RAF	Royal Air Force
RFI	request for information
RFP	Request for proposal. Contractor bid request for aircraft project.
S	surveillance
SA	Soviet SAM designation (see below)
SAB	Scientific Advisory Board
SAC	Strategic Air Command
SAF-OII	Secretary of the Air Force, Office of Internal Information
SAM	surface-to-air missile
SAR	search and rescue
SASC	Senate Armed Services Committee
SECDEF	Secretary of Defense
SLUF	Short little ugly fellow, an A-7 nickname
SPO	System Program Office
STOL	short field takeoff and land
STOVL	short field takeoff and vertical land
Su	Sukhoi
SVN	South Vietnam

T	aircraft trainer, or Soviet tank, depending on context
TAC	Tactical Air Command
TASVAL	joint test of tactical aircraft effectiveness in close air support anti-armor operations
TD&E	tactics development & evaluation
TF	Turbofan
TFS	Tactical Fighter Squadron
TFTW	Tactical Fighter Training Wing
TFW	Tactical Fighter Wing
TFWC	Tactical Fighter Weapons Center
TFX	Designation for F-111 project
TOW	Tube-launched, optically-aimed, wire-guided missile
TRADOC	Training and Doctrine Command
TTW	Tactical Training Wing
U	utility
UCAV	unoccupied combat aerial vehicle
UH	utility helicopter
USA	U.S. Army
USAAD	*United States Army Aviation Digest*
USAAF	U.S. Army Air Force
USAF	U.S. Air Force
USAFAWC	U.S. Air Force Air War College
USAFOHP	U.S. Air Force Oral History Program
USAFR	U.S. Air Force Reserve
USASODP	U.S. Army Senior Officers' Debriefing Program
USAWC	U.S. Army War College
USDR&E	U.S. Directorate of Research & Engineering (part of OSD)
USMC	U.S. Marine Corps
USMCR	U.S. Marine Corps Reserve
USN	U.S. Navy
USSF	U.S. Special Forces
VAL	Designator for Navy light attack plane project
V/STOL	vertical and short takeoff and land
VTOL	vertical takeoff and land
WPAFB	Wright-Patterson Air Force Base, Ohio
WSEG	Weapons System Evaluation Group
ZSU	Soviet radar AAA designation

NOTES

Chapter 1. Introduction
1. "Laughter."
2. Maj. Gen. Jay Garner, U.S. Army assistant chief of staff for operations and plans, cited in Mason, *Centennial Appraisal*, 237.

Chapter 2. The Origins of Air Support
1. Hallion, *Strike*, 23; Reinburg, "Close Air Support"; GAO, *Close Air Support: Principal Issues and Aircraft Choices*, 62–63.
2. Greenhous, "Counter"; Kennett, "Developments," in *Case Studies in the Development of Close Air Support*, ed. Cooling, 16–17; Lee, *No Parachute*, 261 (quote).
3. Lee, *No Parachute*, 252 (quote).
4. Hallion, *Strike*, 17, 21, 24; Kennett, "Developments," 23–24; Muller, "Close Air Support," in *Military Innovation*, ed. Murray and Millett, 164; Smith, *Close Air Support*, 7–9.
5. Corum, *Luftwaffe*, 29–34; Corum, *Luftwaffe's Way*, 63–75; Hoeppner, *Germany's War*, 150; Muller, 146–49; Smith, *Close*, 8–9.
6. *Air Power Showdown, Part III*; historian Peter Grosz's review of records indicated that Germany's dedicated ground attack planes endured relatively fewer losses than others flying the mission. Smith, *Close*, 9, implies same. Hallion, *Strike*, 16–25, 37–38, also implies same but still prefers fighter planes for CAS work. Muller, 148–49, apparently cited Hoeppner, who did not detail ground

support losses.

7. Kennett, "Developments," 42–52; Muller, 172–75.

8. Muller, 163–72; Slessor, *Air Power and Armies*, 90–110 (quote, 90).

9. Corum, *Luftwaffe*, 6, 166–70, 221–23, 243–49; Murray, "The Luftwaffe Experience," in *Development of Close Air Support*, 71–78.

10. Copp, *Few*, 47, 66, 92–127, 214, 393, 419; Faber, "Interwar," in *Paths of Heaven*, ed. Meilinger, 183–238; Kotz, *Wild Blue Yonder*, passim; Maurer, *Aviation*, 131–48, 299–317, 400–412; Woodward, *Commanders*, 74.

11. Kennett, "Developments," 43–52; Mortensen, *Pattern*, 8–9, 24–28; Muller, 179–80; Smith, *Close*, 29.

12. Greenfield, "Study No. 35," 8; Mortensen, *Pattern*, 17.

13. Beaumont, *Joint Military Operations*, 106.

14. Jacobs, "Battle for France," in *Development of Close Air Support*, 246; McLennan, "Close Air Support History," 16–17; Smith, *Close*, 96.

15. Goldberg and Smith, "Army–Air Force Relations," 3–4; Hallion, *Strike*, 174–77; Syrett, "Tunisian Campaign," in *Development of Close Air Support*, 184–85 ("dismay," 185); McLennan, 19–24 (other quotes, 19 and 23).

16. Greenfield, 76–85; Hallion, *Strike*, 179–87; Mark, *Aerial Interdiction*, 175–78; McLennan, 25; Wilt, "Sicily and Italy," in *Development of Close Air Support*, 204–20.

17. Bradley, *Soldier's Story*, 250 (quote); McLennan, Table C, Annex A, 1–2; Hallion, *Strike*, 175–99; Hughes, *Over Lord*, 240–42; Jacobs, "France," 237–46, 267–68.

18. Hughes, *Over Lord*, 119, 182–87 ("simmering" and "laborious," 183); Jacobs, 254–73; Schlight, "Quesada," in *Makers*, ed. Frisbee, 196–97; Spires, "Patton and Weyland," in *Airpower and Ground Armies*, ed. Orange and others, 147–60 (quote, 153).

19. Gooderson, *Air Power at the Battlefront*, 125–56; Hallion, *Strike*, 196, 206–14, 224; Jacobs, 281–82.

20. Brulle, *Angels Zero*, 38 (quote); Gooderson, 107–19; Hallion, *Strike*, 203–6, 214–23; Hughes, *Over Lord*, 24; Jacobs, 246–51, 277–84.

21. Brulle, 69–70; Gooderson, 103–23, 239–41; Greenhous, "Aircraft versus Armor," in *Men at War*, ed. Travers and Archer, 104–7; Hallion, *Strike*, 225–27; Hughes, *Over Lord*, 247; Jordan, *Jarvis*, n.p.

22. Brulle, 93; Colgan, *Fighter-Bomber Pilot*, ix–xi; Gooderson, 201–9, 235–36; Hallion, *Strike*, 225; Jacobs, 279–80. The case study essays in Cooling's *Air Superiority* highlight the attritional nature of World War II air combat.

23. Bergerud, *Fire in the Sky*, 534–38; Gooderson, 165–94, 229–38; Greenfield, 29–36. Representative sources for the U.S. firepower preference: Ambrose, *Citizen-Soldiers*, 489 (quote); Hadley, *Straw Giant*, 195; Krepinevich, *Army*, 6–8; Lynch, "Close Air Support," 72–78; Mrozek, *Air Power*, 5; Weigley, *American Way of War*, xxii, 347, 359.

24. Beaumont, *Joint*, 114–15; Bergerud, 598–621; Hallion, *Strike*, 165–67; Taylor,

"Southwest Pacific," in *Development of Close Air Support*, 301–23; Sherrod, *Marine Corps Aviation*, 150–51, 189–91, 207–11, 250–58, 290–313, 408–11; Smith, *Close*, 115–16.

25. Dick, *Momyer*, 38 (quote); Barnes, "Concept," 16–22; Brulle, 31; Colgan, 98; Hallion, *Strike*, 201–3; Jordan, *Jarvis*, n.p.; Wilt, 212–13; Worden, *Rise*, 12–13.

26. Dick, "Disosway," 181; "Sweat" (USAF Oral History), 119; Futrell, *Ideas*, 1:174, 492; Hallion, "Battlefield Air Support," 19–21; Hallion, *Strike*, 46–51, 264, 267; Col. Budd Peaslee, USAF (ret.), cited in Caidin, *Black Thursday*, 159 (quote).

27. Futrell, *Ideas*, 1:146–47; GAO, *Principal Issues*, 72–74; Gooderson, 74–76; Hughes, *Over Lord*, 127–31; Dick, *Momyer*, 38 (quote).

28. Bergerud, 294–96, 580–84; Sherrod, 290–312; Smith, *Close*, 101–5; Tillman, *Dauntless*, passim.

29. Sources criticizing, praising, or neutral on Stukas include the following: Bavaro, "Tank Busters"; Boyne, *Clash of Wings*, 34–35, 174–79; Futrell, *Ideas*, 2:174, 492; Green, *Famous Bombers*, 1:37–46; Hallion, *Strike*, 49, 146, 239; Higham, *Air Power*, 101, 106, 138, 143; Kross, *Military Reform*, 89, 157; Murray, "Luftwaffe Experience," 71–104; Ratley, "Comparison," passim; Rudel, *Stuka Pilot*, passim; Scott, "Rise and Fall"; Smith, *Close*, 47–48, 63–91.

30. Bavaro, 4; Boyne, *Clash*, 142–43, 162–64, 384; Buckley, *Air Power*, 146; Hallion, *Strike*, 234, 244–45; Higham, 95; Hogg, *Tank Killing*, 204–7; Kennett, "Developments," 51; Whiting, "Soviet Air-Ground Coordination," in *Development of Close Air Support*, 125–40; Smith, *Close*, 79; Whiting, "Soviet," in *Air Superiority*, ed. Cooling, 200–201. Buckley, Hogg, and Whiting mention that Shturmovik was no air fighter either (Hallion ironically blames its losses on Soviet fighters' inability to protect it).

Chapter 3. Air Support in the Atomic Era

1. Barlow, *Revolt*, 86–95; Brown, *Blind*, 66–86; Finletter, "Survival," 3–10; Futrell, *Ideas*, 1:168–72, 214–30; Martin, "Reforging," 32–53; Millett and Maslowski, *Common Defense*, 498–501; Smith, *Plans for Peace*, 17 (quote).

2. The sources listed in note 1, above, also support the statements made in this paragraph. See also Boyne, *Beyond*, 6–19; Builder, *Masks*, 18–20, 32–33, 67–73; Crane, *Airpower*, 14–22; Gorn, *Genie*, 17–19 (quote on 18), 183–86; Hadley, *Straw Giant*, 76–83; Hallion, "Troubling Past"; Meilinger, *Vandenberg*, 63–66; Mrozek, *Air Power*, 16; Rearden, "Doctrine," in *Case Studies in Strategic Bombardment*, ed. Hall, 384–86; Worden, *Rise*, 36.

3. Hadley, 76–83; Hallion, *Naval*, 9–14; Meilinger, *Vandenberg*, 80–82; Millett and Maslowski, 501–4.

4. Barlow, 44–63, 193–268; Futrell, *Ideas*, 1:237–59; Martin, "Reforging," 130–31; Meilinger, *Vandenberg*, 126–53; Crane, 6 (quote); Watson, *Secretary*, 38–39, 56–58, 62; Worden, 30–32.

5. Barlow, 263–67; Bergerson, *Army*, 37–38, 43–46; Davis, *31 Initiatives*, 9; Futrell, *Ideas*, 1:171–78, 256; Hadley, 82–87; Head, "A-7 Attack Airdraft Program," 104; Martin, 37, 47, 56–60.

6. Wolf, *Basic Documents*, 63–80, 151–65 (CAS definition, 165), 237.

7. Grey, "Korea," in *War in the Air*, ed. Stephens, 144; Hadley, 112–13; Martin, 60–63; Millett, "Korea," in *Development of Close Air Support*, 347–50; Smith, *Close*, 136.

8. Boyne, *Wild Blue*, 97–110; Bright, *Jet Makers*, 13–18, 31–39, 103–17; Brown, *Blind*, 68–160; Futrell, *Ideas*, 1:229–31; Rhodes, *Dark Sun*, 340–51.

9. Boyne, *Wild Blue*, 46; Crane, 21–22; MacIsaac, "Evolution," in *Third Dimension*, ed. Mason, 15; Futrell, *Ideas*, 1:224, 244–45; Hadley, 92–93; Hallion, "Troubling Past," 6; Hughes, 310–14; Martin, 39–41; Schlight, "Quesada," in *Makers*, ed. Frisbee, 198–203; Smith, *Peace*, 23–24; Worden, 38–40.

10. Boyne, "Evolution," 36–37, 42; Bright, 10–17; Futrell, *Ideas*, 1:307; Hadley, 113; Hallion, "Troubling Past," 4–7; Head, "Program," 110–14; Huenecke, *Design*, 27–28; Lester, *Mosquitoes*, 16; Millett, "Korea," 349; Mischler, *A-X*, 4.

11. Hadley, 101; Lester, 19–20; Meilinger, *Vandenberg*, 160; U.S. Congress, Senate, *Military Situation in the Far East*, 1378–1500.

12. Crane, 7–8; Futrell, *Korea*, 84–98; Grey, "Korea," 144; Hallion, *Korea*, 75–76; Kropf, "Problems"; Lewis, "Almond," 51–88; Lynch, "Close"; McLennan, "Close Air Support History," 44–50; Millett, "Korea," 353–61; Reinburg, "Close"; Smith, *Close*, 137–39; U.S. GAO, *Principal Issues*, app. 1, 55. Maj. Gen. Ned Almond, USA, who served first as U.N. Command's chief of staff and then as commander of its X Corps, was the most prominent complainer.

13. Armitage and Mason, *Nuclear Age*, 24; Crane, 24–30, 61–62; Dorr, *Skyraider*, 86–91; Futrell, *Ideas*, 1:308; Futrell, *Korea*, 84–98, 391; Grey, "Korea," 144; Hallion, *Korea*, 72–87 (quote, 78), 151–54; Isenberg, *Shield*, 266–67; Lynch, "Close"; Millett, "Korea," 351–71; Reinburg, "Close Air Support," 60–61; U.S. GAO, *Principal Issues*, app. 1, 55; Worden, 41.

14. Bradin, *Hot Air to Hellfire*, 83–85; Crane, 60–62, 171, 177; Davis, *31 Initiatives*, 11–12; Drew, "Low-Intensity Conflict," in *Paths of Heaven*, ed. Meilinger, 327; Goldberg and Smith, "Army–Air Force Relations," 9–12; Grey, 145–47; Martin, 84–85; McLennan, 47–50; Meilinger, *Vandenberg*, 170–72; Millett, "Korea," 377–97; U.S. Congress, Senate, *Close Air Support*, 92d Cong., 186.

15. Carlson, "Close"; Crane, 24 (the Air Force explored using P-47s, but they lacked spare parts), 60–62, 171–77; Davis, *MiG Alley*, 11, 15, 23; Dorr, 88; Grey, 145–47, 159; Futrell, *Korea*, 700–708 (early war desperation CAS said to be a "severe expense to American taxpayers," 707), 816 (index contains four times as many "praise" references as "criticism" references); Hallion, *Korea*, 50; Kropf, "MR Letters"; Lester, 33, 70–72; Martin, 75–76; Meilinger, *Vandenberg*, 172; Millett, "Korea," 362–95; Paulsen, "Air Force," 74–75; Werrell, *Archie*, 75–81. Lester's Air Force FAC history reflected the lukewarm Air Force attitude to CAS. He seemed overly concerned about cost versus benefits of CAS

flown during the stalemate period in Korea, but did not address the interdiction campaign's difficulties of that time.

16. Crane, 136; Dorr, 88; Hallion, *Korea*, 287, 307; Isenberg, 265; Millett, "Korea," 395–96 (the "later" study); Werrell, 75–81.

17. Crane, 171–80; Drew, "Wilderness"; Futrell, *Ideas*, 1:346–49, 351 (report quotes); Grey, 146–51; Hallion, "Retrospective Assessment"; Isenberg, 269–72; Kropf, "MR Letters"; MacIsaac, "Evolution," 16–17; Mark, 287–319; Tilford, *Setup*, 21 (Finletter quote).

18. Futrell, *Ideas*, 1:447 (Ridgway quote); Jones, "Flashblind," in *Air Support: Proceedings of Aerospace Symposium*, passim; Summers, *Strategy*, 41 ("pique" quote).

19. Ahmann, *McNickle*, 56, 60, 69–70; Boyne, *Wild Blue*, 97; Buhrow, "Requirements," 22–23; Drew, "Wilderness," 6; Futrell, *Ideas*, 1:419–23; Hadley, 103, 132–33; Jones, "Flashblind," 352; Lester, 77; Martin, passim; Meilinger, *Vandenberg*, 181, 182 (quote), 194, 206; Millett and Maslowski, 544–51; Mrozek, 19–20; Tilford, *Setup*, 24–28, 39; Weigley, *American*, 399–410; Worden, 81.

20. Ahmann, *McNickle*, 70–71; Crane ("Faustian" quote), 172, 175; Dick, "Disosway," 197, 246; Emme, "Fallacies"; Futrell, *Ideas*, 1:447–68 (intelligence director quote, 464); Head, "Program," 122–23; Hasdorff, "Hildreth," 42–63; Jones, "Flashblind," 346–53; MacIsaac, 17–18; Martin, 153–335; Sbrega, "Southeast Asia," in *Development of Close Air Support*, ed. Cooling, 411 (McConnell quote); Tilford, *Setup*, 20–34; U.S. GAO, *Principal Issues*, app. 1, 55–56; Worden, 75–85; Ziemke, "Shadow of the Giant," passim.

21. Barnes, "Concept," 16–19, 22, 36–38; Bright, 16–17, 68, 112; Buhrow, 19–22; Futrell, *Ideas*, 2:288; Gunston, *Fighters*, 138–214; Hallion, "Troubling Past"; Head, "Program," 114–16, 157–58; Jones, "Flashblind," 353–55; Lambeth, "Pitfalls," 11–14; MacIsaac, 18–19; Martin, 2–3, 353–79; Neufeld, "Sprey," 26; Nihart, "Unresolved"; Tilford, *Setup*, 31–34.

22. Anderegg, *Sierra Hotel*, 5–9; Goforth, "New Aircraft?" 18–20, 23, 33–34; Deatrick, interview by author.

23. Barnes, 45–46; Buhrow, 24–25; Dick, "Disosway," 246–47; Jones, "Flashblind," 354–56; Lambeth, *Transformation*, 36; Mason, *Centennial*, 92; Moody and Trest, "Containing," in *Winged Shield*, ed. Nalty, 1:151–52; Neufeld, "Sprey," 20–27; Sights, "Lebanon," 42 (quote); Tilford, *Setup*, 36.

24. Head, "Program," 126–33, 156–60; Holmquist, "Carrier-Based"; Isenberg, 311–57; Neufeld, "Sprey," 27.

25. Asprey, "Close," 33–37 (quotes, 35, 36).

26. Agnew, *Johnson Papers*, 3:35–36, transcript of interview #14, tape #2; Andreson, *Seneff*, 71; Asprey, 35 (quote); Bergerson, *Army*, 44; Builder, 118–42; Crane, 176; Hadley, 84–87; Kirkpatrick, "Army and the A-10," 4–5; Martin, 341.

27. Andreson, *Seneff*, 71; Asprey, 35–36; Coates and Killian, *Heavy Losses*, 138–39; Crane, 177; Goldberg and Smith, 45–46; Howze, "Growing Vacuum," 3–4; Howze, "Howze Board," 10, 24, 25; Lester, 75; McLennan, 68–70; Ridgway and Walter, *Hamlett*, 56–57; Worden, 85. Both Howze pieces are from Howze-Hawkins Family Papers.

28. Goldberg and Smith, 45, 47; Dixon, Williams, interviews by author. Retired U.S. Army general Robert Williams was the source for the analogy.

29. Bahnsen, "New?"; Bergerson, 72–75, 101–2; Galvin, *Air Assault*, 254–57; Gavin, "Cavalry"; Weinert, *Army Aviation*, 181–87.

30. Bergerson, 70–110 ("bureaucratic insurgency"); Bradin, 92–99; Davis, *31 Initiatives*, 31; Weinert, 185–214.

31. Bergerson, 55, 70; Davis, *31 Initiatives*, 13–14; Dupuy and Dupuy, *Encyclopedia*, 1331; Goldberg and Smith, 12–14; Head, "Program," 120; Ridgway, *Soldier*, 315; Wolf, *Basic Documents*, 317–23.

32. Asprey, 36–37; Bergerson, 90–91; Bradin, 95–96 (Vanderpool quote); Buhrow, 23, 26–29; Gonseth, "Tactical Air Support," 3–16; McLennan, 54–55; Moon, "Joint Doctrine," 8–13; Weinert, 214–17.

33. Bahnsen, 64; Bergerson, 79–80, 102–10.

34. Bergerson, 73, 78, 89, 185; Bradin, 104; Galvin, 264, 315; Rosen, *Winning*, 90–91; Weinert, 203–5.

35. Millett and Maslowski, 548–52; Weigley, 399–440.

36. Bergerson, 63; Bradin, 97, 102–5; Davis, *31 Initiatives*, 23; Futrell, *Ideas*, 2:174; Hadley, 136; Sbrega, 413.

CHAPTER 4. VIETNAM AND THE A-7 CHOICE

1. Millett and Maslowski, *For the Common Defense*, 554; McNamara, *Retrospect*, 4, 24; Shapley, *Promise*, 99–104, 112–13, 192–96; Watson, *Secretaries*, 206–24; Weigley, *American*, 445–48.

2. Hadley, 142–56; McMaster, *Dereliction*, 18–21; Shapley, *Promise*, 202–23; Summers, *Strategy*, 27–32; Watson, *Secretaries*, 228–34; Sprey, interview by author (quote).

3. Bergerson, *Army*, 191; Coulam, *Illusions*, 35–82, 98–100; Dick, "Disosway," 297; Neufeld, "Georgi," 18, 21.

4. Builder, *Masks*, 19, 104–19; Dick, "Disosway," 205–7; Drew, "Air Power Wilderness"; Futrell, *Ideas*, 2:48, 56–57; McMaster, 42–45; Neufeld, "Agan," 18–20; Tilford, *Setup*, 48–50; Worden, 107–55.

5. Barnes, "Concept," 46–47; Brown, *Blind*, 213–26; Coulam, 45–56, 90–96; Futrell, *Ideas*, 2:1–252; Gunston, *Attack Aircraft*, 170–80; Lambeth, "Pitfalls," 13; McLennan, 63; McNamara, Memorandum, SecDef Cont. No. C-595, 7 June 1961 (from A-10 monograph supporting document copies; henceforth called "supporting documents"); Shapley, 202–23; Tilford, *Setup*, 47–52; Sprey, interview by author.

6. Gunston, *Attack*, 234–36, 257; Head, "Program," 157–58 (Graham quote, 158); Holmquist, "Carrier Based Aircraft," 205–7; McNamara Memorandum, 7 June 1961 (quote).

7. Bugos, *Phantom II*, 115–21; Head, "Program," 159–65, 216; Tilford, *Setup*, 50–53; U.S. Congress, House, *DoD Appropriations for 1963, Part IV*, 321; U.S. News Productions, *MiGs versus America* (pilot quote).

8. Gunston, *Attack*, 234–37; Head, "Program," 171, 216; Holmquist, "Carrier Based Aircraft."

9. The sources listed here also apply to the following paragraph. Asprey, 37; Bergerson, 108–9; Bradin, 103–5 (first quote, 105); Davis, *31 Initiatives*, 15; Futrell, *Ideas*, 2:174–75, 468–69; Goldberg and Smith, 17–19; Kirkpatrick, "Army," 11 (Decker quote); McLennan, 62 (Decker quote), Tab F to Annex A; Millett and Maslowski, 558–60; Schlight, *Offensive*, 88–90; Tolson Papers, "Vietnam Diary, 1960–1968"; Weinert, *Army Aviation*, 214.

10. Brown, "Importance," 1; Goldberg and Smith, 19–20; Head, "Program," 216; Kirkpatrick, 11–15; McLennan, 62–66.

11. Bergerson, 110–11 (quote); Davis, *31 Initiatives*, 15–17; Krepinevich, *Army*, 119; Kramer and Powell, "Powell," 47–48, transcript, tape 1, side 1.

12. Bergerson, 111–14 (quote, 111); Bradin, 108–11; Galvin, *Air Assault*, 274–79; Howze, "Howze Board," 10–46.

13. Ahmann and Gallagher, "McNickle," 91–92; Booda, "Roles"; Davis, *31 Initiatives*, 17–18; Dick, "Disosway," 176–78; Futrell, *Ideas*, 2:182, 188; Goldberg and Smith, 20–21; Head, "Program," 221–23; Howze, "Howze Board," 39, 56–57; Mrozek, 16.

14. Bradin, 111; Futrell, *Ideas*, 2:182; Galvin, "Cavalry"; Head, "Program," 220–23; Kirkpatrick, 14–15; McLennan, Tab F to Annex A; McMaster, 81–82, 113–14.

15. Army Aviation School, "Use of Aerial Vehicles"; Agnew, "Johnson," 36–40, transcript interview 12, tape 12, 28–36, and transcript interview 14, tape 2, Johnson Papers; Bergerson, 118; Futrell, *Ideas*, 2:183–87; Goldberg and Smith, 23– 24; Head, "Program," 224–26; Hessman and others, "New Weapons"; Kirkpatrick, 16–21; McLennan, 72; McNamara, "Close Air Support," Memorandum, 16 February 1963, supporting documents; Reeves, "AH-56," 10–11.

16. Bergerson, 114–16; Bradin, 111–12; Davis, *31 Initiatives*, 18–19; Futrell, *Ideas*, 2:182–89; Goldberg and Smith, 21–27; Reeves, 12; Harrison, interview by author.

17. McMaster, 114 (quote).

18. Crane, *Korea*, 177–78; Drew, "Confusion," in *Paths of Heaven*, ed. Meilinger, 327–35; Lester, 84–230; Sbrega, "Southeast Asia," in *Development of Close Air Support*, ed. Cooling, 420–25, 441; Schlight, 3–5; Sheehan, *Bright*, 64–65; Tilford, *Setup*, 63–67, 76–77; U.S. Congress, House, *Close Air Support*, Report, 1 February 1966, 4863. Lester took a lukewarm attitude toward airplanes such as the A-1, emphasizing their vulnerabilities.

19. Andreson, "Seneff," 43–48; Bradin, 111–15; Brownlow, "Burgeoning," 28– 29; Buhrow, 40–42; Dick, "Disosway," 179, 192; Easterbrook, "Oblivion," in *More Bucks, Less Bang*, ed. Rasor, 52–60; Futrell, *Ideas*, 2:299; Krepinevich, 121–26; Moore and Galloway, *Soldiers*, 121, 305–7; Mrozek, 31–32, 76; Rosen, 94; Sbrega, 423, 454–55; Schlight, 16; Sheehan, 214–38.

20. Dick, "Disosway," 203; Mrozek, 17–20; Sbrega, 418–31; Tilford, *Setup*, 68–70, 79 (staff officer quote), 93–98, 114. Some accounts that discuss Air Force interdiction and North Vietnam air campaign frustrations are Broughton, *Going Downtown*, passim; Clodfelter, 73–146; Futrell, *Ideas*, 2:288–94; Gropman, "Air

War," in *Third Dimension*, ed. Mason, 70–76; Hosmer, "Psychological," 129–39; Lambeth, *Transformation*, 21–23; Mark, *Interdiction*, 324–26, 406–9; Sheehan, 676–79; Tilford, *Setup*, 89–163; Worden, 157–83. The soldiers' gratitude for air support, or belief that they came to rely too heavily upon CAS, are from Beaumont, *Joint Military Operations*, 152–53; Dick, "Disosway," 195; Hadley, 182; Krepinevich, 170–71; Moore and Galloway, 305–7; Sbrega, 469; Schlight, 216–17; U.S. GAO, *Principal Issues*, app. 2, 57–60.

21. Anderegg, 14–15; Broughton, 67 ("road recce" quote); Dick, "Disosway," 192; Momyer, *Air Power*, 270; Neufeld, "Sprey," 27; Sbrega, 441; "Talbott," 137 ("toothpicks" quote); Tilford, *Setup*, 217.

22. Ahmann and Gallagher, "Griffin," 62–64 (quotes); Resor and Seamans, "Systems for the Air Delivered Fire Support of Ground Forces," Attachment 3, Annex A, 1–3, supporting documents; "Talbott," 144.

23. Bowers and Rowley, "Deatrick," 55–62 (quotes); U.S. GAO, *Principal Issues*, 63; Deatrick, interview by author.

24. Ahmann and Gallagher, 64–65; Bennett, "Replace?" 14; Bowers and Rowley, 49–50, 56; Brownlow, "Burgeoning," 29–30; Gallagher and Officer, "Costin," 10–12 (SAR quote, 10), 24–25, 31–40; Hasdorff, "Hildreth," 31; Moore and Galloway, 96–97, 189–90, 227, 303–7; Officer, "Officer," 26–28, 31–32; Porter, "Second Defense," iv, 5, 6; Schlight, 99–100; Tilford, *Search*, 66; Deatrick, interview by author.

25. Ahmann, "McNickle," 114; Dorr, 57–60, 93–128; Gallagher and Officer, 6 (quote), 13–14, 27–29, 33; Gunston, *Attack*, 250–51; Hasdorff, "Hildreth," 68–78; Momyer, *Airpower*, 263, 270; Officer, "Officer," 68–69; Sbrega, 441; Deatrick, interview by author.

26. Ballard, *Fixed-Wing*, passim; Dorr, 110–11; Futrell, *Ideas*, 2:306–7; Gallagher and Officer, 14; Mark, 335–38; Mrozek, 125–32; Nalty, *Air War*, 26–30, 67, 371–72, 389–94; Sbrega; 444–46; Schlight, 199, 238–40; Tilford, *Setup*, 175–77; Zeybel, "Truck Count," 36–45; McGrath, interview by author. B-52s also performed air support in Vietnam. Their huge bomb loads delivered maximum shock effect and helped defeat the siege of Khe Sanh and communist attacks in 1972. However, as in World War II, they required much more mission preparation time, and their high-altitude, unguided-bomb attacks were better suited to area targets free of friendly troops. See Hallion, "Retrospective"; Mrozek, 139–45; Sbrega, 445–46, 456–58; West, "Decisive Role," in *Air Support: Proceedings*, 307–41.

27. Buhrow, 1–2, 39; Bergerson, 116; Goldberg and Smith, 27; Hasdorff, "Hildreth," 32; U.S. Congress, House, *Close Air Support*, 89th Cong., 4639–41; U.S. Congress, House, *Close Air Support*, Report, 1 February 1966, 4859; Pike, interview by author.

28. U.S. Congress, House, *Close Air Support*, 89th Cong., 4653–4752.

29. Hasdorff, "Hildreth," 62; U.S. Congress, House, *Close Air Support*, 89th Cong., 4765–97 (Agan-Pike quotes, 4790). The Army top leadership was at this time on record in support of Air Force resistance to OSD's proposal that the service procure the Navy A-7 attack plane; Futrell, *Ideas*, 2:120.

30. U.S. Congress, House, *Close Air Support*, 89th Cong., 4836–57; Schriever, interview by author. Later, answering an Air Force interviewer who considered the gunship a doctrinal violation, Schriever said, "Air Force doctrine. I don't know what the hell you're talking about. What I want is a weapon system that can do the job at hand in the most effective way." See Hasdorff and Officer, "Schriever," 71–72. Schriever told this author that he argued with LeMay for an attack plane because he felt that Air Force air superiority made it possible.

31. Bright, 21, 149; McNamara, "Close Support and SAW Aircraft," 7 January 1965, supporting documents; Mischler, 6–8; Neufeld, *F-15*, 8–9; Pentland, "Evolution of A-10 Mission Requirements."

32. Futrell, *Ideas*, 2:120, 470; Head, "Program," 228–31, 246 (Lemay quote, 228; Enthoven quote, 246).

33. Head, "Program," 243–47; McConnell, "Hq USAF Study," 15 March 1965, supporting documents; Neufeld, "Burns," 20–25; Neufeld, *F-15*, 9–10; Zuckert, "Close Support and SAW Aircraft," 2 February 1965, supporting documents.

34. Head, "Program," 251–53 (staff officer quote, 251; Enthoven quote, 252).

35. Futrell, *Ideas*, 2:174, 492; Hasdorff, "Hildreth," 30, 59; Head, "Program," 284; Neufeld, "Agan," 18–20 (quote, 20); Neufeld, "Georgi," 20 ("killed" quote); Sweat, 119.

36. Hasdorff, "Hildreth," 60 (quote); Head, "Program," 257.

37. Andrews, "Rugged Tests," 17; Bradin, 111–21; Futrell, *Ideas*, 2:189.

38. Davis, *31 Initiatives*, 19; Head, "Program," 273; McMaster, 224–26, 265; Schlight, 21, 61, 88–90, 114; U.S. Congress, Senate, *Air Force Tactical Operations*, 2–3, 37.

39. Hasdorff, "Hildreth," 60; Head, "Program," 258–71, 276–79.

40. Head, "Program," 280 (second quote), 281–82; U.S. Congress, Senate, *DoD Appropriations for FY 1970, Pt. 4*, 122 (first quote).

41. Harold Brown, letter to author; Head, "Program," 281 (quote).

42. Neufeld, "Burns," 25–26; Neufeld, "Georgi," 19; Ruegg, 139.

Chapter 5. The A-10's Improbable Conception

1. U.S. Congress, House, *Close Air Support*, Report, 1 February 1966, 4864, 4872 (quotes), 4873; Deatrick, interview by author.

2. Andrews, "No Red Flags"; Bradin, *From Hot Air to Hellfire*, 115 (first quote), 116–17; Gunston, *Apache*, 3 (second quote); Hessman and others, "New Weapons System"; U.S. GAO, *Principal Issues*, 12–14.

3. Bradin, 118–22.

4. DDR&E, DCP No. 23, 2, 11, supporting documents; Goldberg and Smith, "Army–Air Force Relations," 33; Head, "Program," 337, 415; U.S. Congress, Senate, *Close Air Support*, 92d Cong., 194; U.S. Congress, Senate, *DoD Appropriations for FY 1970, Pt. 4*, 122; Kay and Sprey, interviews by author.

5. Bergerson, *Army*, 117–19; Davis, *31 Initiatives*, 19–20; Futrell, *Ideas*, 2:312–13; Goldberg and Smith, 29–31; Head, "Program," 335–37; "Look but Don't

Shoot," 10–14; Schlight, *War*, 122–24; Williams and Yudkin, interviews by author. Yudkin said that Johnson and McConnell created a discrete joint staff project to accompany their talks, called New Focus. Yudkin chaired the Air Force team, and later surmised that the service chiefs were under pressure to resolve interservice differences, though he stated he did not know the source of the pressure. The Air Force also successfully forbade Navy special operations OV-10s to drop bombs in Vietnam. See Lavell, *Black Ponies*, 82–83, 153; U.S. Congress, House, *Roles and Missions of Close Air Support*, 15.

6. The citations in note 5 also support this paragraph. See also Andrews, "Rugged Tests," 17; Bradin, 122; Davis, *31 Initiatives*, 23.

7. Dick, "Disosway," 180 (quote); Goldberg and Smith, 31–33; Mischler, *A-X*, 15–16; Kay, Lancaster ("monstrosity" and "take over" quotes), Sprey, and Yudkin, interviews by author. Perhaps deliberately, the Air Force's professional journal, *Air University Review*, published differing tactical plane views in its May–June 1966 issue. In "The Rise and Fall of the Stuka Dive Bomber," Air Force Col. William Scott warned his service to avoid overemphasizing air support missions in wars like Vietnam, lest it end up with poorly performing, Stuka-like planes. Hans Multhopp, in his "Challenge of the Performance Spectrum for Military Aircraft," observed that the current fast fighters incurred penalties such as increased takeoff/landing rolls, fuel inefficiency, and poor agility.

8. Goldberg and Smith, 33 (review findings quote); Kaplan, "Beast of Battle"; Neufeld, "Boyd," 25; Kay, Lancaster, Schlesinger, Sprey, and Yudkin (quote), interviews by author. Others praised Yudkin's courage, and some even portrayed him confronting Air Force leaders. But in the author's interviews Yudkin was modest about his role, and declared that his office was supposed to handle such issues.

9. Fulghum, "A-7F," 4 (Mathis quote). Oral history sources for air leader contempt: Ahmann, "McNickle," 114 ("ridiculous" quote; Air Force deputy chief of staff for research and development, 1969–1970); Dick, "Disosway," 180; Hasdorff, "Hildreth," 65 (chief of the Tactical Support Division, Headquarters U.S. Air Force Studies and Analysis Office, 1969); Neufeld, "Georgi," 17–19, 22, 23–25 (TAC's fighter requirements staff); Sweat, 119 ("stupid plane" quote; deputy chief of staff for operations at headquarters, U.S. Air Forces in Europe, 1968–1969, and TAC vice commander, 1972).

10. Andrews, "Weapons Ship," 13 (cartoon); Bergerson, 126, 131; Bradin, 123; Dick, "Disosway," 180; Harrison, "Gift," 36–39 ("illegitimate" quote, 39); Dixon, Sprey (quote), and Williams, interviews by author.

11. Yudkin, interview by author (quote).

12. McConnell, "New Orders," 121–23 (quote, 122). See also Brown, "Southeast Asia," 25–34; Clay, "Shaping"; Worden, *Rise*, 169–70.

13. Brown, "Tactical Air Power," 66–68; Futrell, *Ideas*, 2:472–73; Harold Brown, letter to author.

14. Harrison, "Gift," 37–38; Neufeld, "Sprey," 37–38; Kay, Lancaster, Sprey, and Yudkin, interviews by author. Kay and Yudkin said that Disosway received

coolly their A-X briefing in 1967. Sprey said that A-X concept designers invited TAC to participate, but TAC refused.

15. Burton, *Pentagon Wars*, 24, 241; Fallows, *National Defense*, 95–106; Stevenson, *Pentagon Paradox*, 100; Christie, Fredericksen, Kay, Sprey (quote), Yudkin, interviews by author.

16. Dempster, "Requirements Action Directive RAD 7-69-(1)," 1–6 (quote, 1), supporting documents; Mischler, 20–24; Pentland, "Evolution"; Christie, Kay, Lancaster, Pentland, Sprey, and Welch, interviews by author. Sprey was aware of early 1960s CAS plane articles by U.S. Marine Corps Col. J. H. Reinburg, but felt that Reinburg's proposed design was too heavily armored. The articles coincided with the Army's CAS plane interests, but Sprey was the only source who acknowledged Reinburg.

17. Welch, interview by author (quote).

18. Andrews, "Weapons Ship"; "AF Concerned over AX Opposition"; Avery, "AH-56," para. 2b; DCP 23, 12; Greenhous, "Israeli Experience," in *Development of Close Air Support*, ed. Cooling, 503–9; Gunston, *Attack*, 259; Hessman, "New"; Smallwood, *Warthog*, 11–12; Sweetman, *A-10*, 8; Volz, "A-X"; Wetmore, "Israelis' Air Punch," 21–23; Christie, Kay, Lancaster, and Sprey (quotes), and Stolfi, interviews by author. DCP 23 cites Israeli gun success, but the author believes A-X advocates included it in order to sell the A-X gun. Stolfi, who inspected wrecked Arab armor at war's end as part of a U.S. study, found that ground fire destroyed 97% of Arab tanks. Israeli fighters lacked armor-piercing gun rounds, and Israeli pilots told him that they did not practice tank attacks.

19. Dörfer, 4; Futrell, *Ideas*, 2:470–74; Hasdorff, "Hildreth," 61; Head, "Decision," 223–24; Mischler, 24; Neufeld, "Boyd," 27–29; Neufeld, *F-15*, 11–27; Neufeld, "Sprey," 20; Stevenson, *Paradox*, 27–28; Sweetman, *A-10*, 7; Christie and Sprey (quote), interviews by author.

20. Andrews, "Rugged Tests"; DCP 23, 4–5; McNamara, "Selected Statements, January 1–June 30, 1968," 5; Mischler, 37, 43–57; Wilson, "Fairchild"; Sprey, interview by author (quote). IOC meant the first plane delivered to an operational squadron.

21. Mischler, 51–54; Pentland, "Evolution"; Reeves, "AH-56," 27; Wall, *GAU-8A*, iv–v, 1–6; Christie, Sprey, and Welch, interviews by author.

22. Buchta, Christie, Dilger, Kishline, Oates, McMullen, and Sprey, interviews by author. Gun design decision making occurred in a study group set up between Systems Command's main armament laboratory and the Eglin AFB arms lab. Christie worked at the Pentagon and also in OSD's Joint Technical Coordinating Group (JTCG), which had some oversight authority over the project.

23. Bradin, 117; Mischler, 49–57; Reeves, 45–48; Christie, Sprey, and Welch, interviews by author. Sprey told the author that "the only thing that counts is effectiveness. . . . We're not in this to field technology."

24. Armament Systems Division (ASD), "News Release," 18 May 1970, 1, supporting documents; Gunston, *Attack*, 256–57; McDermont, "Engineering for Survivability," 13–14 (put loss statistics, derived from both Vietnam and the

Six-Day War, at 62% due to fire/explosions, 20% due to loss of control, and 18% due to pilot incapacitation); Mischler, 54–55; Sweetman, *A-10*, 7, 16, 24–25; U.S. Congress, House, *Cost Escalation*, 1336; U.S. Congress, Senate, *DoD Appropriations for FY 1970, Pt. 4*, 41; Christie, Foster, Fredericksen, Spinney, and Sprey, interviews by author.

25. Sprey, interview by author (quote).

26. Mischler, 25–31, 39; Sweetman, *A-10*, 27–30; Sprey, interview by author (quote).

27. Gunston, *Attack*, 254–56; Mischler, 44; and Sprey, interview by author.

28. Andrews, "'Known'"; Andrews, "No Red Flags"; Mischler, 43.

29. DCP 23; Mischler, 57–59; Foster, interview by author.

30. DCP 23; Mischler, 59–64; Wilson, "Fairchild"; Foster and Sprey, interviews by author.

31. Mischler, 62, 66, 82–83.

32. Bradshaw, "Close Air Support," CFP suppl., 1.2–2.5, supporting documents; DCP 23, 13; Reeves, 13; Wolf, *Basic Documents*, 180.

33. BDM Corp., *Statistical Lessons*, 6:18–19; Nophsker, "Perceptions of Fighter Strikes," 14–40, 80–89.

34. Andrews, "Weapons Ship," 12–13; Avery, "AH-56," para. 2b; Davis, *31 Initiatives*, 21; Kirkpatrick, 27–30; Sbrega, 454–55.

35. Andreson and Kramer, "Goodhand," 1–2, 7, transcript, tape 2; Bradin, 117, 123–24, 142–43; Court and Powell, "Williams," 21–22, transcript, side 1, tape 3; Kramer and Powell, "Powell," 26–27, transcript, tape 2, side 1; Reed, "Howze," 58–59, transcript; Volz, "AX vs. AH-56," 25; Bahnsen and Williams, interviews by author.

36. The citations given here also apply to the next paragraph. "AF Concerned"; Bright, *Jet Makers*, 70–73, 149–67; Barnes, "Concept," 21–22, 36; DCP 23, 14; Fallows, *National Defense*, 36–37; Head, "Program," 416, 463, 476–520; Ludvigsen, "Up from the Ground"; "Stopping the Incredible Rise in Weapons Costs," 60–61 (quote); U.S. Congress, Senate, *DoD Appropriations, FY 1973, Pt. 4*, 691–95; U.S. Congress, Senate, *Tactical Air Program*, 92–116; Volz, 25–26.

37. "AF Concerned"; Frankosky, "Systems Management Directive #SMD-0-379-329A(1)," supporting documents; Gunston, *Attack*, 250; Mischler, 91–92; Ulsamer, "AX"; Volz, 25; Christie, interview by author.

38. "AX Fighter Paved Way for Prototyping"; "Competitive Prototype Program Planned for USAF AX Aircraft"; Gunston, *Attack*, 254–67; "Fly Two before Deciding"; Mischler, 72–82; Odgers, "Design-to-Cost," 1–5; Packard, "Improvement in Weapons Systems Acquisition," supporting documents; Stevenson, 99–100; Sweetman, *A-10*, 8; Ulsamer, "A-10 Approach"; Christie and Sprey, interviews by author.

39. Burton, *Pentagon Wars*, 13–19; Gunston, *Attack*, 260; Hansen, "Address," in *Supplement*, 2–7; Mischler, 66–67; Neufeld, *F-15*, 32–36, 66–68; Neufeld, "Sprey," 5–9, 18–19 ("best plane" comment); Stevenson, 1–50, 152–58; Volz, 26. Gunston, Hansen, Neufeld, and Stevenson are the "observers."

40. DDR&E, "Minutes of the Defense Acquisition Review Council," 31 December 1969, 1–2, supporting documents; "HASC Restores Most Senate Weapons Cuts"; Harrison, "Gift"; Mischler, 66–67; Volz, 25.

41. Dick, "Disosway," 181–82 ("rid" quote); Harrison, 38 ("remarkable" quote); Neufeld, "Sprey," 19–20.

42. The citations given here also apply to the next paragraph. Davis, *31 Initiatives*, 22 ("scholar"; "agree" quote); Futrell, *Ideas*, 2:521–22; Goldberg and Smith, 34–35; Resor and Seamans, "Systems for the Air Delivered Fire Support," supporting documents; Seamans, *Aiming*, 174 ("angels" quote), 175; Volz, 25–26; Seamans, interview by author. Interestingly, an attachment to the Seamans-Resor Agreement stated that armed helicopters in Vietnam were not collocated with infantry units and had a 20- to 30-minute ground request response time.

43. Futrell, *Ideas*, 2:519–20; Goldberg and Smith, 34–36 (Westmoreland quote, 35; Ryan quote, 36); Harrison, "Gift," 38–39; Seamans, *Aiming*, 174 (first and second quotes), 175; Seamans, interview by author.

44. Mischler, 67–71, 79–84; Pentland, "Evolution."

45. Bell, "Close Air Support for the Future," 34–36, 93–99; Crawford, "Low Level Attacks," 11–16; Gunston, *Attack*, 271; Haynes and Jones, "Air Support of the 'Close-In Battle,'" in *Air Support: Proceedings*, 110–11 (neither IR nor radar can guarantee identification at normal weapons firing ranges); Rasmussen, "A-10," 27–28; Reis, "Close Air Support Systems," 29–35; Ritchie, "Fixed-Wing," 102–10 (no technological device provides complete insurance against target misidentification); Sweetman, *A-10*, 8; Wessely, "Limiting Factors," 18–21; Westenhoff, "Close Air Support," in *Air Support: Proceedings*, 59–63; Sprey, interviews by author.

46. Ahmann, "Manor," 93; Field, "A-10 Test," 47; Gunston, *Attack*, 256; Mischler, 83–84; Sweetman, *A-10*, 28–32; Christie, Sprey, and Welch, interviews by author.

47. Aeronautical Systems Division, "A-X Specialized Close Air Support Aircraft, Request for Proposal," 1–2, supporting documents; Mischler, 83–102. The twelve companies were Beech Aircraft, Boeing, Cessna Aircraft, Fairchild Hiller (later Fairchild-Republic), General Dynamics, Grumman Aerospace, Lockheed Aircraft, LTV Aerospace, McDonnell Douglas, North American Rockwell, Northrop, and Textron. Fairchild Hiller, Boeing, Northrop, Cessna, General Dynamics, and Lockheed were the six semifinalists. During RFP formulation, TAC tried to replace the CFP and DCP design concept with one resembling the F-5.

48. Brown, *Blind*, 332; Carleton, "A-10 Design-to-Cost," 3; Mischler, 98; Field, "A-10 Test," 49; Odgers, "Design-to-Cost," 1–5; Wilson, "Fairchild."

49. Mischler, 98; Odgers, 6; Wall, GAU-8A, 3–9; McMullen, interview by author.

50. "Mahon Rites Friday," 1A, 2A; U.S. Congress, House, *DoD Appropriations for 1971*, Pt. 1, 554, 636–40; U.S. Congress, House, *DoD Appropriations for 1971*, Pt. 6, 101–3 (quote, 102); U.S. Congress, Senate, *DoD Appropriations for FY 1970*, Pt. 4, 41; Volz, 25.

51. Bergerson, 125; Braddon, "Know about Harrier"; Goldberg and Smith, 36–37; Gunston, *Attack*, 96–99; Gunston, "Harrier," in *Great Book of Modern Warplanes*, ed. Bonds and others, 349–51; U.S. GAO, *Principal Issues*, 19–28;
52. "Close Air Support: Cheyenne vs. Harrier vs. A-X?"; "Fly Two before Deciding"; Ludvigsen, "Up from the Ground"; New, "Perspectives"; Nucci, "Minutes, DSARC Review," 17 December 1970, 2, supporting documents; Schemmer, "Packard Personally Heads New Group"; U.S. Congress, Senate, *Close Air Support*, Report, 18 April 1972, 1; U.S. GAO, *Principal Issues*, 6 (quotes).

Chapter 6. CAS Aircraft on Trial

1. Schemmer, "Packard" (quote). The Packard study's formal name was the Close Air Support Review Group. The other members were Dr. John Foster, DDR&E; Gardiner Tucker, assistant defense secretary for systems analysis; and Navy Vice Adm. John Heinel.
2. Bergerson, 123; Everett-Heath, *Helicopters*, 100–102; Nalty, *Air War*, 250–73; Seibert, "Memo for Record," in Seibert Papers; Weiss, "'Gunships'"; Johnson, interview by author. Nalty added that cloud conditions forced helicopters along heavily defended routes.
3. Ahmann, "McNickle," 93–94; Loye and others, "Lam Son 719," xii–xvii; 104–15; Nolan, *Into Laos*, 358; Seamans, *Aiming*, 174; Tilford, *Setup*, 201; Werrell, *Archie*, 112–14; Seamans, interview by author.
4. Bergerson, 129–30; Futrell, *Ideas*, 2:283–84; Goldberg and Smith, 37; Johnson and Winnefield, *Joint Air Operations*, 70–73; Mrozek, 82; Sbrega, 456–64; U.S. Congress, Senate, *Close Air Support*, 92d Cong., 12–13, app. A, 413.
5. Nihart, "Packard Review Group," 20, 39–40; U.S. Congress, Senate, *Close Air Support*, 92d Cong., app. A.
6. U.S. Congress, Senate, *Close Air Support*, 92d Cong., app. A, 418–36; Fredericksen, interview by author.
7. U.S. Congress, Senate, *Close Air Support*, 92d Cong., app. A, 409 (quote), 410–11. The Packard Report was an appendix to the hearings.
8. "Each Service Has Its Own Views"; U.S. Congress, Senate, *Close Air Support*, 92d Cong., app. A, 411–13.
9. Bergerson, 125–26 ("observer"); Goldberg and Smith, 36–39; Packard, letter to Cannon, 13 May 1970, and Packard, letter to Stennis, 11 September 1970 (both in supporting documents); U.S. Congress, Senate, *Close Air Support*, 92d Cong., 32–33, 58–59, 104 (Cannon quote); U.S. Congress, Senate, *Close Air Support*, Report, 18 April 1972, 1; U.S. GAO, *Principal Issues*, 5–6.
10. Gunston, *Attack Aircraft*, 254–55; U.S. GAO, *Principal Issues*, passim; Christie, interview by author.
11. U.S. Congress, Senate, *Close Air Support*, 92d Cong., 1–55.
12. Bradin, *From Hot Air to Hellfire*, 125–31; U.S. Congress, Senate, *Close Air Support*, 92d Cong., 71–172 (Cannon quote, 71).
13. U.S. Congress, Senate, *Close Air Support*, hearings, 92d Cong., 80–84 (Cannon, Goldwater, and McIntyre exchange), 112–31.

14. Ibid., 173–49 (quote, 189). This citation also applies to the following paragraphs. Sources for Momyer's military attitude are Broughton, 105–7, 265; Dick, "Disosway," 162; Hasdorff, "Hildreth," 32–34; Sweat, oral history, 118–20; Adams, Christie, Deatrick, and Yudkin, interviews by author.

15. U.S. Congress, Senate, *Close Air Support*, 92d Cong., 251 (quote), 252–330.

16. Ibid., 307–29.

17. Ibid., 331–64.

18. Ibid., 1, 71, 173, 251, 312, 320, 331, 474. Sources for Texas's military contract success are Coulam, 63; Dörfer, *Arms Deal*, 42; Gunston, 247–69; Hadley, 154–55; Hance, Mahon oral history; Adams, Battista, Fredericksen, Giaimo, McMullen, Pike, Price, Seamans, interviews by author.

19. U.S. Congress, Senate, *Close Air Support*, 92d Cong., 365–88, and app. B.

20. Ibid., 372–400 (quotes, 399–400). The critical *AFJ* pieces are Weiss, "They Call It SLUF"; Schemmer, "Bum Dope." None of the major U.S. print news organs covered the story. Perhaps this was because they knew that it would confuse the lay public, and because it involved relatively low-cost weapons. The most expensive CAS aircraft contestant cost $3.8 million per copy—much less than the cost of such planes as the B-1 and C-5, which generated so much media interest at the time.

21. U.S. Congress, Senate, *Close Air Support*, Report, 18 April 1972, 1–36 (helicopter quote, 26; Hughes quote, 27). Actually, the report recommended a flyoff among the A-7, A-X, and A-4. The Cheyenne's demise is in Bergerson, 140; Bradin, 124; Futrell, *Ideas*, 2:527.

22. Bright, *Jet Makers*, 189–91; Hansen, "A-10 Prototype"; "Northrop Streamlines"; Watson, *A-10*, 13–22. The first flights of the A-9 and A-10 took place on 30 May 1972 and 10 May 1972, respectively. The A-9's maximum weight was 41,000 pounds, while the A-10's maximum was nearly 45,600 pounds. The F102 fan developed 7,500 pounds of thrust and the TF34 generated 9,275 pounds of thrust. The A-9's dimensions were fifty-eight feet wingspan, fifty-four feet length, and sixteen feet height. The A-10's dimensions were fifty-seven feet wingspan, fifty-three feet length, and nearly fifteen feet height.

23. Gunston, *Attack*, 263–64; "Northrop Streamlines"; Watson, *A-10*, 13–16.

24. Gunston, *Attack*, 264–65; Hansen, "A-10 Prototype"; Sweetman, *A-10*, 16–33; Watson, *A-10*, 16–19; Sanator, Sprey, interviews by author.

25. Bridges, "Realism," 28–35 ("stringent" quote, 30); DDR&E, DCP 23B, 5, 15; Field, "A-10 Test," 56–61; "History of the Air Force Systems Command, FY 73," 195–202; Robinson, "A-10A Future"; Sweetman, *A-10*, 11–12 ("unprecedented" quote, 11); Watson, *A-10*, 22–23. The pilots routinely overstressed the A-9 during dive bomb pullouts until Northrop engineers reduced its pitch control sensitivity. The A-10's engines surged during high angles of attack (the angle between the aircraft and the relative wind). This problem was rectified shortly after the flyoff.

26. Sweetman, *A-10*, 11–12; Watson, *A-10*, 23–24, Sprey, interview by author.

27. "History of the Air Force Systems Command, FY 73," 202 (quote); Seamans, *Targets*, 175; Smallwood, *Warthog*, 13–17; Watson, *A-10*, 24–29; Wilson, "A-10"; Seamans, Sprey, interviews by author.

28. Gunston, *Attack*, 265; Sweetman, *A-10*, 11.

29. Powell, "A-10 Source Selection"; Seamans, *Aiming*, 175; Fredericksen, Pike, Seamans (quotes), interviews by author.

30. The citations given here also apply to the next paragraph. Brindley, "Fairchild's A-10"; "Fairchild Can-Opener," 267–82 (TAC boast, 271); Minutes, DSARC Review of A-X Aircraft—Milestone II, 5–6, supporting documents; Robinson, "USAF Unveils War Game"; Smallwood, 120 (Sanator quote); U.S. Congress, House, *DoD Appropriations for 1974, Pt. 6*, 1351; U.S. Congress, Senate, *DoD Appropriations for FY 1974, Pt. 4*, 68–69; Wilson, "A-10"; Fredericksen, Kishline, Sprey, interviews by author.

31. "A-10/A-7 Flyoff Rules Drafted"; "A-7/A-10 'Fly-Off' Suggested"; Brownlow, "Senate Unit"; Futrell, *Ideas*, 2:527–29; Geddes, "A-10—USAF Choice"; Geiger and others, "History of the Aeronautical Systems Division, July 1973–June 1974," 1:80–84; Minutes, DSARC Milestone II, 7, supporting documents; Powell, "A-10 Source Selection"; Robinson, "USAF Unveils"; "Senate Unit Cuts A-10A Funds; Recommends Flyoff"; "USAF Agrees to Flyoff for A-10, A-7D"; U.S. Congress, House, *Cost Escalation*, 1339–40; "Washington Roundup: Strike Delay"; Watson, *A-10*, 38–40; Fredericksen, interview by author. The official name of Air Force studies was the SABER ARMOR series. No available source mentioned why the A-4 did not participate in the flyoff, per the Senate report, but reasons are apparent. The flyoff was complex enough with only two jets. McDonnell-Douglas's A-4 had a wide domestic and international market. The A-7's and A-10's respective strengths, such as weapons carriage and loiter, made them better candidates. DCP 23B (10) scored the A-4 for its comparative weaknesses in these areas as well as its greater vulnerability to battle damage.

32. "A-X Takes Flak Head On," 22 (Uhl quote); Cleary, "Dixon," 301; Furlong, Memorandum for DDR&E, 17 September 1973, supporting documents; U.S. Congress, House, *DoD Appropriations for 1974, Pt. 6*, 1346–48; U.S. Congress, House, *DoD Appropriations for 1974, Pt. 7*, 999–1004 (Mahon quote, 1000); U.S. Congress, House, *DoD Appropriations for 1975, Pt. 7*, 1052–60; U.S. Congress, House, *Cost Escalation*, 1330–36; U.S. Congress, House, *Military Posture and H.R. 12564, Pt. 1*, 614; U.S. GAO, *A-10 Close Air Support Aircraft*, 1–9; Adams, Battista, Pike, Price, Schlesinger, interviews by author. Adams observed that LTV pushed the issue continually. Battista said the flyoff seemed like a battle between Mahon and Pike.

33. Andradé, *Trial*, 28–40, 87–88, 156–57, 510–37; Brownlow, "Senate Unit"; Doglione and others, "1972 Spring Invasion," in *Air War—Vietnam*, 97–206; Dorr, *Skyraider*, 150–51; Futrell, *Ideas*, 2:482–83, 529; Geiger and others, "History, July 1973–June 1974," 80–81; McAdoo, "A-7D/A-10A Debate"; Nalty, *Air War*, 359–95; Nordeen, *Air Warfare*, MP 48; *Watson*, A-10, 38–40; Werrell, 115–16; McGrath, Price, interviews by author.

34. Cordesman and Wagner, *Lessons*, 1:90–97; Futrell, *Ideas*, 2:484–87; Greenhous, "Israeli Experience," in *Development of Close Air Support*, ed. Cooling, 493–511.

35. These citations also apply to the next paragraph. Alberts, "Tactical Air Power in NATO"; Cockburn, *Threat*, 132–33; Coleman, "Israeli Air Force Decisive"; Crabtree, *Air Defense*, 152–57; Greenhous, "Israeli Experience," 510–27 (quote, 515); Herzog, *War*, 251–61; Hogg, *Tank Killing*, 166; Mason, *Centennial*, 72–76; Nordeen, *Air Warfare*, 131–46; Werrell, 139–46; Stolfi, interview by author. The SA-7 was a small missile that often damaged only a jet's tailpipe. However, the SA-6 and ZSU-23-4 were quite lethal. Both sides suffered high helicopter casualties. The Arabs salvo-fired their missiles regardless of who was airborne, and downed as many of their own planes as they did Israeli aircraft. Coleman claimed that the IAF was effective against tanks, but Dr. Stolfi also did tank battle damage assessment after this war, and his findings matched those of his research on the Six-Day War. Hogg, Cordesman, and Wagner reached similar conclusions. Greenhous notes IAF air support successes but gives only moderate credit to its antitank claims.

36. Astor, "World's Toughest Air Force"; Futrell, *Ideas*, 2:529; U.S. Congress, Senate, *DoD Appropriations for FY 1975, Pt. 4*, 603–4; U.S. Congress, Senate, *FY 1975 Authorization for Military Procurement, Research and Development, Pt. 8*, 4310–47; U.S. GAO, *A-10*, 8. The following are among the many sources discussing the war's significance: Deitchman, "Implications"; Dupuy, *Evolution*, 282–83; "Five Lessons of the War"; Futrell, *Ideas*, 2:485–90; Hotz, "Mideast Surprise"; Macksey, *Technology*, 194–200; O'Rourke, "TacAir Dead?"; Rasmussen, "Central"; Rich and Janos, *Skunk Works*, 16–18. Many sources described the Israelis as masters of war. Americans have cited Israeli opinions to make their own points; the 1993 U.S. Air Force Air War College Associate Programs essays on the October War are a good example. Apparently wanting to reinforce the proper place of CAS in Air Force doctrine, two of the four readings featured Israeli complaints about the mission.

37. Cleary, "Dixon," 301, 303; U.S. Congress, House, *DoD Appropriations for 1975, Pt. 4*, 860; U.S. Congress, House, *Cost Escalation*, 1337; U.S. Congress, Senate, *FY 1975 Authorization for Military Procurement*, 4183, 4319.

38. Cleary, "Dixon," 274–76, 299–300 (quote, 300); Davis, *31 Initiatives*, 24–29; Futrell, *Ideas*, 2:539–46; Momyer, "Close Air Support"; Puryear, "Brown," 212; Sweat, 120; Ulsamer, "A-10," 41–45; Ulsamer, "New Muscle," 22–23; Winton, "Partnership," 100–106; Adams, interview by author.

39. "A-10/A-7 Flyoff Rules Drafted"; Bradin, 133–56; Futrell, *Ideas*, 2:490–510, 546–48; Geddes, "A-10"; Gunston, *AH-64*, 4–8, 41–44; Kelly, "King," 230–31; Lambeth, *Transformation*, 62–63; Romjue, *Active Defense to AirLand Battle*, 4–7 (quote, 6), 82–86; Worden, 222–23; Adams, McMullen, Schlesinger, interviews by author.

40. Dörfer, 14–15; Mintz, "Maverick Missile," reprinted in *More Bucks*, 183 (quote); Stevenson, *Pentagon*, 161, 176; Christie, Jones, Schlesinger, Sprey, interviews by author. Schlesinger believed that Jones becoming chief cemented the agreement because Jones was more supportive than Brown.

41. The following citations apply to the next paragraph as well. Bridges, 39; Freck and others, "Comparison," 1–7, 27–38; Geiger and others, "History," 90–91; "USAF Agrees to Flyoff," 23 (Cannon quote); U.S. Congress, Senate, *FY 1975 Authorization, Pt. 8*, 4321–28; Watson, *A-10*, 40–46; McMullen, interview by author. Supporting document sources are Brown, "A-10/A-7"; Currie, Memorandum for the Secretary of Defense, 2 October 1973; Headquarters USAF, "Program Management Directive [PMD] for the A-10/A-7D Fly-Off," 11 February 1974; Helfman, "A-10/A-7D Comparative Flight Evaluation," 4 April 1974.

42. Wall, *GAU-8/A*, ix–xi, 12–23, 32; Christie, Dilger, Fredericksen, Kishline (chairman of the gun selection board), Oates, and Sprey, interviews by author. Fredericksen, who was CAS programs monitor in DDR&E at the time, said he welcomed Sprey's Oerlikon input because it guaranteed competition against what he considered GE's almost unbeatable weapon. Citing corporate officials' comments, he was convinced that a third competitor kept gun costs down. Sprey proudly acknowledged his role to the author, and his assertion about strafing is valid (Walker, "GAU-8/A," 2–5).

43. Suydam, "GAU," 132–33; Wall, *Gau 8/A*, 20–27, 30–31; Christie, Fredericksen, Kishline, McMullen, Oates, Sprey, interviews by author. Oerlikon advocate Sprey conceded that the competition was fair. The other interviewees mentioned the Eglin Armaments Lab's preference for Philco-Ford's gun.

44. Wall, *Gau 8/A*, 18–45; McMullen, interview by author.

45. Bridges, 41–44; 76; Brown, "A-10/A-7"; Currie Memo, 2 October 1973, supporting documents; Freck and others, "Comparison," 4–22, 36–53; Geiger and others, "History," 94; Headquarters, USAF, "PMD for the A-10/A-7D Fly-Off," supporting documents; "USAF Agrees to Flyoff"; Watson, *A-10*, 49; Adams, McMullen, Sprey, Tabor, interviews by author. The planes carried bombs during the missions to compare performance, but the pilots did not drop them. Postulations can create suspicions that the parameters are rigged; for example, Sprey obviously liked the A-10, but he dismissed this flyoff as biased toward it.

46. Cleary, "Dixon," 301–4 (quote, 302); McMullen, Tabor, interviews by author. Dixon added, "I also had to call in the contractor[s] and tell them that if they got down there and . . . interfered with the operation . . . I would have thrown them off the field" (303). Tabor, one of the four pilots, recalled that isolation procedures were stringent. The pilots also carried a prepared statement in case these measures failed to thwart enterprising reporters.

47. Freck and others, "Comparison," 40–66; Geiger and others, "History," 95–96; Watson, *A-10*, 49–53; Fredericksen, McMullen, Tabor, interviews by author. Firing the Edwards A-10's GAU-8/A impressed the pilots. The test excited LTV's interest in a 30-mm gun pod for the A-7. The Air Force rejected it as too big and costly. The OSD criticized the A-7's survivability faults, but LTV and others answered that it would fly fast enough to avoid hits. They asserted that the A-10 might be durable, but its slow speed meant that A-10s would

spend a war in hangars receiving battle damage repairs. In response, Fredericksen sanctioned a study that found the A-10's design envisioned mostly overnight repairs. The Israelis achieved this repair time with their A-4s and F-4s during the October War.

48. "A-10A Capabilities Set Basis for Flyoff"; Freck and others, "Comparison," app. A, A-1 through A-22; Geiger and others, "History," 94; Adams, interview by author.

49. "Decision on A-10 Production May Sway Action in Congress"; "Flyoff Decision Favoring A-10 Expected"; "Lack of Clear Flyoff Decision Could Postpone Funds for A-10"; U.S. Congress, House, *Full Committee Consideration of H.R. 8591, H.R. 11144, H.R. 15406*, 16–50 (quote, 29); McMullen, Schlesinger, Tabor, interviews by author. McMullen and Tabor say that the first time the pilots learned the results and the other pilots' opinions was when the results were presented to Congress. The test did not determine the planes' ability to defeat fighter attacks, but the Air Force produced a study concluding that the A-10 could outmaneuver and evade enemy fighters. The formal announcement had to wait for certain arrangements. The Air Force still wanted both planes, and was trapped between two congressional factions: those who wanted one plane, and the Texas delegation which threatened to block A-10 funds in favor of more A-7s. With congressional compromise established in July 1974, the Air Force declared the A-10 the official flyoff winner and retained some A-7 units (later moved to Air National Guard and Reserve units).

Chapter 7. Developmental and Operational Challenges

1. U.S. Congress, House, *DoD Appropriations for 1975, Pt. 7*, 1050–58; U.S. Congress, House, *H.R. 8591*, 49–50; Battista, Dickinson, Giaimo, Lloyd, Pike, and Price, interviews by author.

2. Porth, "Enforcer"; Robinson, "Flight Test Program"; USAF, Project PAVE COIN, "FAC and LSA Evaluation Report," iii–xv, 153–204.

3. U.S. Congress, House, *Briefing on the Military Airlift Capability during the Middle East Conflict and Enforcer Aircraft*, 93d Cong., 2–40; U.S. Congress, House, *Military Posture and H.R. 3689*, 5151; U.S. Congress, Senate, *FY 1978 Authorization for Military Procurement, Pt. 7*, 5141–65.

4. Anderson, "Secrets Leak"; Bryant and Dapore, "Stewart," 95–96; "House Unit to Urge Enforcer Flight Test"; U.S. Congress, House, *Enforcer Aircraft*, 95th Cong., 3–4; U.S. Congress, House, *Military Posture and H.R. 3689*, 5150–69; U.S. Congress, House, *Middle East Conflict and Enforcer Aircraft*, 8–39; U.S. Congress, Senate, *FY 1978 Authorization, Pt. 7*, 5141–5251; Battista, Christie, Creech, Fredericksen, Lloyd, McMullen, Myers, Pike, and Sanator, interviews by author. After Helms's Enforcer briefing in *FY 1978 Authorization* hearings, Thurmond told him, "I have never heard a more convincing or more effective presentation . . . in the Congress of any defense weapon, Mr. Helms, than you made this morning" (5204).

5. USAF Project PAVE COIN, 157–71; U.S. Congress, House, *Enforcer Aircraft*, 140–53; U.S. Congress, Senate, *FY 1978 Authorization, Pt. 7*, 5140, 5199, 5205–51; Allen, interview by author.

6. Orr, letter to Thurmond; Baso, Battista, Christie, Creech, McMullen, Orr, interviews by author. Battista and Christie were the two "observers."

7. Odgers, "Design-to-Cost," 24; U.S. Congress, House, *DoD Appropriations for 1975, Pt. 7*, 1050–58; U.S. Congress, House, *DoD Appropriations for 1977, Pt. 3*, 121–23 (Giaimo quote, 123); U.S. Congress, House, *DoD Appropriations for 1977, Pt. 5*, 104–12 (Slay quote, 110); U.S. Congress, House, *H.R. 3689*, 5154; U.S. Congress, House, *Middle East Conflict and the Enforcer Aircraft*, 9; U.S. Congress, Senate, *DoD Appropriations for FY 1976, Pt. 4*, 324–25; U.S. Congress, Senate, *FY 1978 Authorization, Pt. 7*, 80; U.S. GAO, *Review of the Department of Defense's Design-to-Cost Concept*; Kishline, interview by author.

8. The following sources are from supporting documents: Clements, deputy secretary of defense, "A-10 Production Decision," memorandum for the secretary of the Air Force, 31 July 1974. Other sources are Systems Command, "History, 1 July 1974–30 June 1975," n.d., 255–57; Systems Command, "History, 1 July 1975–31 December 1975," n.d., 79.

9. Hails, "Report Relative to Production Readiness Posture of the A-10 Program," n.d., 3–4, pt. 1, 19–35 (henceforth called the Hails Report), supporting documents.

10. Hails Report, pts. 1–3; Watson, *A-10*, 67–89; McMullen, Sanator, and Toomay, interviews by author.

11. Watson, *A-10*, 67–89; Wetmore, "A-10 Program"; Downey, Kishline, McMullen, and Sanator, interviews by author. The Air Force initially cited comparative labor costs as another reason for relocating production to Hagerstown, but the congressionally driven review of its figures indicated that savings were minimal.

12. Bridges, "Realism," 44; Carleton, "A-10," 6–21; Hails Report, 12–13, 65, 68; U.S. Congress, House, *DoD Appropriations for 1975, Pt. 7*, 1050–53, 1057–58; U.S. Congress, House, *DoD Appropriations for 1977, Pt. 5*, 109; U.S. Congress, Senate, *DoD Appropriations for FY 1976, Pt. 4*, 72 (quote), 325; U.S. GAO, *Design to Cost*; Welch, interview by author. Welch said that a radar ranging device for the gun was another budget casualty, but because the A-10 was a low-priority project with several enemies, the omission stuck.

13. Dilger, "Tank Killer Team," 190–91 (has a picture of the GAU-8A sitting next to and dwarfing a Volkswagen Beetle); Wall, *GAU 8/A*, 30, 37, 63–64.

14. Dilger, "Team"; Elko and Stuelpnagel, "30-mm GAU-8A Ammunition"; Geiger, and others, "Aeronautical Systems Division," 101; Robinson, "USAF Plans A-10 Deployment"; Suydam, "GAU the Giant Killer"; Sweetman, *A-10*, 14; Wall, *GAU 8/A*, 35–45; Watson, *A-10*, 94; Buchta, Casey, Christie, Dilger, Kishline, Lieberherr, Oates, and Yates, interviews by author. Gun residue also accumulated on the windscreen and engine blades, and required special windshield and engine washes. Lieberherr recalled how secondary gun gas ignition caused an A-10 dual-engine flameout in 1978.

15. Burton, *Pentagon*, passim; Buchta, Burton, Dilger, O'Bryon, and Yates, interviews by author.

16. Ericson, "Recycling"; Geiger and others, 98; Gilson, "Can A-10 Thunderbolt II Survive?"; Wall, *GAU 8/A*, 49–57 (quote, 55); Dilger, Kishline, and Oates, interviews by author. Environmentalists claimed that depleted uranium ammunition usage in Desert Storm caused some veterans' postwar maladies. Also, one anonymous interviewee who was associated with gun testing opined that depleted uranium in large quantities could be hazardous. The author sides with the government study, which was conducted during the 1970s, when there was skepticism about military weapons in general and nuclear test safety in particular. Congress had a post-Watergate, post-Vietnam Democratic majority that probably would have challenged any suspicious findings.

17. DDR&E, "DCP 23B," 1, 4, supporting documents; Dilger, 190–91; Elko and Stuelpnagel, 52; Kaplan, "Beast," 12; Wall, *GAU 8/A*, 66–67; Dilger, Kishline, Oates, O'Bryon, Stolfi, and Yates, interviews by author. A-10s made twenty two passes against fifteen tanks, immobilizing all fifteen and destroying eight. A couple of interviewees differed over whether the Air Force "spiced" the targets with gasoline to create a spectacular display. I believe not. GAU eight. A rounds penetrate armor outright with pyrophoric effect or cause spalling (pushes armor inward, causing it to liquefy and separate at roughly the same velocity as the impacting round). If fuel or ammunition is contacted, it can detonate with even greater dramatic effect.

18. Coates and Killian, *Heavy Losses*, 154–55; Riley, "Thunderbolt"; Wall, *GAU 8/A*, 17–20, 36–37, 69–70, 78–80; Buchta, Casey, Dilger, Kishline, and Oates, interviews by author.

19. "AFSC History, 1 January 1976–31 December 1976," n.d., 168–69; DCP 23B, 1; Watson, *A-10*, 96–97.

20. Burton, 24–25, 58–59; Coates and Killian, 154–55; "Cost Cutter"; Greve, "Career Cut Short," reprinted in *More Bucks*, ed. Rasor, 284–88; Kaplan, 12; Mintz, "Maverick Missile," in *More Bucks*, ed. Rasor, 176.

21. Drendel, *A-10*, 3–5; Famiglietti, "A-10"; "Remarks of General Dixon at 100th A-10 Ceremony"; Smallwood, *Warthog*, 9, 16–19; Sweetman, *A-10*, 12, 15; Adams ("wing commander") and Lieberherr, interviews by author. The Thunderbolt II name apparently received official sanction in 1978 at a ceremony honoring an A-10 production milestone. Sweetman cited Eglin AFB Tactical Air Warfare Center [TAWC] Maj. Michael Major with the first recorded use of the "Warthog" nickname in a 1974 *TAWC Review* article. Lieberherr recalled that TAC Headquarters initially banned the "Warthog" nickname due to fears about A-10 unit morale and negative public image, but Air Force aviation painter Keith Ferris's picture *Warthog at Work* illustrated, so to speak, the ban's futility.

22. Anderegg, 168–70; "First A-10 Squadron Operational Ahead of Schedule Despite Transition Difficulties," 208 (quote); Mekkelsen, "Close"; Mintz, "Maverick," 177; "Talbott," 144; Bass, Chaleff, Lieberherr, Lloyd, and Sanator,

interviews by author. Both Lloyd and Sanator heard informal Air Force complaints about the Hog. Maj. Gen. James Hildreth, who commanded most of Nellis's operations from 1977 to 1979, doubted the plane's survivability.

23. "A-10: A NATO Preview"; "A-10 Program Unaffected by Crash at Paris Airshow"; "A-10 Reliability Expands Crew Training"; USAF, "Instructional Text: Manual Weapons Delivery," 3-7, 3-8, 5-23 through 5-26, Drendel, 3; "First A-10 Squadron Operational"; Gulick, "A-10"; Halbert, Lucas, and Sherman, "Warthog"; Schlitz, "Lethal"; Smallwood, *Warthog*, 1–9, 16–19 ("sack" quote, 17); Tabor, "Open Letter"; Wacker, "A-10," 76–77; Bass, Croker, Fox, George, Hennigar, Henry, Houle, Jenny, Koechle, Lieberherr, Swift, Wilson, interviews by author.

24. Winton, "Partnership."

25. "Air Force Studies Fighter Pilot Workload"; "A-10: A NATO Preview"; Brown, "A-10 Pilots"; Carleton, 5–11, 22–29; "Center Seeks New Combat Techniques"; DCP 23B, 13–17; Crawford, "Low Level," 11–12; "First A-10 Squadron Operational"; Halbert, Lucas, and Sherman, "Warthog"; Helbling, "Spinich"; Robinson "USAF Plans"; Sweetman, 42–44, 56–57. Some proof of Hog driver emphasis upon pilot skills is in Maj. Rod Schraeder's "Augustine's Law." Writing after A-10s received INS devices, he warned about the pitfalls of overreliance on INS, and urged Hog drivers to maintain their fundamental skills in navigation and weapons delivery. Interestingly, Schraeder flew F-117s in Desert Storm and was one of the Stealth wing's weapons officers.

26. AFTEC, summary of internal letter, 30 September 1976; Schaeffer, "A-7 Strikefighter"; Oates, Stenner, Whitley, Yates, interviews by author.

27. Drendel, 4–6; "First A-10 Squadron Operational"; Gilson, 189; Dixon, Yates, interviews by author.

28. Anderegg, 89–94; Fink, "A-10 Survivability"; Fink, "Joint Combat"; Neubeck, *A-10*, 18; Houle, interview by author.

29. "A-10 Program Unaffected by Crash"; "A-10 Testing to Continue in Europe"; Drendel, 4–6; "Fairchild Can-Opener," 267 (Luftwaffe general quote), 268–87; Futrell, *Ideas*, 2:498–99; Gilson, 184–89; Gunston, *Attack*, 270–71; Halbert, Lucas, and Sherman, "Warthog"; Harvey, "Aircraft for Export"; Kocivar and Rhea, "A-10"; Starry, letter to ALFA, and Starry, letter to Canby; Sweetman, *A-10*, 46–48, 52–53; Adams, interview by author.

30. "A-10, A NATO Preview"; "A-10 at Exercise Reforger"; "A-10 Program Unaffected by Crash"; Brown, "A-10 Pilots"; Drendel, 5; Winton, "Partnership"; "6 A-10 'Tank Killers' Win NATO Acclaim in Europe."

31. Bradin, 133–54; Davis, *31 Initiatives*, 27–30; Dixon, speech draft for Langley AFB Officers' Wives Club, 14 October 1975; Dixon, 100th A-10 ceremony speech; Evans, "Thinking 'We,'"; "Fairchild Can-Opener"; Futrell, *Ideas*, 2:548; Geddes, "AH-64," 1249; Gilson, 189; Gunston, *AH-64*, 5–8; Hackett, *Untold Story*, xiii–xv; Hackett and others, *Third World War*, passim; Lambert, "A-10s on the German Front Line"; Mekkelsen, 1–2; Powell, "A-10," 43; Schemmer, "Army's YAH-64"; Schemmer, "Getting Ready"; U.S. Congress, House, *Close*

Air Support, 101st Cong., 16 (cartoon quote and Air Force study); "Where USAF's A-10's Fit in NATO's Defense"; Winton, "Partnership," 103–6.

32. "AF, Army Agree on Role of Copters"; Berry, "TAC and TRADOC"; "CAS Panel, Investigations Subcommittee, HASC [House Armed Services Committee]," September 1976; Chapman, *National Training Center*, 30–39; Cleary, "Dixon," 287–300; Currie, "Tactical Air Programs," 24; Evans, "Thinking 'We,'"; "Journal Interviews . . . BG Sidney L. Davis"; Quinn, "Deadly"; Starry, letters to ALFA and Canby; "Tactical Airpower Subcommittee of SASC [Senate Armed Services Committee]," March 1976, 5640 (Dixon quote); Wolf, *Basic Documents*, 401–5.

33. "A-10/AAH Together Pack More Punch"; Kocivar and Rhea, 1193 (quote); Ray, "JAAT," 42–47; Bahnsen, interview by author.

34. Alder, "Airborne Anti-Tank, Part I"; Beach, "JAWS," 44–47; Berry, "JAWS," 32–34 (kill ratios quote, 32); Drendel, 36–38; Gilson, 187–88; Lambeth, *Transformation*, 84–85; Lieberherr, "JAAT," 42–43; Mullendore, "Future of the Joint Air Attack Team"; Sweetman, *A-10*, 62–63; Bahnsen, Lieberherr, Whitley, interviews by author.

35. Bahnsen, "New Army Air Corps?" 70; Bunyard, "Tomorrow's War," 64; Dayton, "57 TTW TASVAL Detachment after Action Report," 1-1 (quote); Hartman, "Airborne Anti-Tank, Part I," 53–54; LaRue, "TASVAL"; Levens and Schemmer, "Loyalty Down," 28; Riley, "Organic Asset"; U.S. GAO, *Aerial Fire Support Weapons*, i–iv.

36. Bunyard, 64–66; Dayton, "57 TTW TASVAL," 1-1 through 4-4; DCP 23B, 4–8; "Eastern Bloc Augments Attack Force," 57–61; Hartman, "Part I," 53–62; Hartman, "Airborne Anti-Tank, Part II," 54–64; LaRue, "TASVAL," 26–29.

37. The citations given in this note also apply to the following paragraph. Berry, "TAC and TRADOC," 28, 30, 36; Dayton, "57 TTW TASVAL," 4-4 through 7 5; Drendel, 7; Hubbert, "JAAT"; Adams, Bahnsen, Henry, Houle, Jenny, Lieberherr, and Pentland, interviews by author.

38. Futrell, *Ideas*, 2:559.

39. The following sources also apply to the subsequent paragraph. "A-10 Assembly Line Geared for 1982 Peak"; Barnard, "Feasible?" 33–34; Brown, "Bill Expands," WB3A; "CBO Views Europe Defenses," 12; "Fairchild Can-Opener," 287; Futrell, *Ideas*, 2:560–61; Harvey, "Aircraft for Export," 45–46; Harvey, "Dangerous Midnight," 12–15, 43; "Korea Eyeing A-10"; Lopez, "Night/Adverse Weather A-10"; Neubeck, 17, 20–23; Ropelewski, "Night/Adverse Weather Version," 46–48, 53–56; Sweetman, *A-10*, 48–51; Wessely, "Limiting," 1–21; Allen, Russ, interviews by author.

40. "A-10 Deployment," 53; "A-10s Set for European Deployment in January"; Engle and Patton, "Basing," 82–85; Sweetman, *A-10*, 46–48; "Where USAF's New A-10s Fit," 18; Christie, Schlesinger, interviews by author.

41. Rasmussen, "A-10"; Rasmussen, "Central"; Sweetman, *A-10*, 52–53, 60–62.

42. "First A-10 Squadron Operational"; Neubeck, 17–19; Pentland, "A-10 Completes FOL Validation," 18–20; Henry, interview by author.

43. "A-10 Program Unaffected by Crash"; DCP 23, 12–13; Dilger, 191; Hogg, *Tank Killing*, 8, 20; Kocivar and Rhea, 1191; Pentland, "Evolution"; Pivarsky, "30-mm Gun"; Stolfi, Clemens, and McEachin, "A-10/GAU-8," 1–2; Christie, Dilger, Jenny (observer), Oates, Pentland, Stolfi, interviews by author.

44. Elko and Stuelpnagel, 52; McDermont, "Avenger," 13; Ratley, "A-10"; Ratley, "Comparison"; Stolfi, Clemens, and McEachin, "A-10/GAU-8A," passim; Whitley, "A-10 OT&E, TD&E," 34; Dilger, Henry, Oates ("save" quote), Ratley, Stolfi (other quotes), Whitley, interviews by author. The test used T-62 tanks in the early stages and U.S. M-47s later on, due to limited availability of T-62s. The test team did not put live ammunition in the tanks' magazines, but instead used dummy rounds. The effectiveness rating used three classifications: Mobility Kill (M-Kill), when a tank cannot execute controlled movement; Firepower Kill (F-Kill), when the tank cannot direct controlled fire from its main gun; Catastrophic Kill (K-Kill), the tank is destroyed and unrepaired. The 1978 T-62 LAVP test featured K-Kills for all rear-aspect attacks, and a combination of M-Kills and F-Kills for side attacks. Another directorate officer, Maj. Richard Hackford, thought up LAVP, and Dilger assigned Ratley as project officer.

45. Coates and Killian, 154–55, 172–73; "Cost Cutter"; Elko and Stuelpnagel, 52–54; Greve, "Career Cut Short," 284–88; Buchta, Dilger, Oates, Ratley, interviews by author.

46. Anderegg, 135–41; Blohm, "AGM-65D," 10–14; Coates and Killian, 155–57; Cordesman and Wagner, *Lessons*, 1:95; Futrell, *Ideas*, 2:559–60; Isherwood, "Combat Hammer"; Mintz, "Maverick," in *More Bucks*, ed. Rasor, 135–85; Nordeen, *Missile Age*, 168; Reis, "Close Air Support Systems," 29–35; Simpkin, *Antitank*, 212; Sweetman, *A-10*, 39–42; TAC Headquarters, "A-10 AGM-65D IR Maverick Training Program"; Allen, Battista, interviews by author.

47. Berry, "81st"; Brown, "A-10 Crews"; Drendel, 7; Lambert, "A-10s," 1146; Schlitz, "Getting A-10 Firepower"; Schraeder, "Using the AGM-65D"; Sweetman, *A-10*, 47–62; Wacker, "A-10"; Warwick, "Europe's Anti-Tank Vanguard." The squadrons occasionally swapped base deployments to anticipate combat contingencies.

48. Brown, "A-10 Crews," 37; Sweetman, *A-10*, 47; "Targets Lacking"; Hennigar, interview by author.

49. Drendel; 6–7; Lambeth, *Transformation*, 64; Neubeck, 17–19; Spencer, "Army Field Manual 100-5 vs. Sun Tzu," 4–19; Sweetman, *A-10*, 47–48; Wilson, "Impossible to Shoot Down," A3; George, Hennigar, Klein, McPeak, Wilson, interviews by author. The A-7s went to Air National Guard (ANG) and Air Force Reserve units, and the Texas congressional delegation dictated Air Force A-7 purchases through the early 1980s. A-7s flew in the Guard and Reserves through the early 1990s. A-10s started entering the Guard and Reserve in 1979.

50. Armaments Systems Division, "Are ZSU-23/4 Drivers Really Ten Feet Tall?"; Brown, "A-10 Crews"; Builder, *Masks*, 120–21; Lambert, "A-10s in the Ger-

man Front Line"; Price, *Air Battle*, 92–101; Schlitz, "Getting A-10 Firepower"; Sweetman, *A-10*, 46–48, 52–63; Wacker, 74–77; Allison, Bass, Condon, George, Hennigar, Kemp, Koechle, Lieberherr, Wilson, Wickstrom, Yohe, interviews by author.

51. The *Air Force Magazine* articles are Beavers, "Monstrosity"; Schlitz, "Getting A-10 Firepower"; Schlitz, "Lethal"; Schlitz, "Thunderhog II"; Wacker, "A-10 Operations." Other articles are Harris, "Non-Stop Knuckle-Busters"; Harris, "Wart Hogs!"; Katzman, "Groundwork for the Future"; McMullen, "TACAIR Outlook"; Rubinstein, "Ugliest Plane"; Taylor, "Mean Machine." Observers are Davis, *31 Initiatives*, 25–33; and Winton, "Partnership," 100–111.

Chapter 8. Fire in the Minds of Men

1. Bingham, Roberts, and Wormser, "Fire, Counterfire,"; Canby, "Tactical"; Luttwak, "American Style"; Madelin, "Emperor's Close"; Ralph and others, "Commentary"; Saye, "Close."

2. De Czege, "Army Doctrinal Reform," in *Defense Reform Debate*, ed. Asa Clark and others, 101–19; Kelly, *King*, 230–41; Romjue, *Active*, 15–33; Romjue, "Evolution."

3. Davis, *31 Initiatives*, 29–31; Futrell, *Ideas*, 2:546–54; Romjue, *Active*, 63–64; Stein, "Development," 8–35 (quote, 33); Tuttle, "AirLand?" 4–11; Winton, "Partnership"; Creech and Russ, interviews by author. The NATO air support doctrinal document was Allied Tactical Publication (ATP) 27(B).

4. Samples of reformer works: Barlow, ed., *Reforming*; Burton, *Pentagon Wars*; Clark and others, eds., *Debate* (essays both for and against the movement); Coates and Killian, *Losses*; Cockburn, *Threat*; Etzold, *Defense or Delusion?*; Fallows, *National Defense*; Gabriel, *Military Incompetence*; Hadley, *Straw Giant*; Rasor, ed., *More Bucks*; Stevenson, *Pentagon Paradox*. Critiques of the reformists include the following: Brown, *Thinking about National Security*; Dupuy, "Why Deep Strike Won't Work"; Kelly, *King*, 242–45; Kross, *Military Reform*; Strode, "Soviet Design Policy," 46–61.

5. Bradin, 133–56; Easterbrook, "All Aboard Air Oblivion," in *More Bucks*, ed. Rasor, 50–66; Gunston, *Apache*, 5–8.

6. De Czege, "Army Doctrinal Reform," 108 (quote); Fulghum, "Doubts Cast," 2; Romjue, *AirLand Battle*, 25–26; U.S. Congress, House, *National Defense Authorization Act for FY 1989—H.R. 4264*, 290–91; Richardson and Wickham, interviews by author.

7. The citations given in this note also apply to the following paragraph. Dörfer, *Arms Deal*, 3–28; Fallows, 95–106; Hallion, *Storm over Iraq*, 41–42; Herzog, *Defense Reform and Technology*, 1–55; Kaplan, "Little Plane," in *More Bucks*, ed. Rasor, 186–90; Richardson, *F-16*, 4–17; Stevenson, passim.

8. Burton, 100–102; Dörfer, 3–28; Futrell, *Ideas*, 2:562, Hallion, *Storm*, 43; Hasdorff, "Hildreth," 62 (quote); Richardson, *F-16*, 10–17, 48–54; Stevenson, 188–98; Taylor and Young, "Gallery" (May 1982), 155–56; Battista, Christie, Fredericksen, Loh, Pentland, and Ringo, interviews by author.

9. Burton, 57–61, 99–101; Futrell, *Ideas*, 2:561; U.S. Congress, House, *Military Posture and H.R. 5968*, 470–75; Adams, Christie, Orr, Pentland, Russ, and Welch, interviews by author. Writing in 1984, Sweetman observed that the F-16 was the "A-10's real problem" (*A-10*, 46). The A-10 FAC idea had appeared as early as 1978 in a staff college research paper; see Profitt, "FAC."

10. Davis, *31 Initiatives*, 30–61 ("rear area" quote, 53); Hallion, *Storm*, 74–81; Lambeth, *Transformation*, 85–90; Machos, "Air-Land Battles"; McPeak, "TacAir Missions"; U.S. Congress, House, *H.R. 4264*, 308–11 ("analysis" quote, 311); U.S. Congress, Senate, *DoD Authorizations for Appropriation for FY 1989, Pt. 4*, 48; U.S. GAO, *Close Air Support: Status of the Air Force's Efforts to Replace the A-10*, 14–16; Orr, Russ, Welch, Wickham, interviews by author. Orr said that while he was Air Force secretary in the early 1980s, he found little enthusiasm for the A-10 among senior leaders because it was considered vulnerable to defenses.

11. Davis, *31 Initiatives*, 153–54 (agreement quotes , 154); Greeley, "USAF Reviewing," 16–18 (staff quote, 16); Pentland, "Warfighting View"; Ropelewski, "USAF Seeks"; Sweetman, *A-10*, 46–47; U.S. Congress, Senate, *DoD Authorizations for Appropriation for FY 1989, Pt. 4*, 48; U.S. GAO, *Replace the A-10*, 19–20; Orr, Russ, Welch, interviews by author.

12. Barger, "What the Air Force Has to Do"; Beyers, "Russ' Support of F-16"; Morrocco, "USAF Plans"; Ropelewski, "Affordability"; Ropelewski, "Fighter Derivative"; Fredericksen, Orr, Russ, interviews by author. Barger wrote that Air Force Headquarters staff considered traditional CAS a "myopic" approach to war-fighting. Air Force secretary Verne Orr also wanted a quick resolution, which would demonstrate his service's air support dedication and preempt any interservice roles-and-missions fight. The industry study request's formal title was Request for Information (RFI). The project justification's formal title was Justification for Major System, New Start (JMSNS).

13. Isaacson, "Winds of Reform"; U.S. Congress, House, *Military Posture and H.R. 5968*, 470–75 (Russ/Byron quotes, 472–73); U.S. Congress, Senate, *DoD Authorization for Appropriations for FY 1983, Pt. 4*, 2595–96 (Goldwater quote 2595); Russ, interview by author.

14. James Blackwell, "American Close Air Support"; Burton, 57–68, 99–101, 242 ("symbol" quote); Neubeck, 39 (Sprey quote); Burton, Christie (quote), Fredericksen, Spinney, Sprey, interviews by author. To Christie, Air Force leaders seemed wary of an OSD CAS plane initiative because it reminded them of Sprey and others' A-10 labors: "It was like, 'Here we go again. We've got to stop this.'"

15. "Battle Brews over Follow-On Close Air Support Aircraft"; Blackwell, 59; Sweetman, *A-10*, 46–47; U.S. Congress, House, *H.R. 4264*, 252, 310–12; U.S. GAO, *Replace the A-10*, 19–20; Christie, Fredericksen, Pentland, and Spinney, interviews by author.

16. "Defense Dept. Asks USAF to Broaden Design Options," 28–29 (requirement quote, 28); Morrocco, "Study," 29–30; U.S. Congress, House, *H.R. 4264*,

253–55, 312; U.S. Congress, Senate, *DoD Authorizations, FY 1989, Pt. 4*, 56–57; U.S. GAO, *Replace the A-10*, 20–22; Fredericksen, Pentland, and Spinney, interviews by author.

17. Barger, "What the Air Force Has to Do," 58 ("enthusiasm" quote); Bingham, "Close-In Battle," in *Air Support: Proceedings*, 5–49; Carlson, "Close"; Greeley, "USAF Reviewing," 17; Russ, "No Sitting Ducks"; Stevenson, 17–20; Lieberherr, Spinney, Welch, interviews by author.

18. Canan, "Flak"; Correll, "Budget Wars"; Dudney, "Cuts"; Foglesong and Shirron, "Close"; Leibstone, "Halcyon"; Morrocco, "USAF Plans"; Pentland, "Warfighting View"; Ropelewski, "Affordability"; Pentland, Russ (quote), Welch, and Yates, interviews by author.

19. "Battle Brews," 19; Burton, *Pentagon Wars*, 242; Morrocco, "USAF Plans," 23; U.S. Congress, House, *H.R. 4264*, 250–51; U.S. Congress, Senate, *DoD Authorizations for FY 1989, Pt. 4*, 48, 54–55; Christie (quote), Fredericksen, Myers, and Spinney, interviews by author.

20. Adkin, *Urgent Fury*, 225; Builder, *Masks*, 138–40, 185–90; Currie, "Operational Lessons," ii, 9–27; de Czege, "Army Doctrinal Reform," 107; Gabriel, *Incompetence*, 160–79; Harding, *Air War Grenada*, 22–52; Lind, "Grenada Operation"; U.S. Congress, House, *Roles and Missions of Close Air Support*, 23–25; U.S. GAO, *Replace*, 14; Swift, interview by author.

21. Menetrey, "U.S. Military Posture," 14 (quotes); Spencer, "Army Field Manual 100-5 vs Sun Tzu," 16–18; Hennigar, Klein, Wickham, Wilson, interviews by author.

22. Cordesman and Wagner, *Lessons*, 1:203–5, 3:316–17, 388 (quote); Ethell and Price, *Air War, South Atlantic*, 55–59, 201–16. Coordination problems also occurred in the Grenada invasion. Radio incompatibility plagued interservice communications, and there were too many units trying to participate. The A-10s waited for an opportunity that did not come.

23. Backlund, "Can the Army Take Over CAS?" 24; Bahnsen, "New Army Air Corps?"; Brown, "Attack Helicopter Operations," 2–11; Collins, "Another Look," 26–33; Ferrell and Reynolds, "Apache Thunder"; Greenhous, "Israeli Experience," 525; Kruzel, ed., *American Defense Annual, 1987–1988*, 102; Mullendore, "Future," 27–44; Riley, "Organic Army Asset," 16; RisCassi, "Army Aviation"; Saint and Yates, "Deep Operations"; Terrien, "Close Air Support," 81 (quote); White, "Case," 13–14; Richardson, Wickham, interviews by author.

24. Bahnsen, "New?" 78 (quote); Garrett, "Fuss," 41–43; Littlejohn, "Fourth Dimension," 39–40; Riley, "Asset," 127–29; U.S. Congress, Senate, *DoD Authorization for Appropriations for FY 1990 and 1991, Pt. 5*, 219–21; Bahnsen, Richardson, Vuono, and Wickham, interviews by author. Former TRADOC commander Richardson agreed that there were overaggressive younger helicopter officers, but noted that generals existed to rein such people in.

25. Adkin, *Urgent Fury*, 245; Coates and Killian, 137–38; Currie, "Operational Lessons," 17–18; Ethell and Price, *South Atlantic*, 200 (quote); Gabriel, *Incompetence*, 180–81; Greenhous, "Israeli Experience," 525; Harding, *Grenada*, 51–52;

Lind, 5; U.S. Congress, House, *Roles and Missions of Close Air Support*, 23; Werrell, 154, 164. Lind observed that helicopters might not do well on a confused battlefield, which was what the Army's AirLand Battle doctrine anticipated.

26. Armour, "Soviet Aviation," 16–18; Cockburn, 133; Cockerham, "Fighting Hind"; Coyne, "Frontal"; Everett-Heath, *Helicopters*, 120–44; Mason, *Centennial*, 96–98, 188–89; Mason and Taylor, *Aircraft*, 68–96; Nordeen, *Missile Age*, 165–68; Schneider, "Soviet Frontal Aviation," in *Soviet Air Forces*, ed. Murphy, 145–47; Sweetman, "New Shturmovik"; Sweetman, "STOL Striker"; Whiting, *Soviet Air Power*, 196–202; "B" (senior intelligence analyst), Isby, Pentland, interviews by author.

27. Cordesman and Wagner, *Lessons*, 3:174–201; Everett-Heath, *Helicopters*, 144–56; Mason, *Centennial*, 188–89; Nordeen, *Missile Age*, 169–70; Rodman, *More Precious*, 336–40 ("cosmonauts" quote, 339); "Sukhoi Incorporates Changes to Su-25"; U.S. Congress, House, *Roles and Missions of Close Air Support*, 28, 42; Werrell, 165–66; Isby, interview by author. The "Sukhoi" article mentioned a combat problem with the Frogfoot's fuselage-mounted engines— a concern that led some to prefer the A-10 over the A-9 in the A-X flyoff. When one engine was hit, flames and/or engine parts often affected other aircraft components and the other engine, causing numerous aircraft losses.

28. Ethell, "Close," 97–100 (quote, 100); Foster, "Defense Forum."

29. Heitman, *War in Angola*, 310–30 ("impossible" quote, 313; "remarkable," 328). The passivity was not only because of the war's imminent end, but also because of the lack of effective countermeasures equipment on both sides' planes. Also, neither side had air defense suppression units. Thus, both air arms assiduously avoided defenses, and the most active South African attack squadron's losses for this period were very low (0.3%).

30. Lynch, "Close"; Myers, "Air Support"; Fredericksen, Russ, interviews by author.

31. The citations listed here also apply to the next paragraph. Amouyal, "Gen. Galvin," 21 (quote); "Army's Analysis of LHX Helicopters," 18; Builder, *Masks*, 24; Canan, "Sorting," 53 (officer quote); Correll, "Battle Damage," 41; Morrocco, "USAF Plans," 23; U.S. Congress, Senate, *DoD Authorization, FY 1990 and 1991, Pt. 5*, 218–19; Bahnsen, Fredericksen, Myers (other debate source), Spinney, Vuono, Williams, interviews by author.

32. Anderson and van Atta, "Air Force Plan"; Correll, "Battle Damage," 41 ("substantial faction" quote); "Defense Dept. Asks USAF to Broaden Design Options"; Garrett, "Fuss," 3; Greeley, "USAF, Army Grapple"; Littlejohn, "Fourth Dimension," 5–6; Fredericksen, Yates, interviews by author.

33. Two works about selling airpower or weapons are Kotz, *Wild Blue Yonder*; Meilinger, "de Seversky," in *Paths of Heaven*, ed. Meilinger, 239–78.

34. The "battlefield of the 90s" mantra or similar terms can be found in Amouyal, "Pentagon Expected to Assign Role"; Barrett, "FAC's Perspective"; Canan, "Sorting"; Chapman, "Technology, Air Power"; Daskal, "Adapt Tactical Air"; Hinds, "Replacing," 3; Pentland, "View"; Russ, "Battlefield of the 1990s."

35. *Air Support: Proceedings*, passim.
36. Barrett, "Perspective"; Bingham, "Bad Idea"; Carlson, "Close," 51–59 (quote, 54); Chapman, "Technology," 42–51 (quote, 43); Foglesong and Shirron; Kropf, "MR Letters," 90; Pentland, "View," 92–96; Pentland, interview by author. Pentland recalled that he wrote his article out of frustration with the F-16 opponents. Kropf rebutted Carlson's article, pointing out that the F-16's runway requirements limited its utility. He also criticized Carlson for misleading comments about Korean War CAS.
37. The Russ articles are "Battlefield"; "Open Letter"; "No Sitting Ducks." Rebuttals are "Darts," 95; Haynes, Brown, and Flynn, "Airmail." In the author interview, Russ said that he could not recall the articles, but added that these were often ghostwritten and disseminated as general policy statements.
38. Bloom, "Attack Aircraft"; Daskal, "Adapt Tactical Air," 17–20 (quotes, 17); Fulghum, "Defense Trends," 1–15 ("battlefield" quote, 2).
39. Canan, "Flak"; Canan, "Sorting"; Correll, "Battle Damage"; Gorton, "Mudfighters"; Rhodes, "Improving the Odds"; Ulsamer, "Roadmap."
40. Brown, *Blind*, 233–34, 310–26; Cockburn, *Threat*, 139–56; Fallows, *National Defense*, 35–75; Greeley, "Speed"; Haynes, Brown, and Flynn, "Airmail"; Kotz, *Yonder*, passim; Nordeen, "Debate"; Prados, *Soviet Estimate*, 41–50.
41. Hallion, *Strike*, xviii–xx. Air University papers include the following: Gonzales, "Tactical Air Support"; Greene, "Close Air Support"; Hinds, "Replacing."
42. The citations given here also apply to the following paragraph. Hallion, "Assessment," 11–22 ("peculiar circumstances" quote, 12); Hallion, *Strike*, 263–69 ("mirror image" quote, 267).
43. Hain, "Letters," 7 (quote).
44. Gorn, *Harnessing the Genie*, passim; Toomay, interview by author.
45. "USAF Panel Urges Separate CAS, Interdiction Aircraft"; Toomay interview.
46. Burton, *Pentagon Wars*, 40, 48, 56–57, 58–59; "USAF Panel Urges Separate CAS, Interdiction Aircraft"; Fredericksen, Myers, Stenner, Toomay, Welch (quote), interviews by author. Myers said that Air Force leaders recalled the known copies of the report and destroyed them. Fredericksen also claimed that the service suppressed the report. Burton made clear that Toomay did not like the Blitzfighter idea. Besides Chairman Toomay, the SAB CAS committee consisted of the following: Dr. Paul Chrzanowski (Lawrence Livermore Laboratory), Dr. William Heiser (director, Propulsion Research Institute, Aerojet General), Eugene Kopf (president, Litton Optical Systems), Dr. F. Robert Naka (vice president, Engineering and Planning, GTE), Elbert Rutan (president, Scaled Composites), Dr. A. Richard Seebass (dean, College of Engineering, University of Colorado), Richard Shevell (Department of Aeronautics and Astronautics, Stanford University), and Major General Loh (Air Force liaison). The committee visited TRADOC Headquarters for joint doctrine briefings, Eglin AFB for weapons briefings, and Fort Irwin to watch air support missions during NTC exercises. Their activities concluded with a wrap-up at Lawrence Livermore Laboratory.

47. Greeley, "TAC to Modify"; Morrocco, "Pentagon to Review"; "Washington Roundup: Thunderbolt Retirement"; Fredericksen, Pentland, Phillips, Ringo, Russ, interviews by author.
48. "AFTI F-16 Testbed"; Dryden, "Comparisons," 22–23; Wanstall, "Squabbles"; Pentland, Russ, Yates, interviews by author.
49. "A-16 CAS Demonstration," draft copy; "F-16/A-16 Close Air Support Demonstration, Final Report"; "Grumman, Rockwell Team to Bid," 19; Headquarters Air Force, "PMD for Class II Modification for A/OA-10"; Phillips, "Integration of the F-16"; Rhodes, "Close Support Test Bed," 56–59; Scott, "AFTI F-16," 38–40; Shifrin, "TAC Demonstration," 49–50 (Vuono quote, 49); Stout, "CAS"; "USAF to Begin Evaluating FLIR Candidates for A-10"; Robert "Muck" Brown, letter to author; Chaleff, Hennigar, Pentland, Russ, Yates, "Z," interviews by author. Pentland claimed that one month before the test began, he quashed an already written demonstration report quoting Army sources who praised the F-16's performance. Brown said it was more an infomercial production than a test.
50. Fulghum, "Equipping F-16 Squadron"; Hughes, "Syracuse Wing"; "Special Avionics, Weapons Needed"; Muck Brown letter; Chaleff, Kuebler, Pentland, Spinney, Stenner, Yates, interviews by author. The Sidewinder air-to-air, heat-seeking missile was one Syracuse A-10 initiative adopted by the Air Force in the very late 1980s for Hog drivers' self-defense.
51. Canan, "Flak," 79 (quote).
52. Dikkers, "Overcoming," 23, 61 ("hearts" quote, 61;); Foglesong, "Defense Forum"; Icarus, "Choice"; Loida, "Viper"; McCarthy, "Human Factors"; U.S. Congress, Senate, *DoD Authorization, FY 1989, Pt. 4*, 53; Fredericksen, McGrath, Russ, interviews by author. F-16 weaknesses generated their own jokes, such as "Careful, bad guys . . . I'm carrying BOTH bombs today!"
53. Burton, *Pentagon Wars*, 57–68, 99–101; "Close Air Charade," 9; Myers, "Airmail," 9; Myers, "Air Support," 46–47; Scott, "Rutan's Scaled Composites," 27; Fredericksen, Myers, and Spinney, interviews by author. Fredericksen observed that other companies had designs, but that Rutan was not afraid of Air Force reprisals for pressing his case.
54. Myers, "Air Support," 47; Fredericksen, Myers, Spinney, and Sprey, interviews by author. In a letter to the author, Myers wrote that he saw that the A-10 was "dead." It was a nice first effort to build a dedicated CAS plane, he thought, but a more agile replacement was better.
55. "Defense Dept. Asks USAF"; Pentland, "View"; U.S. Congress, House, *H.R. 4264*, 253–55, 312; U.S. Congress, Senate, *DoD Authorizations, FY 1989, Pt. 4*, 56–57; U.S. GAO, *Replace the A-10*, 20–22; Fredericksen, Pentland, Ringo, Spinney, interviews by author.
56. Carlson, "Close"; Gorton, "Mudfighters," 106–7; Myers, "Air Support"; von Spangenberg, "Fighter Mafia"; Fredericksen, Myers, Pentland, Russ (quote), Spinney, interviews by author.

57. "Close Air Charade"; Corley and Flanders, "Letters"; Foglesong and Shirron, 34; Garrett, "Which Way"; Gorton, "Mudfighters"; Greeley, "Let the Army"; Myers, "Air Support"; Reynolds, "Letters"; Bahnsen, Myers, interviews by author.

58. Anderson and van Atta, "Air Force Plan"; Coram, "Case," 17 ("written extensively" quote); Kruzel, ed., *Annual, 1989–1990,* 14 ("reserves" quote).

59. Anderson and van Atta, "Plan"; Coram, "Case," 17–24 (citation applies to next two paragraphs).

60. Barger, "What the USAF Has to Do," 64 ("observers" quote); Brown, "Real Hog War"; Covault, "Alaskan A-10s"; Fulghum, "Study Touts A-10"; Whitcomb, "Hogs in the Rear"; Muck Brown, letter; Fredericksen, interview by author. Brown was told not to publish his article in *Fighter Weapons Review* because it was too "politically contentious."

61. Smallwood, "Old 'Hog,' New Tricks"; author recollection; Muck Brown letter; Fredericksen, Ringo, Spinney, interviews by author.

Chapter 9. The Bureaucratic War, the Political War, and the Real War

1. The citations in this note also apply to the first paragraph. Morrocco, "Pentagon to Review"; Morrocco, "Unable to Agree," 84, 86 (quote); Ropelewski, "Congress Stirs Pot"; U.S. GAO, *Replace the A-10,* 15–18; Fredericksen, Pentland, Ringo, Russ, Spinney, "Z," interviews by author. The service increased classification restrictions because it claimed that CASADA was considering the highly classified Stealth technology for CAS.

2. "USAF, Defense Dept. Reach Compromise"; Fredericksen, Pentland, and Spinney, "Z," interviews by author. Fredericksen said he wanted resolution before he left office. The compromise and the Air Force's CASADA security classification tactic led some OSD staff to leave the CASMARG. CASADA remains classified, and the information derived is from other sources.

3. "Close Air Charade"; U.S. Congress, Senate, *DoD Authorization FY 1989, Pt. 4,* 57; U.S. GAO, *Replace,* 21–22; "Washington Roundup: Seeing Red," 15 (quote); "Z," interview by author. The six companies were Boeing, General Dynamics, Lockheed, McDonnell-Douglas, Northrop, and Rockwell International.

4. "Defense Dept. Asks USAF," 28; Mann, "Senate Girds"; Maze, "Aspin"; Battista, Christie, "D," Fredericksen, Highbush, Ireland, Pentland, Phillips, Ringo, Rosa, Russ, Welch, Wheeler, Yates, "Z," interviews by author. Fredericksen, Pentland, and Welch thought that Dixon's motivation for pushing the Harrier seemed to be the McDonnell-Douglas constituency.

5. Brown, *National Security,* 233–34; Ginovsky, "GAO Report," 54; U.S. Congress, House, *H.R. 4264,* 257–63, 298; U.S. Congress, Senate, *DoD Authorization for FY 1989, Pt. 1,* 17–64; U.S. GAO, *Close Air Support: Airborne Controllers,* 2–13; U.S. GAO, *Close Air Support: Comparison of Air Force and Marine Corps Requirements,* 2–5; U.S. GAO, *Replace,* 2–4; Pentland, Phillips, Rosa, Russ, interviews by author. General Russ described another inadvertent pitfall

    of his FAC A-10 plan. He had to stop overzealous TAC Headquarters staff from removing the GAU-8A gun, since they wanted to save money and believed a FAC A-10 did not need a gun.

6. Bloom, "Attack Aircraft"; Bond, "Defense Dept."; Carlson, "Close"; Dörfer, *Arms Deal*, 42; Fulghum, "A-7F"; Greeley, "TAC to Modify"; Hallion, *Strike*, 267; Mann, "Senate Girds"; Nordeen, "Debate"; Schaeffer, "A-7 Strike-fighter"; Schaeffer, "LTV YA-7F"; Shifrin, "LTV Receives"; U.S. GAO, *Upgraded A-7*, 2–4, 15–18; Wilson, "Opponents Find"; Fredericksen, Russ, Toomay, Welch. interviews by author.

7. "Fairchild Tightens Procedures"; "Grumman: Moving Beyond" (does not mention Grumman's A-10 responsibility); "Grumman Team Evaluates Buy"; Neubeck, 24; "New Pilot Tries to Pull Fairchild Out"; "New York Legislators Ask for Fairchild Reprieve"; Shifrin, "Sale"; "T-46A—Dead or Alive?"; "T-46 Termination Will Force Closure"; "USAF Urges Canceling Fairchild T-46"; U.S. Congress, House, *H.R. 4264*, 254–55; "Washington Roundup: Eternal Trainers"; Dixon, Downey, Hennigar, McMullen, Russ, Sanator, Stenner, Toomay, interviews by author. TAC commander Dixon was Fairchild's president from 1978, when he retired from the Air Force, to 1982. He apparently did not get along with the civilian management. See Cleary, "Dixon," 328–29.

8. Charlton, "Heating"; Eggers, "Freefall"; Greeley, "TAC to Modify"; "Grumman, Rockwell Team to Bid"; Rennspies, "LASTE"; Schaeffer, "A-7 Strike-fighter"; Ulsamer, "New Roadmap"; Adams, Chaleff, Fredericksen, Hennigar, Henry, Kuebler, Pentland, Stenner ("ANG officer"), interviews by author. All interviewees believed that without the flight safety contribution, the TAC leadership would not have accepted LASTE. Pentland added that budget concerns delayed modifications for other fighters. Hogs did receive AIM-9 heat-seeking, air-to-air missiles, thanks in part to the Syracuse A-10 unit's efforts, and there were plans to incorporate some avionics changes.

9. Mann, "Senate Girds," 21 ("hardened" quote); U.S. Congress, House, *H.R. 4264*, 254–88 ("battlefield" quote, 279); U.S. Congress, Senate, *DoD Authorization FY 1989, Pt. 4*, 8–64. Pentland stated that one reason the Marines praised their Harrier was to achieve a larger buy and thus reduce the plane's per-unit cost. He also observed in a letter to the author that Congress complained about the CAS F-16, but "it was also Capitol Hill that was forcing 120 F-16s/year down the Air Force's throat!"

10. Garrett, "Close Air Support," 1; Mann, "Senate Girds"; U.S. GAO, *Comparison*, 11–12; White, "Divestiture Planning," 1, 4; Pentland, interview by author.

11. Baker, "Dispute"; Dudney, "Cuts"; Leibstone, "Halcyon"; Morrocco, "Pentagon Proposes"; OSD, "Operational Test Plan Concept"; "USAF to Accept Plan on Modifying F-16, A-10"; U.S. Congress, House, *Close Air Support*, 101st Cong., 33–40; Christie, "D," Ringo, Rosa, "Z," interviews by author. Fredericksen's compromise F-16s would be a mix of older models modified with survivability features and brand-new models built to CAS specifications. "Z" said that Sen. Carl Levin wanted to scare the Air Force into line. Ringo

cited Sen. Sam Nunn's and Sen. John Warner's exasperation with the Air Force, but added that they were not serious about the Army assuming the CAS mission. Rosa thought that the congressionally directed test marked the point when the Air Force started to lose control of the CAS plane dispute.

12. Baker, "Dispute," 10 ("high-speed aircraft" quote); Morrocco, "Pentagon Proposes," 97 ("not happy" quote); U.S. Congress, House, *Close Air Support*, 101st Cong., 37–40; Christie, "Z," interviews by author.

13. Anderson, "Close Air Support"; U.S. Congress, House, *Close Air Support*, 101st Cong., iii, 1–5; Christie, Ringo, "Z," interviews by author.

14. U.S. Congress, House, *Close Air Support*, 101st Cong., 5–10 (quote, 6), 41–43.

15. Durden and others, *Falklands*, 26–27, 95–109; Ethell and Price, *Air War*, passim; U.S. Congress, House, *Close Air Support*, 101st Cong., 10–15 (quote, 10).

16. U.S. Congress, House, *Close Air Support*, 101st Cong., 29–61 (cartoon, 61). The GAO report containing the cartoon is *Replace the A-10.*

17. U.S. Congress, Senate, *DoD Authorization for Appropriations for FY 1990 and 1991, Pt. 5*, 213–48.

18. Amouyal, "Galvin"; Bond, "Budget"; Ropelewski, "A-16"; Shifrin, "TAC Demonstration"; U.S. Congress, House, *Close Air Support*, 101st Cong., 47; Fredericksen, Pentland, interviews by author.

19. Amouyal, "DAB"; Bond, "Defense Dept.," 32 (staff member quote); Bond and Morrocco, "Close"; Smallwood, "Old Hog"; Christie, Pentland, Rosa, and Spinney, "Z," interviews by author. Although, Christie, Spinney, and "Z" did not admit to any specific contacts with the press and Congress, Pentland and Rosa saw them as the hard core of resistance.

20. Amouyal, "Powell," 1, 43; Bond, "Congress"; Crowe, "Roles," app. B to encl. B, 10–11; Littlejohn, "Fourth Dimension," 1–5; Rosa, Vuono, and Welch, interviews by author. Welch said little about this incident, but Vuono thought Crowe's assessment was too simplistic and risked creating interservice rancor. Army officer Littlejohn thought congressional attention helped drive the rebuttal, and claimed that reliable sources told him that TAC Headquarters pressed for a response that guaranteed the service's right to the mission.

21. The citations given here also apply to the preceding paragraph. Amouyal, "DAB," 26; Bond and Morrocco, "Close," 22 ("Khomeini" quote), 23, 24 ("jerk" quote); Fulghum, "Congress," 25 ("shapes up" and "lobbying" quotes), 26 ("creeps" and "officials" quotes); Ropelewski, "Congress," 22–24 (Russ quote, 24); Pentland (Air Force staffer quote), "Z," interviews by author. Congress's specific demands appeared in an F-16 procurement authorization bill. "Z" recalled that the Navy did not want the F/A-18 to compete in the test, lest results force the Navy to accept some other plane.

22. Harrison, "Gift," 36–37; Schemmer, "Congress Mandates," 20; McPeak (quote), Pentland, Phillips, Ringo, Rosa, Russ, Welch, and "Z," interviews by author.

23. Bond, "Congress," 32; Fulghum, "Air Force Calls," 21–22; Christie and Hennigar, interviews by author.

24. Amouyal, "DAB to Settle"; Burley, *Joint*, 5-19 through 5-23; Rosa and "Z," interviews by author.

25. Amouyal, "AF," 1, 51; Amouyal, "Tactical," 6, 22; Fukuyama, "End?" 3; Miller and others, "Peace," 44–45.

26. Amouyal, "Pentagon Expected"; Amouyal, "Prospects"; Bond, "Defense"; Brown, "LTV," 19–21; Fulghum, "Air Force Calls," 22; Fredericksen, Phillips, Russ, Welch, and "Z," interviews by author. Phillips was at one time the Air Force A-7F program spokesman, and thought that the program existed because of Texas congressional interest and the desire to create CAS plane selection competition. To him, even the Air National Guard seemed unenthusiastic. Fredericksen recalled that the A-7F also lost support because its vulnerability was greater than earlier models. The Senate rejected John Tower's nomination for defense secretary amid allegations of wild living and Texas defense industry influence peddling. Wright stepped down as speaker due to ethics violations.

27. The citations given here also apply to the following paragraph. Amouyal, "AF," 1, 51; Amouyal, "Pentagon Panel," 3, 44; Morrocco, "Air Force Drops," 28 (quote); "Washington Roundup: Endless Saga"; Adams, Rosa, Russ, and "Z," interviews by author. Adams said that the service also recognized the political importance of not contesting the DAB's A-10 decision.

28. Anderson and van Atta, "Hero"; Clancy and Horner, *Every Man*; Fialka, "A-10 'Warthog'"; Hallion, *Storm*, 210; Schwarzkopf and Petre, *It Doesn't Take a Hero*, 311–12, 331; Muck Brown, letter to author; Cheney, Fox, Lieutenant Colonel Henry, General Henry, Pentland, Swift, and Welch, interviews by author. Anderson and van Atta made the claim about Cheney, and Pentland said that the Air Force did not initially plan to deploy A-10s in Desert Shield. Fialka claimed that Schwarzkopf ordered Horner to accept A-10s. Brown, Fox, Lieutenant Colonel Henry, and Swift recalled hearing the story, and Hallion called it a rumor. Horner and Schwarzkopf were unable to respond to this author's queries about the story and other matters, and they did not mention it in their books. Welch contacted Horner on the author's behalf, and related that Horner remembered no such confrontation. Cheney recalled nothing about ordering anyone to deploy A-10s, though he knew that Air Force leaders were unenthusiastic about the Hog. General Henry served on Horner's staff and dismissed the story as false.

29. Atkinson, *Crusade*, 60–62; Clancy and Horner, 208, 263; Gordon and Trainor, *Generals' War*, 92–94; *Gulf War Air Power Survey* (henceforth cited as *GWAPS*), *Statistical Compendium*, 58–64; Hallion, *Storm*, 116–18, 142–43, 150–51; Adams, Fox, and General Henry, interviews by author.

30. Gordon and Trainor, 63–65; *GWAPS*, *Effects and Effectiveness*, 18–19, and *Logistics*, 81–140; Smallwood, 20–35; Henry, interview by author. The postwar accounts are Mann's *Thunder and Lightning* and Reynolds's *Heart of the Storm*.

31. Anderson, "Close Air Support," 28; Christie, Ireland, and Isby, interviews by author. Christie thought that the panel members wanted him to describe Air Force CAS F-16 bureaucratic skullduggery. Isby believed that the panel was

reacting to the 1989 hearings' Air Force witnesses, and that some of its members feared a loss of Air Force CAS expertise if A-10s were retired. He also thought that the committee saw him as an independent witness.

32. Bradin, 157–72; Smith, "Army Aviation and Just Cause," passim; U.S. Congress, House, *Roles and Missions of Close Air Support*, 3–29 ("permit" quote, 6; "spectrum" quote, 9); Muck Brown, letter. Isby alluded to another conflict not so far covered in this work: the December 1989 Operation Just Cause invasion of Panama. The fighting was not extensive enough to require much fixed-wing air support, though Air National Guard A-7s as well as Air Force A-37 FAC planes and AC-130s flew patrols. As for helicopter fire support, some Apaches were deployed via C-5 transports, and the initial fire support demand forced their pilots to navigate in confused combat conditions without defense suppression. Small arms fire seriously damaged one Apache, but its survivability features allowed it to return to base safely. Other helicopter types, especially transports, suffered more serious damage.

33. U.S. Congress, House, *Roles and Missions of Close Air Support*, 30–40 ("fifties" quote, 39).

34. Ibid., 41–55 (quote, 44).

35. Austen, and others, "Lessons," 36 (first quote), 38 (second quote); Ewing, "Reprieve," 14, 16.

36. Smallwood, 38–41; Fox, Koechle, Swift, and Wilson, interviews by author. Fox recalled that resistance initially accompanied the decision to attack from higher altitudes. The minimum altitude was ten thousand feet AGL.

37. Smallwood, 66–79. Glosson told Smallwood that the ground commanders wanted the Hogs held in reserve in case the Iraqis launched a ground attack when the air campaign started. Glosson and Horner overrode their concerns.

38. Smallwood, 114–21; Yohe, interview by author (Yohe flew A-10 night missions in the war).

39. Neubeck, 27, 33; Koechle, Swift and Wilson, interviews by author. Wilson, a proponent of the gray paint scheme, was told that gray would stand out on the black pavement during an air attack (however, F-15s and F-16s were light gray), and also that allied troops might shoot different-colored Hogs. Swift and Koechle recalled that the A-10 leaders believed that repainting would be impossible given the increasing tempo of war preparation. The 81st TFW had since split up into two wings; the 511th TFS was part of the recently formed 10th TFW.

40. "Desert Storm Performance Sparks Program," 44; U.S. Dept. of Defense, *Conduct of the Persian Gulf War* (henceforth cited as *CPGW*); Frostic, "Air Campaign," 48–51; *GWAPS, Statistical Compendium*, 323, 642–54; Smallwood, 206. None of the A-10s carried the LASTE targeting avionics in the war; the system was not developed in time. Although A-10 claims were devalued by two-thirds, conservative BDA estimates affected other planes, too. See Andrews, *Airpower*, 46–67; Frostic, 37–38; *GWAPS, Effects and Effectiveness*, 209–20, 260–64; Smallwood, 88–91.

41. Andrews, *Airpower,* 42–50; "Fahd Squad A-10 Briefing"; Fulghum, "A-10s," 23; *GWAPS, Summary Report,* 229, and *Weapons, Tactics, and Training,* 48–49, 55, 82; Smallwood, 80–159; Condon, Fox, Houle, Kemp, Koechle, and Swift, interviews by author. In addition to the list given in the text, the Hogs' typical bomb load included a radar-jamming pod, a full complement of chaff and flares, and two Sidewinder air-to-air, heat-seeking missiles. The two forward bases were Al Jouf (about one hundred miles from the Iraqi border) and King Khalid Military City (roughly fifty miles from Kuwait). King Fahd airbase was nearly two hundred miles from Kuwait.

42. Clancy and Horner, 394–97; Frostic, 14–18; *GWAPS, Operations,* 266–71, and *Weapons, Tactics, and Training,* 53–54; Hallion, *Storm,* 246–47; Smallwood, 91–110 (quote, 96); Tilford, *Search and Rescue,* passim; Fox, Kemp, Koechle, and Yohe, interviews by author. Earl Tilford recounted the A-1's SAR escort prowess. He concluded that combat situations may allow SAR, but questioned its viability in certain areas, including the Middle East. Early in its existence, the A-10 was regarded as the A-1's successor: Anderegg, 169; Bennett, "A-X"; Furtak, "A-10"; Kasbeer and Lyon, "A-10 Pilot SAR Handbook." Lyon was a Desert Storm A-10 deputy wing commander. Hallion observed that SAR deficiencies were one of Desert Storm's "lessons."

43. Andrews, *Airpower,* 46 ("qualified" quote); Frostic, 35–36; *GWAPS, Chronology,* 184, and *Weapons, Tactics, and Training,* 53–54, 70, 221; Hughes, "USAF Firing"; Jamieson, *Lucrative Targets,* 78–79; Neubeck, 27–33; Smallwood, 68–81, 96 (Horner quotes; Horner's praise appeared in the A-10 Battle Staff Directive), 117–75; Fox, Kemp, Koechle, Swift, and Wilson, interviews by author. *R* stood for "reconnaissance," *F* for "air superiority fighter," *O* for observation (FAC); *G* recalled the F-4G Wild Weasel defense suppression planes.

44. "Air Force Pilot Tests A-10's Toughness"; Andrews, *Airpower,* 44–45; Frostic, 43–44; Fulghum, "A-10s"; *GWAPS, Chronology,* 160–210, *Operations,* 280, *Statistical Compendium,* 642–47, and *Summary Report,* 21; Henderson, "A-10"; Jamieson, 80–81, 93; Neubeck, 33–34; "Pentagon Gave Short Shrift"; Smallwood, 160–202; Yohe, interview by author. The new restriction was between four thousand and seven thousand feet AGL. One A-10 pilot died attempting to land his crippled plane; thus, not all of the A-10 battle damage stories ended well. Yohe recalled that the restrictions did not apply to night operations.

45. Atkinson, 310–14; Clancy and Horner, 345 (quote), 479–82; Frostic, 36, 44; *GWAPS, Operations,* 279–80, and *Statistical Compendium,* 651; Hallion, *Storm,* 210–11; Smallwood, 124–26, 176–202. A-10s had suffered three battle damage incidents before Glosson's altitude restriction change, and suffered six more between then and the 15 February shootdowns. Two of the A-10's six Gulf War losses were FAC planes. These aircraft operated alone, which denied them a wingman's mutual support. Smallwood's account has the 15 February flight dueling with a SAM site that had shot at them—a sign of zeal. He also wrote that Captain Johnson was so upset with the over-aggressiveness that led to his Hog losing part of its wing that he slammed his helmet on the ground after

landing. Although the shootdown and some other battle damage accounts reveal a rising level of aggressiveness among the Hog drivers, this author cannot impugn these pilots' combat airmanship. First, he was not there. Second, aggressiveness is a necessary part of war that can yield mixed results. Third, the A-10 pilot interviewees felt uneasy about criticizing any fallen comrade's tactics and abilities. They believed that Capt. Steve Phyllis justifiably risked his life attempting to save his downed wingman in the 15 February double shootdown. (Phyllis was shot down and killed.) However, they agreed that over-aggressiveness was a likely factor in many of these cases. Two interviewees considered the paint scheme a partial cause, and one cited the Iraqi term *black jet* for the A-10 as proof.

46. Clancy and Horner, 425; *GWAPS, Effects and Effectiveness*, 234–42; Jamieson, 102–7; Smallwood, 128–31, 188–205 (Horner quote, 188). During the war, there were three other A-10 friendly fire incidents; the worst occurred when two A-10s attacked a British armored unit and killed some soldiers. Confusion existed between the ground FAC and pilots about how far the British troops had advanced. Of the twenty-eight Gulf War friendly fire cases, sixteen were ground-to-ground incidents. One notable event involved an Apache that shot a U.S. Army vehicle, killing its occupants. Fratricide has been a factor in warfare since ancient times, and it is one of the CAS mission's inherent risks. For Gulf War and general fratricide treatments: *AH-64 Apache*; Atkinson, 314–20; Congress, *Who Goes There?* 27; Frostic, 41–43; Hillman, "Task Force"; Ritchey, "Technological Solutions"; Shrader, "Friendly Fire."

47. Battlefield and Armor/Anti-Armor Assessment Teams, "Armor/Anti-Armor Operations," 34 ; Clancy and Horner, 344; Frostic, 24–25; *GWAPS, Effects and Effectiveness*, 242–48, *Operations*, 296–307 (quote, 306), and *Summary Report*, 72; Jamieson, 150–51; Neubeck, 38; Fox, Kemp, Koechle, Swift, Wilson, and Yohe, interviews by author.

48. Atkinson, 2; "Fahd Squad" (debrief quote); *GWAPS, Operations*, 306 ("wonderful day" quote).

49. Anderson and van Atta, "Hero"; "Desert Storm Performance Sparks Program"; Fialka, "A-10 'Warthog'"; Henderson, "A-10," 42–43; Higgins, "Tank–Killing 'Warthog,'" 1A (Clifton quote); Kaplan, "Beast"; Warrick, "Air Force Steps Up Attacks," 8 (Clifton and "day in the sun" quotes).

50. Burton, *Pentagon Wars*, 242; U.S. Congress, Senate, *Operation Desert Shield/Desert Storm*, 275–76 (quotes, 276). Commenting about Horner's apparent change of heart, Burton wrote, "After the war was over, however, the Air Force A-10 lobotomy took effect" (242).

51. *CPGW*, 138–39.

52. Andrews, *Airpower*, 46–67; Battlefield and Armor/Anti-Armor Assessment Teams, "Armor/Anti-Armor," 18; *CPGW*, 664–65; Frostic, 51; *GWAPS, Operations*, 155 (loss of accuracy); Smallwood, 88–89, 110–11, 195–206; U.S. Army, "Battle Damage Assessment, Volume I," 1, 16, 19; Condon, Fox, Kemp, Koechle (quote), Swift, Wilson, and Yohe, interviews by author.

53. *CPGW,* 665; Smallwood, 70–71, 114–21; Koechle and Yohe, interviews by author.

54. Hallion, *Storm,* 210–11 (quotes, 210), 284–85.

55. Frostic, 47–48; *GWAPS, Operations,* 205, 277, and *Weapons, Tactics, and Training,* 44; Hallion, *Storm,* 211–13, 287–88.

56. Frostic, 52–57; *GWAPS, Chronology,* 221, *Operations,* 155, 170–77, 261, *Statistical Compendium,* 646–47, and *Weapons, Tactics, and Training,* 49; Hallion, *Storm,* 287–88; Hallion, *Strike,* 268; Jamieson, 90–91; Lambeth, *Transformation,* 119, 143; Neubeck, 39; Fredericksen, Houle, and O'Bryon, interviews by author. Fredericksen observed that the F-16's speed made Maverick usage tough.

57. Lambeth, *Transformation,* 85–90, 131–37 ("scholar"); Scales, *Certain Victory,* 368–70 (quotes, 369–70).

58. Bradin, 173–232; Frostic, 49; Goodman, "Army Aviation," 16; *GWAPS, Summary Report,* 112; Hallion, *Storm,* 77–79, 243, 252; Nelson, "Combat Use," 48–65; U.S. GAO, *Apache,* 2–11. Works discussing the war and AirLand Battle are Atkinson, 448–83; Gordon and Trainor, 375–432; Hackworth, *Hazardous,* 58–60; Mann, "Beyond," 31; Mann, "Operation?" 27, 61; Winton, "Persian," 23.

59. Frostic, 15–60. Examples of *GWAPS* A-10 passages are *Operations,* 260–61, 277–78, 305–6, and *Weapons, Tactics, and Training,* 52–55. The AC-130 enjoyed some success early in the war doing night attacks, especially in conjunction with A-10s. However, AC-130 Scud hunt missions deeper into enemy territory led to one gunship suffering serious damage and subsequent removal of AC-130s from these efforts. Later, an AC-130 flying air support in the Battle of Khafji was shot down with the loss of fourteen members of the crew. See *GWAPS,* "Operations," 187, and "Weapons, Tactics, and Training," 118; Hallion, *Storm,* 221.

60. Winnefield, Niblack, and Johnson, "League," 250–51 (all quotes, 251), 259–85.

61. U.S. GAO, *Operation Desert Storm: Evaluation of the Air Campaign,* 19–37, 128–29, 144–45; Wheeler, interview by author. The GAO investigators talked to approximately one hundred A-10 pilots, and they found that the Hog drivers had to press even their Maverick attacks before achieving target lock-on.

62. Battlefield and Armor/Anti-Armor Assessment Teams, "Armor/Anti-Armor," 34 (quote); Frostic, 17; Fulghum, "Senate Panel," 44–45; *GWAPS, Weapons, Tactics, and Training,* 60; Hallion, *Storm,* 211–13, 287–88; "Pentagon Gave Short Shrift," 45; Fredericksen, Myers, interviews by author.

CHAPTER 10. CONCLUSION

1. Oliveri, "Warthog Round," 32–37; Henry, "LASTE Employment," 20–22; Lieutenant Colonel Henry, interview by author. Henry was the 175th's weapons officer, and many consider him the "brains" behind the 175th's success.

2. Bell, "Close Air Support for the Future," 92–93.

3. "A-10s Take on the Night"; "A-10 Thunderbolt Ground Attack"; Boatman, "USAF and the DAB"; Fulghum, "Night-Fighting"; Fulghum, "Signature"; Goodman, "Close"; Grossman, "USAF"; Kolcum, "USAF," 54; Neubeck, 40–44; Smallwood, "Old Hog"; Muck Brown, letter to author; Chaleff, Fredericksen, Lieutenant Colonel Henry, Koechle, Kuebler, and Swift, interviews by author.

4. "Air Combat Command"; Dudney, "Battle"; Goodman, "Juggling"; Hudson, "Fewer," 11; Y'Blood, "Metamorphosis," in *Winged Shield*, ed. Nalty, 2:553; McPeak, Ringo "Z," interviews by author.

5. Fulghum and Wall, "Joint Strike," 27; Gamboa, "Lockheed"; Scott, "Joint Strike," 51–54; Tirpak, "Scoping," 36–41.

6. Bryan, "Close"; Callero and Veit, "Close"; Ford, "Future"; Fredericksen, "Close," in *Selected Papers*, 112; Hall, "Viewpoint"; Harrison, "Gift"; Hebert, "Smaller Bombs"; Hudson, "AF Said to Slight"; Jones, "Close Air Support: A Doctrinal Disconnect"; Meyer, "For the Guys"; Otis, "Shuffling Assets"; Sweetman, "Close"; Welch, "Fixed-Wing CAS." Some Bingham articles are "Air Interdiction"; "Let the Air Force Fight"; "Revolutionizing Warfare." After retirement, Bingham worked for Northrop-Grumman as a JSTARS spokesman.

7. Boatman, "ACC," 1081; Boatman, "USA Redefines," 17; Hudson, "AF Said," 11; Lambeth, *Transformation*, 280–321; McPeak, "Presentation," 88–122; McPeak, "Roles," 32–34; Winnefield, "Moint Air Warfare," in *Air Power Confronts*, ed. Hallion, 151–52; McPeak, interview by author. McPeak told me that CAS had a "shrinking market share" of mission tasking, and that the A-10 was a "disappointment" whose pilots made it work.

8. Coniglio, "Close"; Dean, "Correspondence"; Faletti, "Close"; Fedorchak, "Close"; Hedberg, "Role"; Maduras, "Correspondence"; Meyer, "For the Guys"; Wright, "Urban"; Croker, Richardson, Russ, Saint, Vuono, Welch, and Wickham, interviews by author.

9. Bowden, *Black Hawk Down*, 335–41; Hackworth, *Hazardous Duty*, 173–83; Jordan and Compart, "'Global Mission,'" 22; Stevenson, *Losing Mogadishu*, 100–105, 128–29. Bowden concluded that heavy CAS firepower would have been irrelevant, Hackworth believed the opposite, and Stevenson thought that it might have helped.

10. Hackworth, "Air Power"; Hackworth, *Hazardous*, 257–83; Owen, ed., *Deliberate Force*, 23–29, 134–413; "Plan for Action"; Wall, "USAF Eyes"; Wall, "USAF Steps"; Condon, Koechle, Swift, and Wickstrom, interviews by author. Swift commanded the A-10 Air National Guard unit that flew the 1995 Bosnian attack missions and recalled that the scenario suited A-10s well. There was a high minimum altitude during the air campaign, but the LASTE-equipped A-10s were so accurate that they were the only non-laser-guided, bomb-equipped planes allowed to expend ordnance (though they also used Mavericks). Wickstrom flew some of the missions and said that on certain days, the weather was such that only the maneuverable A-10s could strike the targets.

11. Aubin, "Newsweek"; Fulghum, "Air War Pays"; Cordesman, *Kosovo*, 255; Fulghum, "Isolated"; Fulghum, "Pentagon Dissecting"; Halberstam, *War*, 446–73; Lambeth, "Kosovo," 122–23; Lambeth, *Transformation*, 196–98, 207–13; McPeak, "Kosovo"; Tirpak, "Short's View"; Wall, "Airspace Control"; Wall, "Joint STARS"; Wall, "NATO Endgame"; Wall, "NATO Shifts"; Koechle, letter to author.
12. Baker, "Afghans Strengthen"; Loeb and Graham, "Seven Soldiers"; McDaid and Oliver, *Smart Weapons*, 50–155; "Pentagon Taps Odd Mix of Weapons in Afghan Campaign"; "Report: Armed UAVs Fire First Missiles"; Ricks and Graham, "U.S. Special Forces."
13. The sources cited in note 12, above, also apply here. Dougherty, "Bombers" (quote); Goodman, "Airborne Goliath"; Loeb, "Technology Changes"; Ricks, "Bull's-Eye War"; Sherman, "Out of the Loop"; Trimble, "Air Campaign."
14. Bacon, "Saving."
15. Callero and Veit, "Close Air Support Evaluation"; Cordesman, *Campaign in Kosovo*, 355; Harrison, "JSF," 60–61; Lambeth, "Kosovo," 96–97; McDaid and Oliver, 33–34 and 60; "E.," interview by author.
16. Lieven, "World Turned," 43 (quote).
17. Cassidy, *Dot.Con*; Cross and Szostak, *Technology*, 312–13; Marcus and Segal, *Technology*, 352; Pursell, *Machine*, 301–19; Tenner, *Why Things Bite Back*, 206.
18. Levine, "Some Things," 120–67.
19. I recall that the Air Force's officer professional education courses—Squadron Officer School, Air Command and Staff College, and Air War College—all contained large sections on the budgeting and procurement process.
20. Some air power works have recognized the wide spectrum of warfare possibilities in the post–Cold War world, and the need for the Air Force to address diverse scenarios. Examples are several of the essays in Hallion, ed., *Air Power Confronts*, especially Meilinger, "Air Targeting Strategies"; Lambeth, *Transformation*, 307–10, 313–14; and Mason, *Air Power*, xiii, 240.

BIBLIOGRAPHY

Government, Published and Unpublished Works

AFTEC. Summary of internal letter. 30 September 1976. AFHRC (no holding number).

Agnew, James, Lt. Col., USA. *The Harold K. Johnson Papers.* Vol. 3. Carlisle Barracks, Pa.: AMHI, 5 November 1973. HAAC.

Ahmann, Hugh. "Interview with Lt. Gen. Leroy J. Manor." USAFOHP, 26–27 January and 9 May 1988. AFHRC Holding K239.0512-1799.

Ahmann, Hugh, and Lt. Col. Vaughn Gallagher, USAF. "Interview with Major Richard L. Griffin." MAFB: 24 March 1972. AFHRC Holding K239.0512-542.

———. "Lt. Gen. Marvin L. McNickle." USAFOHP, 21 October 1985. AFHRC Holding K239.0512-1679.

"Air Combat Command." Air Force News Factsheet. Database on-line. Updated February 2002, cited 9 March 2002. http://www.af.mil/news/factsheets/Air_Combat_Command.html.

Air Support of the Close-in Battle: Proceedings of the United States Air Force Aerospace Power Symposium, 9–10 March 1987. MAFB: USAFAWC.

Anderegg, C. R. *Sierra Hotel: Flying Air Force Fighters in the Decade after Vietnam.* Washington, D.C.: Air Force History and Museums Program, 2001.

Andreson, Ronald, Lt. Col., USA. "Lt. Gen. George Seneff, USA (ret.)." USASODP. Carlisle Barracks, Pa.: AMHI, 1978. HAAC.

Andreson, Ronald, Lt. Col., USA, and Col. Bryce Kramer, USA. "BGEN O. Glen Goodhand, USA (ret.)." USASODP. Carlisle Barracks, Pa.: AMHI, 1978. HAAC.

Andrews, William, Lt. Col., USAF. *Airpower against an Army.* MAFB: AUP, 1998.

Army Aviation School and Center. "Army Views on the Use of Aerial Vehicles." Briefing. 24 October 1968. Carlisle Barracks, Pa.: AMHI, Fort Rucker Papers.

"A-16 CAS Demonstration Project Plan, Introduction." Draft copy. 57 FWW, n.d.

"A-10 Thunderbolt Ground Attack, USA." Database on-line. Cited 9 March 2002. http://www.airforce-technology.com/projects/a-10/.

Avery, James T., Col., USA. "AH-56 Outline Plan of Test." Army Combat Developments Command letter of transmittal. Carlisle Barracks, Pa.: AMHI, Fort Rucker Papers.

Ballard, Jack. *Development and Employment of Fixed-Wing Gunships, 1962–1972.* Washington, D.C.: OAFH, 1982.

Battlefield and Armor/Anti-Armor Assessment Teams. "Armor/Anti-Armor Operations in Southwest Asia." Marine Corps Research Center Research Paper #92-0002. Quantico, Va.: Marine Corps Research Center, 1991.

Bowers, Ray, Col., USAF, and Ralph Rowley, Maj., USAF. Interview with Col. Eugene Deatrick (USAF), C.O. 1st Air Commando Sqdn, Pleiku, SVN, Feb 66–Feb 67, and Col. Eleazar Parmly (USA), Spec. Forces Adviser to Laos, May 60–May 61, USSF Det C.O. I and III Corps, Aug 66–Jul 67. Washington, D.C.: USAFOHP, 9 March 1972. Transcript, AFHRC Holding K239.0512-705.

Brown, Harold. "The Air Force Role in Southeast Asia." *Supplement to the Air Force Policy Letter for Commanders* 3-1968. Washington, D.C.: SAF-OII, March 1968, 25–34.

———. "Secretary Brown Speaks about the Importance of Tactical Aviation." *Supplement to the Air Force Policy Letter for Commanders* 12. Washington, D.C.: SAF-OII, December 1965, 1.

Bryant, Lloyd, Maj., USAF, and Maj. Thomas Dapore, USAFR. *A Man for the Time: An Oral History Interview with Lieutenant General James T. Stewart.* Fairborn, Ohio: Aeronautical Systems Division Historical Office, 16 October 1983. AFHRC Holding K239.0512-1927.

Burley, Boyce, Lt. Col., USAF, ed. *The Joint Staff Officer's Guide.* Armed Forces Staff College Publication 1. Washington, D.C.: GPO, 1993.

Callero, Monte, and Clairice Veit. *Close Air Support Evaluation: Preliminary Insights for ASCIET.* Report #PM-464-JS. Santa Monica, Calif.: RAND, 1995.

"CAS Panel, Investigations Subcommittee, HASC." Declassifed transcript of classified September 1976 congressional hearing, no other information. AFHRC Holding K417.168-123.

Chapman, Anne. *The Origins and Development of the National Training Center, 1976–1984.* Fort Monroe, Va.: Office of the TRADOC Historian, 1992.

Cleary, Mark, Capt., USAF. "General Robert Dixon." USAFOHP, 18–19 July 1984. AFHRC Holding K239.0512-1591.

Cooling, Benjamin Franklin, ed. *Case Studies in the Achievement of Air Superiority.* Washington, D.C.: OAFH, 1994.

————. *Case Studies in the Development of Close Air Support.* Washington, D.C.: OAFH, 1990.

Court, Philip, Lt. Col., USA, and Col. Ralph Powell, USA. "Gen. Robert Williams, USA (ret.)." USASODP. Carlisle Barracks, Pa.: AMHI, 1978. Transcript, HAAC.

Crawford, Natalie. "Low Level Attacks of Armored Targets." Research Study #P-5982. Santa Monica, Calif.: RAND, August 1977.

Crowe, William, Adm., USN. "Roles and Functions of the Armed Forces." DoD report. Washington, D.C.: Chairman of the Joint Chiefs of Staff, 1989.

Currie, James, Maj. Gen., USAF. "Tactical Air Programs." *Supplement to the Air Force Policy Letter for Commanders* 9-1977. Washington, D.C.: SAF-OII, September 1977, 24.

Davis, Richard. *The 31 Initiatives.* Air Staff Historical Study. Washington, D.C.: OAFH, 1987.

Dayton, Hugh, Lt. Col., USAF. "57 TTW TASVAL Detachment after Action Report." October 1979. Air University Library Holding.

Dick, John, Lt. Col., USAF. "Corona Ace Interview with General William Momyer, USAF (ret.)." Washington, D.C.: USAFOHP, 31 January 1977. AFHRC Holding K239.0512-1068.

————. "Gen. Gabriel P. Disosway." Washington, D.C., USAF Oral History Interview, 4–6 October 1977. AFHRC Holding K239.0512-974.

Dixon, Robert, Gen., USAF. "Remarks of General Dixon at 100th A-10 Ceremony on April 3, 1978." AFHRC Holding K417.168-119.

————. Draft of speech before Langley AFB Officers' Wives Club, 14 October 1975. AFHRC Holding K417.168-96.

Evans, William, Gen., USAF. "Thinking 'We.'" *Supplement to the Air Force Policy Letter for Commanders* 8-1978. Washington, D.C.: SAF-OII, August 1978, 9–11.

"Fahd Squad A-10 Briefing." 14 May 1991. AFHRC Holding K417.054-176.

Field Manual 100-5: Operations. Washington, D.C.: Department of the Army, 1982.

Finletter, Thomas. "Survival in the Air Age." U.S. President's Air Policy Commission. Washington, D.C.: GPO, 1 January 1948.

Freck, P. G., and others. "Comparison of Some Aspects of the Relative Operational Effectiveness of the A-7D and A-10 Aircraft Engaged in Close Air Support." Arlington, Va.: WSEG, June 1974.

Frisbee John, ed. *Makers of the United States Air Force.* Washington, D.C.: AFHMP, 1996.

Frostic, Fred. "Air Campaign against the Iraqi Army in the Kuwaiti Theater of Operations." Project Air Force Study #MR-357-AF. Santa Monica, Calif.: RAND, 1993.

"F-16/A-16 Close Air Support Demonstration, Final Report." 57 FWW. Draft copy, n.d.

Futrell, Robert Frank. *Ideas, Concepts, Doctrine.* Vols. 1 and 2. MAFB: AUP, 1989.

————. *The United States Air Force in Korea.* Washington, D.C.: OAFH, 1996.

Gallagher, Vaughn, Lt. Col., USAF, and Lyn Officer, Maj., USAF. "Interview #653 of Major James Costin." USAFOHP, 6 February 1973. AFHRC Holding K239.0512-653.

Geiger, Clarence, and others. "History of the Aeronautical Systems Division, July 1973–June 1974." Vol. 1. WPAFB: History Office, Aeronautical Systems Division, n.d.

Goldberg, Alfred, and Donald Smith. "Army–Air Force Relations." Study #R-906-PR. Santa Monica, Calif.: RAND, October 1971.

Gorn, Michael. *Harnessing the Genie.* U.S. Air Force Air Staff Historical Study. Washington, D.C.: OAFH, 1988.

Greenfield, Kent Roberts. "Army Ground Forces and the Air-Ground Battle Team, Study No. 35." Historical Section, Army Ground Forces, 1948.

Gulf War Air Power Survey. Department of the Air Force. Washington, D.C.: GPO, 1993. *Summary Report* by Eliot Cohen and Thomas Keaney. Vol. 2, pt.1: *Operations* by Barry Watts. Vol. 2, pt. 2: *Effects and Effectiveness* by Barry Watts. Vol. 3, pt. 1: *Logistics* by Richard Gunkel.Vol. 4, pt. 1: *Weapons, Tactics, and Training* by John Guilmartin Jr., Ph.D. Vol. 5, pt. 1: *A Statistical Compendium* by Col. David Tretler, USAF, and Lt. Col. Daniel Kuehl, USAF. Vol. 5, pt. 2: *Chronology* by Col. David Tretler, USAF, and Lt. Col. Daniel Kuehl, USAF.

Hall, R. Cargill, ed. *Case Studies in Strategic Bombardment.* Washington, D.C.: AFHMP, 1998.

Hansen, Grant. Speech to the American Institute of Aeronautics and Astronautics. Reprinted in *Supplement to the Air Force Policy Letter for Commanders* 3-1973. Washington, D.C.: SAF-OII, March 1973, 2–7.

Hasdorff, James. "Interview with Maj. Gen. James R. Hildreth." USAFOHP, 27–28 October 1987. AFHRC Holding K239.0512-1772.

Hasdorff, James, and Lyn Officer, Maj., USAF. "General Bernard A. Schriever." USAFOHP, 20 June 1973. AFHRC Holding K239.0512-1566.

Headquarters Air Force. "PMD for Class II Modification for A/OA-10 Technology Demonstrator Program." 1 December 1989.

"History of Air Force Systems Command, FY 73." WPAFB: HQ/AFSC, n.d.

Hosmer, Stephen. "Psychological Effects of U.S. Air Operations in Four Wars, 1941–1991." Study #MR-576-AF. Santa Monica, Calif.: RAND, 1996.

Howze, Hamilton, Maj. Gen., USA (ret.). Howze-Hawkins Family Papers, Draft Writings Articles Box. Carlisle Barracks, Pa.: AMHI, n.d.

Jamieson, Perry. *Lucrative Targets.* Washington, D.C.: AFHMP, 2001.

Jordan, Frank. "Interview with William Jarvis." USAFOHP, 7 October 1977. AFHRC Holding K239.0512-1124.

Kirkpatrick, Charles, Maj., USA. "The Army and the A-10." Washington, D.C.: U.S. Army Center of Military History, n.d.

Kramer, Bryce, Col., USA, and Col. Ralph Powell, USA. "Gen. Edwin Powell, USA (ret.). USASODP." Carlisle Barracks, Pa.: AMHI, 1978. HAAC.

Kross, Walter, Col., USAF. *Military Reform.* Fort McNair, Washington, D.C.: National Defense University Press, 1985.

Lambeth, Benjamin. "NATO's Air War for Kosovo: A Strategic and Operational Assessment." Project Air Force Study F49642-01-C-0003. Santa Monica, Calif.: RAND, 2001.

———. "Pitfalls in Fighter Force Planning." Study #P-7064. Santa Monica, Calif.: RAND, February 1985.

Lester, Gary. *Mosquitoes to Wolves.* MAFB: AUP, 1997.

Lind, William. "The Grenada Operation." Report to the Congressional Military Reform Caucus, 5 April 1984. Air University Library.

Loye, J. F., Col., USAF, and others. "Lam Son 719, 30 January–24 March 1971." Project CHECO Report. Hickam AFB, Haw.: HQ PACAF, 24 March 1971.

"Lt. Gen. Carlos M. Talbott." USAFOHP, 10–11 June 1985. AFHRC Holding K239.0512-1652.

"Lt. Gen. Dale S. Sweat." USAFOHP, 15 February 1984. AFHRC Holding K239.0512-1572.

"Lt. Gen. Robert G. Ruegg." USAFOHP, 13–14 February 1984. AFHRC Holding K239.0512-1571.

Mann, Edward, III, Col., USAF. *Thunder and Lightning.* MAFB: AUP, 1995.

Mark, Eduard. *Aerial Interdiction.* Washington, D.C.: OAFH, 1994.

Maurer, Maurer. *Aviation in the U.S. Army, 1919–1939.* Washington, D.C.: OAFH, 1987.

McLennan, D. G., Col., USA. "Close Air Support History." Ft. Meade, Md.: CONARC CAS Requirements Board, 29 August 1963.

McNamara, Robert, Secretary of Defense. "Selected Statements by DoD and Other Administration Officials, January 1–June 30, 1968." Washington, D.C.: Office of the Secretary of the Air Force Analysis Division, n.d. AFHRC Holding K168.04-48.

McPeak, Merrill A., Gen., USAF. "Presentation to the Commission on Roles and Missions of the Armed Forces." Washington, D.C.: Chief of Staff, USAF, 14 September 1994.

Meilinger, Phillip, ed. *The Paths of Heaven.* MAFB: AUP, 1997.

Mischler, Edward. *The A-X Specialized Close Air Support Aircraft.* WPAFB: Office of History, AFSC, 1971.

Momyer, William, Gen.,USAF. *Airpower in Three Wars.* MAFB: AUP, 1978.

———. "Close Air Support." *Supplement to the Air Force Policy Letter for Commanders* 6-1973. Washington, D.C.: SAF-OII, June 1973, 13–21.

Mortensen, Daniel. *A Pattern for Joint Operations.* Washington, D.C.: OAFH and U.S. Army Center for Military History, 1987.

Mortensen, Daniel, ed. *Airpower and Ground Armies.* MAFB: AUP, 1998.

Mrozek, Donald. *Air Power and the Ground War in Vietnam.* MAFB: AUP, 1988.

Nalty, Bernard. *Air War over South Vietnam.* Washington, D.C.: Air Force History and Museums Program, 2000.

Nalty, Bernard, ed. *Winged Shield, Winged Sword.* Vols. 1 and 2. Washington, D.C.: AFHMP, 1997.

Neufeld, Jacob ("Jack"). *The F-15 Eagle.* Washington, D.C.: OAFH, 1974.

———. "Interview of Brig. Gen. William F. Georgi." USAFOHP, 5 June 1973. AFHRC Holding K239.0512-964.

———. "Interview K239.0512-857 of Lt. Gen. Arthur C. Agan, USAF (Ret)." USAFOHP, 2 October 1973.

———. "Interview of Maj. Gen. John J. Burns." USAFOHP, 22 March 1973. AFHRC Holding K239.0512-961.

———. "Mr. Pierre Sprey." USAFOHP, 12 June 1973. AFHRC Holding K239.0512-969.

———. "Interview #859 of Col. John Boyd." USAFOHP, 23 May 1973. AFHRC Holding K239.0512-859.

Office of the Secretary of Defense, Office of the Director, Operational Test and Evaluation. "Operational Test Plan Concept for Evaluation of Close Air Support Alternative Aircraft." 31 March 1989. Air University Library holding.

Officer, Lyn, Lt. Col., USAF. "Interview with Maj. Upton D. Officer." USAFOHP, 11–13 July 1973. AFHRC Holding K239.0512-680.

Orr, Verne. Letter to Sen. Strom Thurmond. 27 March 1981. Verne Orr Papers. AFHRC Holding K168.7270-28.

Owen, Robert, Col., USAF, ed. *Deliberate Force.* MAFB: AUP, 2000.

Packard, David. Letter to Senator Howard Cannon. 13 May 1970. AFHRC Holding K168.7094-38.

———. Letter to Senator John Stennis. 11 September 1970. AFHRC Holding K168.7094-39.

Pentland, Pat ("Doc"), Maj., USAF. "Evolution of A-10 Mission Requirements." Headquarters Air Force (Fighter Plans) Research Paper, 10 June 1988.

Phillips, Albert, Maj., USAF. "The Integration of the F-16 into the Close Air Support Environment." Report submitted to 57 FWW Test Directorate, May 1989.

Porter, Melvin, Capt., USAF. "Second Defense of Lima Site 36." Project CHECO Report. Hickam AFB: PACAF, 28 April 1967.

Ratley, Lonnie, Capt., USAF. "The 'A-10 Pilots Coloring Book' Christmas 1977, or 'What You Always Wanted to Know about the T-62 but Were Afraid to Ask.'" A-10 Formal Training Unit pamphlet, n.d.

Reed, Robert, Lt. Col., USA. "Gen. Hamilton Howze, USA (ret.)." USASODP. Carlisle Barracks, Pa.: AMHI, 1972. HAAC.

Reynolds, Richard T., Col., USAF. *Heart of the Storm.* MAFB: AUP, January 1995.

Ridgway, Jack, Col., USA, and Lt. Col. Paul Walter, USA. "Gen. Barksdale Hamlett, USA (ret.)." USASODP. Carlisle Barracks, Pa.: AMHI, 23 January 1976. Transcript.

Romjue, John. *From Active Defense to AirLand Battle.* Fort Monroe, Va.: TRADOC, 1984.

Scales, Robert, Brig. Gen., USA. *Certain Victory.* Washington, D.C.: Office of the Chief of Staff, U.S. Army, 1993.

Schlight, John. *The War in South Vietnam.* Washington, D.C.: OAFH, 1988.

Schraeder, Rod, Capt., USAF. "Using the AGM-65D in Combat." 81st Tactical Fighter Wing training pamphlet, 6 January 1987.

Seamans, Robert. *Aiming at Targets*. Washington, D.C.: NASA History Office, 1996.

Seibert, Donald. Papers. Carlisle Barracks, Pa.: AMHI.

Starry, Donn, Gen., USA. "Close Air Support." Letter to ALFA. 8 August 1977. Gen. Donn Starry Papers. Carlisle Barracks, Pa.: AMHI.

———. Letter to Steven Canby. Gen. Donn Starry Papers. August 1977. Carlisle Barracks, Pa.: AMHI.

Stein, David. "The Development of NATO Tactical Air Doctrine, 1970–1985." Study #R-3385-AF. Santa Monica, Calif.: RAND, 1987.

Stolfi, Ron, Ph.D., John Clemens, Ph.D., and Raymond McEachin. "Combat Damage Assessment Team, A-10/GAU-8 Low Angle Firings, Individual Soviet Tanks." Executive Summary. WPAFB: Armaments Directorate, A-10 SPO, February–March 1978.

Summers, Harry, Col., USA. *On Strategy*. Carlisle Barracks, Pa.: USAWC Strategic Studies Institute, 1981.

Sunderland, Riley. "Evolution of Command and Control Doctrine for Close Air Support." Washington, D.C.: OAFH, 1973.

"Tactical Airpower Subcommittee of SASC, March 1976." 17 March 1976. AFHRC Holding K417.168-122.

Tilford, Earl, Jr. *Search and Rescue in Southeast Asia*. Washington, D.C.: Center for Air Force History, 1992.

———. *Setup*. MAFB: AUP, 1991.

Tolson, John. "Vietnam Diary, 1960–1968." John Tolson Papers. Carlisle Barracks, Pa.: AMHI, n.d.

USA. "U.S. Army Battle Damage Assessment Operations in Operation Desert Storm, Volume I." Report BRL-TR-3408. Aberdeen Proving Ground, Md.: BRL, 1992.

USAF. "Air War College Associate Program Military Studies Course (MS 610AP)." Vol. 1. 4th ed. MAFB: Air University, 1993.

USAF Armaments Systems Division. "Christmas 1978: The A-10 Pilot's Multi-Color Guide to Touring in Central Europe, or Are ZSU-23/4 Drivers Really Ten Feet Tall?" WPAFB, Ohio: ASD, n.d.

USAF Project PAVE COIN. "Forward Air Control (FAC) and Light Strike Aircraft (LSA) Evaluation Report." Task #327, Project 1559, P.E.-647085. September 1971. AFHRC Holding K143.054-2.

USAF Systems Command. "History, 1 July 1974–30 June 1975." WPAFB: HQ/AFSC, n.d.

———. "History, 1 July 1975–31 December 1975." WPAFB: HQ/AFSC, n.d.

———. "History, 1 January 1976–31 December 1976." WPAFB: HQ/AFSC, n.d.

USAF Tactical Air Command. "Instructional Text: Manual Weapons Delivery." Course A1000IDOPN, Part One. Nellis AFB, Nev.: FWS, 1981.

———. "USAF Fighter Weapons Instructor Course, A-10." TAC Syllabus #A1000IDOPN. Langley AFB, Va.: TAC, 1983.

USAF Tactical Air Command Headquarters Training Directorate (HQ TAC/DOT). "A-10 AGM-65D IR Maverick Training Program." July 1984. Air University Library.

U.S. Congress, House. *Briefing on the Military Airlift Capability during the Middle East Conflict and the Enforcer Aircraft.* 93d Cong., 2d sess., 8 August 1974.

———. *Close Air Support.* 89th Cong., 1st sess., 22, 23, 28, 29, 30 September and 6, 14 October 1965.

——— *Close Air Support.* Report. 1 February 1966.

———. *Close Air Support.* 101st Cong., 1st sess., 19 April 1989.

———. *Cost Escalation in Defense Procurement Contracts and Military Posture and H.R. 6722, Pt. 2.* 93d Cong., 1st sess., 21 May 1973.

———. *Crisis in the Persian Gulf.* 101st Cong., 2d sess., 12 December 1990.

———. *DoD Appropriations for 1963, Part IV, Procurement.* 87th Cong., 2d sess., 9 March 1962.

———. *DoD Appropriations for 1971, Pt. 1.* 91st Cong., 2d sess., 19 February 1970.

———. *DoD Appropriations for 1971, Pt. 6.* 91st Cong., 2d sess., 27 April 1970.

———. *DoD Appropriations for 1974, Pt. 6.* 93d Cong., 1st sess., 19 September 1973.

———. *DoD Appropriations for 1974, Pt. 7.* 93d Cong., 1st sess., 27 September 1973.

———. *DoD Appropriations for 1975, Pt. 4.* 93d Cong., 2d sess., 7 March 1974.

———. *DoD Appropriations for 1975, Pt. 7.* 93d Cong., 2d sess., 23 May 1974.

———. *DoD Appropriations for 1977, Pt. 3.* 94th Cong., 2d sess., 1 March 1976.

———. *DoD Appropriations for 1977, Pt. 5.* 94th Cong., 2d sess., 8 March 1976.

———. *Enforcer Aircraft.* 95th Cong., 2d sess., 22 June 1978.

———. *Full Committee Consideration of H.R. 8591, H.R. 11144, H.R. 15406.* 93d Cong., 2d sess., 20 June 1974.

———. *Military Posture and H.R. 3689 [H.R. 6674], DoD Authorization for Appropriations for FY 1976 and 197T, Part 4.* 94th Cong., 1st sess., 9 April 1975.

———. *Military Posture and H.R. 5968, DoD Authorization for Appropriations for FY 1983, Part 3.* 97th Cong., 2d sess., 11 March 1982.

———. *Military Posture and H.R. 12564.* 93d Cong., 2d sess., 4 March 1974.

———. *National Defense Authorization Act for FY 1989—H.R. 4264 and Oversight of Previously Authorized Programs, Title I.* 100th Cong., 2d sess., 9–10 March 1988.

———. *Roles and Missions of Close Air Support.* 101st Cong., 2d sess., 27 September 1990.

U.S. Congress, Office of Technology Assessment. *Who Goes There?* Report OTA-ISC-537. June 1993.

U.S. Congress, Senate. *Close Air Support.* 92d Cong., 1st sess., 22, 26, 28, 29 September, and 1, 3, 8, November 1971.

———. *Close Air Support.* Report. 18 April 1972.

———. *DoD Appropriations for FY 1970, Pt. 4.* 91st Cong., 1st sess., 29 July 1969 and 4 May 1970.

———. *DoD Appropriations for FY 1973, Pt. 4.* 92d Cong., 2d sess., 16 February 1972.

———. *DoD Appropriations for FY 1974, Pt. 4.* 93d Cong., 1st sess., 2 April 1973.

———. *DoD Appropriations for FY 1975, Pt. 4.* 93d Cong., 2d sess., 7 March 1974.

———. *DoD Appropriations for FY 1976, Pt. 4.* 94th Cong., 1st sess., 26 February 1975.

———. *DoD Authorization for Appropriations for FY 1983, Pt. 4.* 97th Cong., 2d sess., 16 March 1982.

———. *DoD Authorization for Appropriations for FY 1989, Pt. 4.* 100th Cong., 2d sess., 21 March and 13 April 1988.

———. *DoD Authorization for Appropriations for FY 1990 and 1991, Pt. 5.* 101st Cong., 1st sess., 4, 16 May, and 6 June 1989.

———. *FY 1975 Authorization for Military Procurement, Research and Development, and Active Duty, Selected Reserve and Civilian Personnel Strengths, Pt. 8.* 93d Cong., 2d sess., 14 March 1974.

———. *FY 1978 Authorization for Military Procurement, Research and Development, and Active Duty, Selected Reserve, and Civilian Personnel Strengths, Pt. 7.* 95th Cong., 1st sess., 16 March 1977.

———. *Military Situation in the Far East.* 82d Cong., 1st sess., 28 May 1951.

———. *Operation Desert Shield/Desert Storm.* 102d Cong., 1st sess., 21 May 1991.

———. *U.S. Air Force Tactical Air Operations and Readiness.* 89th Cong., 2d sess., 9, 10 May 1966.

———. *U.S. Tactical Air Program.* 90th Cong., 2d sess., 14, 17, 28 May, and 6 June 1968.

U.S. Department of Defense. *Conduct of the Persian Gulf War.* Washington, D.C.: GPO, 1992.

U.S. Government Accounting Office (GAO). *Aerial Fire Support Weapons.* Report #PSAD-79-65. 11 June 1979.

———. *Apache Was Considered Effective in Combat but Reliability Problems Exist.* Frank Conahan. Report #NSIAD-92-146, 1992.

———. *A-10 Close Air Support Aircraft.* Report #093514. March 1974.

———. *Close Air Support: Airborne Controllers in High Threat Areas May Not Be Needed.* Nancy Kingsbury. Report #GAO/NSIAD-90-116, 1990.

———. *Close Air Support: Comparison of Air Force and Marine Corps Requirements and Aircraft.* Harry R. Finley. Report #GAO/NSIAD-89-218, 1989.

———. *Close Air Support: Principal Issues and Aircraft Choices.* Report #B-173850, 1971.

———. *Close Air Support: Status of the Air Force's Efforts to Replace the A-10 Aircraft.* Harry R. Finley. Report #GAO/NSIAD-88-211, 1988.

———. *Close Air Support: Upgraded A-7 Aircraft's Mission Effectiveness and Total Cost Unknown.* Report #GAO/NSIAD-88-210, 1988.

———. *Combat Air Power: Assessment of Joint Close Air Support Requirements and Capabilities Is Needed.* Richard Davis. Report #GAO/NSIAD-96-45, 1996.

———. *Operation Desert Storm: Evaluation of the Air Campaign.* Henry Hinton. Report #GAO/NSIAD-97-134, 1997.

———. *Review of the Department of Defense's Design-to-Cost Concept.* R. W. Gutman. Report to Congress, #B-163058, 1978.

Wall, Malcolm. *The Development and Acquisition of the GAU-8/A.* Vol. 1. WPAFB: Office of History, AFSC, 1989.

Watson, George, Jr. *The A-10 Close Air Support Aircraft.* WPAFB: Office of History, AFSC, 1978.

———. *The Office of the Secretary of the Air Force, 1947–1965.* Washington, D.C.: Center for Air Force History, 1993.

Weinert, Richard. *A History of Army Aviation, 1950–1962.* Fort Monroe, Va.: TRADOC Office of the Command Historian, 1991.

Werrell, Kenneth. *Archie, Flak, AAA, and SAM.* MAFB: AUP, 1988.

Wessely, H.W. "Limiting Factors in Tactical Target Acquisition." Study #P-5942. Santa Monica, Calif.: RAND, March 1978.

Winnefield, James, Preston Niblack, and Dana Johnson. "A League of Airmen." Study #MR-343-AF. Santa Monica, Calif.: RAND, 1994.

Wolf, Richard. *The United States Air Force: Basic Documents on Roles and Missions.* Washington, D.C.: OAFH, 1987.

Worden, Mike, Col., USAF. *The Rise of the Fighter Generals.* MAFB: AUP, 1998.

INTERVIEWS BY AUTHOR

Adams, Jimmie V., Gen., USAF (ret.). February and April 1997. Telephone and personal interviews. TAC A-7/A-10 flyoff project officer, A-10 wing commander, Air Force Headquarters senior staff, Pacific Air Forces commander.

Allen, Lew, Gen., USAF (ret.). April 1997. Telephone interview. U.S. Air Force chief of staff, 1978–1982.

Allison, John ("Jack," "Coach"), Maj., USAF. May 1997. Telephone interview. A-10 Weapons School instructor and squadron commander.

"B." Anonymous interviewee for intelligence information on Su-25 Frogfoot. June 1997. Telephone interview. CIA senior analyst of Soviet aviation.

Bahnsen, John ("Doc"), Brig. Gen., USA (ret.). July 1997. Telephone interview. Senior army aviation officer and army director of JAWS exercises.

Bass, Marvin, Col., USAF (ret.). July 1997 and July 1998. Telephone interviews. A-10 squadron commander, tactics/weapons test pilot, and TAC project officer for Enforcer aircraft test.

Battista, Anthony. April 1997. Telephone and personal interviews. Congressional staff member, 1970s and 1980s.

Brown, Harold, Hon. July 1997. Letter correspondence. Secretary of the U.S. Air Force, 1965–1969; U.S secretary of defense, 1977–1981.

Brown, Robert ("Muck"), Lt. Col., USAF. July 1999. E-mail correspondence. A-10 Weapons School instructor and author.

Buchta, Bob. August 1997, September and October 1999. Telephone interviews and letter correspondence. A-10 SPO ammunition program manager in the 1970s.

Burton, James, Col., USAF (ret.). March 1997. Telephone interview. U.S. Air Force officer and military reform author.

Casey, Al, Lt. Gen., USAF (ret.). May 1997 and August 1999. Telephone interview and letter correspondence. Senior A-10 SPO officer in the 1970s.

Chaleff, Robert, Col., USAF. May 1997 and August 1999. Telephone interviews. U.S. Air Force Reserve/Air National Guard A-10 test project officer.

Cheney, Richard, Hon. March 1997. Telephone interview. Secretary of defense, 1989–1993.

Christie, Tom. February, April, and July 1997, March 1998, August and September 1999. Telephone and personal interviews. Occupied several different OSD positions during A-10's development and life.

Condon, John, Lt. Col., USAF. March 1997. Telephone interview. A-10 weapons officer in Desert Storm and squadron commander in 1990s.

Creech, W. L., Gen., USAF (ret.). March 1997 and August 1999. Telephone interviews. TAC commander, 1978–1984.

Croker, Stephen, Lt. Gen., USAF (ret.). May 1997 and January 2000. Personal interview and letter correspondence. ACC vice commander, early 1990s.

Deatrick, Eugene, Col., USAF (ret.). May 1997 and September 1999. Personal interview and letter correspondence. A-1 squadron commander in Vietnam.

Dickinson, William, Hon. March 1997. Telephone interview. Member of Congress (R-AL), 1965–1993.

Dilger, Robert, Col., USAF (ret.). February 1997. Telephone interview. Senior A-10 SPO officer in mid- to late 1970s.

Dixon, Robert, Gen., USAF (ret.). April 1997. Telephone interview. U.S. Air Force Headquarters staff officer at time of A-X decision; TAC commander, 1974–1978.

Downey, Thomas, Hon. March 1997. Telephone interview. Member of Congress (D-N.Y.), 1975–1993.

"E." Anonymous interviewee. February 1998. ASCIET participant.

Foster, John, Hon. December 1997. Telephone interview. Director of defense research & engineering, 1960s.

Fox, Charles ("Chuck"), Col., USAF. April 1997. Personal interview. A-10 weapons officer, Desert Storm.

Fredericksen, Donald. February and April 1997, September and October 1999. Telephone interviews, personal interview, and letter correspondence. Occupied several OSD positions during A-10's development and life; CASMARG chairman.

George, Arthur ("Cub"), Lt. Col., USAF. March and June 1997. Telephone interviews. A-10 Fighter Weapons School instructor.

Giamio, Robert, Hon. March 1997 and September 1999. Telephone interviews. Member of Congress (D-Conn.), 1959–1981.

Harrison, Benjamin, Brig. Gen., USA (ret.). June 1997. Telephone interview. Army aviation officer and Army Headquarters staff officer, late 1960s.

Hennigar, Bruce ("Spanky"), Lt. Col., USAF (ret.). May 1997. Telephone interview. A-10 Fighter Weapons School instructor.

Henry, Larry, Gen., USAF (ret.). March 1997. Telephone interview. Deputy on General Horner's staff during Desert Storm.

Henry, Ron, Lt. Col., Maryland ANG. March 1997, August and September 1999. Telephone interviews and e-mail correspondence. A-10 unit weapons officer.

Highbush, John. February 1997. Telephone interview. Congressional staff member during CAS plane debate.

Houle, Edward, Col., USAF (ret.). June 1997. Telephone interview. A-10 pilot in TASVAL, and A-10 Fighter Weapons School instructor.

Ireland, Andy, Hon. April 1997. Telephone interview. Member of Congress (R-FL), 1977–1993.

Isby, David. September 1997. Telephone interview. Military aviation history author.

Jenny, David ("Bubba"), Col., USAF (ret.). June 1997. Telephone interview. A-10 senior wing officer at Myrtle Beach AFB, late 1970s.

Johnson, Richard, Lt. Col., USA. March 1997. Telephone interview. Army helicopter pilot in Lam Son 719.

Jones, David, Gen., USAF (ret.). March 1997. Telephone interview. U.S. Air Force chief of staff, 1974–1978.

Kay, Avery, Col., USAF (ret.). March 1997 and September 1999. Telephone interviews and letter correspondence. Air Force Headquarters staff officer at time of A-X decision.

Kemp, Dar K. March 1997. Telephone interview. Desert Storm A-10 pilot.

Kishline, Sam, Col., USAF (ret.). April 1997 and September 1999. Telephone interviews and letter correspondence. A-10 SPO deputy director in 1970s; chairman of A-10 gun selection board.

Klein, William, Lt. Col., USAF (ret.). September 1999. Telephone interview. A-10 pilot in Korea.

Koechle, Mark, Maj., USAF. June 1997 and November 1999. Letter correspondence. A-10 weapons officer, Desert Storm, and A-10 squadron operations officer, Kosovo.

Kuebler, Dan. May 1997 and December 1999. Telephone interviews. Syracuse ANG A-10 pilot.

Lancaster, Ray, Col., USAF (ret.). April 1997. Telephone interview. Air Force Headquarters staff officer at time of A-X decision.

Lieberherr, John, Col., USAF. May 1997 and September 1999. Personal interview and e-mail correspondence. Part of initial A-10 Fighter Weapons School instructor cadre and NATO A-10 squadron commander.

Lloyd, Jim, Hon. March 1997. Telephone interviews. Member of Congress (D-CA), 1975–1981.

Loh, John, Gen., USAF (ret.). September 1999. Telephone interview. F-16 SPO director, 1975–1977; TAC/ACC commander, 1990–1995.

McGrath, Thomas, Lt. Col., USAF (ret.). November 1997. Personal interview. Airborne FAC in Vietnam.

McMullen, Thomas, Lt. Gen., USAF (ret.). May 1997 and September 2000. A-10 SPO director, 1973–1974.

McPeak, Merrill A., Gen., USAF (ret.). March 1997 and September 1999. Telephone interview and letter correspondence. U.S. Air Force chief of staff, 1990–1994.

Myers, Charles E. ("Chuck"), Jr. February and May 1997, and September 1999. Personal and telephone interviews, and letter correspondence. OSD staff member and prominent Fighter Mafia and military reform member.

Oates, Richard. July 1997. Telephone interview. Project officer in 1970s A-10 gun program.

O'Bryon, James. May 1997. Personal interview. 1970s BRL staff and later deputy director of OSD Live Fire test issues.

Orr, Verne, Hon. June 1997. Letter correspondence and telephone interview. Secretary of U.S. Air Force, 1981–1985.

Pentland, Pat ("Doc"), Col., USAF. March and June 1997, and September 1999. Telephone interviews and letter correspondence. A-10 pilot, Fighter Weapons School instructor, and Air Force Headquarters staff officer during 1980s CAS plane debate.

Phillips, James, Col., USAF. April 1997. Letter correspondence. A-10 pilot, weapons officer, and Air Force Headquarters staff officer during 1980s CAS plane debate.

Pike, Otis G., Hon. March 1997. Telephone interview. Chairman of 1965 House close air support hearings; member of Congress (D-N.Y.), 1961–1979.

Price, Robert, Hon. March 1997. Telephone interview. Member of Congress (R-TX), 1967–1975.

Ratley, Lonnie O., Col., USAF (ret.). July 1998. Telephone interview. A-10 SPO officer, late 1970s.

Richardson, William, Gen., USA (ret.). May 1997. Personal interview. U.S. Army TRADOC commander, 1983–1987.

Ringo, Durwood W.("Skip"), Jr. February, April, and June 1997, and October 1999. Telephone interviews and e-mail correspondence. Congressional staff member during CAS plane debate.

Rosa, John, Col., USAF. March 1997. Telephone interview. A-10 pilot and Air Force Headquarters staff officer during 1980s CAS plane debate.

Russ, Robert D., Gen., USAF (ret.). March 1997. Telephone interview. TAC commander, 1986–1990.

Saint, Crosbie, Gen., USA (ret.). May 1997 and November 1999. Personal interview and letter correspondence. Commander, U.S. Army, Europe, late 1980s.

Sanator, Robert, Ph.D. May 1997 and August 1999. Telephone interviews. Fairchild-Republic A-10 preliminary design team manager, 1969–1973.

Schlesinger, James, Hon. April 1997 and July 1999. Telephone interviews and letter correspondence. U.S secretary of defense, 1973–1975.

Schriever, Bernard, Gen., USAF (ret.). May 1997. Personal interview. Commander of U.S. Air Force Systems Command, and witness in the 1965 House hearings on close air support.

Seamans, Robert, Hon. June 1997. Telephone interview. Secretary of U.S. Air Force, 1969–1973.

Spinney, Franklin C. ("Chuck"). February and May 1997, and October 1999. Telephone and personal interviews. OSD staff member and prominent military reform movement member.

Sprey, Pierre. February, March, and May 1997. Telephone and personal interviews. OSD staff member and prominent Fighter Mafia and military reform member.

Stenner, Charles, Col., USAFR. May 1997. Telephone interview. Air Force Reserve A-10 Test Project officer.

Stolfi, Russel, Ph.D. March 1997 and September 1999. Telephone interviews. Historian who conducted battle damage surveys after two Arab-Israeli wars as well as A-10 gun evaluations.

Swift, Daniel, Lt. Col., Massachusetts ANG. May 1997. Telephone interview. A-10 weapons officer in Desert Storm and ANG squadron commander in Bosnia.

Tabor, Dale, Maj. Gen., USAF (ret.). February 1997 and September 1999. Telephone interview and letter correspondence. Flew in A-10/A-7 flyoff.

Toomay, John, Maj. Gen., USAF (ret.). March 1997 and September 1999. Telephone interviews. Head of the 1987 U.S. Air Force Close Air Support Science Advisory Board Study.

Vuono, Carl, Gen., USA (ret.). May 1997 and November 1999. Personal interview and letter correspondence. U.S. Army chief of staff, 1987–1991.

Welch, Larry, Gen., USAF. May 1997 and September 1999. Personal and telephone interviews, and e-mail correspondence. Air Force Headquarters staff officer in mid-1960s, and U.S. Air Force chief of staff, 1986–1990.

Wheeler, Winslow. May 1997. Telephone interview. Congressional staff member during CAS plane debate.

Whitley, Alton, Col., USAF (ret.). July 1997. Telephone interview. Pilot in initial A-10 cadre and A-10 Fighter Weapons School instructor.

Wickham, John, Gen., USAF (ret.). March 1997. Telephone interview. U.S. Army chief of staff, 1983–1987.

Wickstrom, Jefry W. ("Rowdy"), Maj., USAF. March 1997. Telephone interview. A-10 weapons officer and pilot in Bosnia air campaign.

Williams, Robert, Gen., USA (ret.). June 1997. Telephone interview. U.S. Army aviation leader.

Wilson, Seth ("Growth"), Lt. Col., USAFR. May 1997. Telephone interview. FAC in Vietnam; A-10 pilot, Korea; and A-10 Air Reserve operations officer, Desert Storm.

Yates, Ron, Gen., USAF (ret.). April 1997. Telephone interview. A-10 SPO officer in 1970s.

Yohe, Kent, Maj., USAF. April 1997. Telephone interview. Desert Storm A-10 pilot.

Yudkin, Richard, Maj. Gen., USAF (ret.). March 1997 and August 1999. Telephone interviews. Director of Headquarters Air Force Doctrine, Concepts, and Objectives Office at time of A-X decision.

"Z." May 1997 and November 1999. Personal interview and letter correspondence. Pentagon OSD staff.

Dissertations, Theses, and Academic Research Papers

Backlund, William, Maj., USA. "Can the Army Take Over CAS with Its Organic Aircraft?" Research report, ACSC, 1985.

Barnard, Howard, Maj., USAF. "Is Tactical Support of an Airborne Battalion Feasible in Adverse Weather?" Master's thesis, C&GSC, 1979.

Barnes, Thomas, Maj., USAF. "A Concept for Tactical Fighter Design: The Case for Single Purpose Aircraft." Research study, ACSC, 1972.

Bell, Steven, Maj., USAF. "Close Air Support for the Future." Master's thesis, C&GSC, 1992.

Bennett, Russell, Maj., USAF. "Can the A-X Adequately Replace the A-1 in the COIN and SAR Roles?" Research study, ACSC, 1973.

Bridges, Roy, Maj., USAF. "Realism in Operational Test and Evaluation for Close Air Support." Research study, ACSC, 1976.

Bryan, William, Col., USA. "Close Air Support Doctrine." Research report, USAFAWC, 1994.

Buhrow, Robert, Lt. Col., USAF. "Close Air Support Requirements." Research paper, USAWC, 1971.

Carleton, Roger, Maj., USAF. "The A-10 and Design-to-Cost: How Well Did It Work?" Research study, C&GSC, 1979.

Collins, John, Lt. Col., Connecticut ANG. "Close Air Support: Another Look." Study project, USAWC, 1989.

Cox, Gary, Maj., USAF. "Beyond the Battle Line." Master's thesis, ACSC, 1996.

Currie, Michael Patteson, Cdr., USN. "Operational Lessons of Urgent Fury." Research paper, Naval War College, 1988.

Field, Edsel, Col., USAF. "A-10 Test and Evaluation." Research report, USAFAWC, 1976.

Frey, Thomas. *F-16 Fighting Falcon Upgraded Cockpit.* Fort Worth, Tex.: General Dynamics, n.d.

Furtak, Ronald, Maj., USAF. "The A-10 in the Rescue Escort Role." Research study, ACSC, 1975.

Garrett, Thomas, Lt. Col., USA. "Close Air Support: Why All the Fuss?" Individual study project, USAWC, 1990.

Goforth, Charles, Maj., USAF. "Do We Need a New Aircraft for Close Air Support?" Master's thesis, ACSC, 1967.

Gonzales, Harold, Lt. Col., USAF. "Tactical Air Support of Ground Forces in the Future." Research report, Airpower Research Institute, 1990.

Greene, Melvin ("Smoky"), Col., USAF. "Close Air Support: Proud Past, Uncertain Future." Research report, USAFAWC, 1988.

Head, Richard, Maj., USAF. "Decision-Making on the A-7 Attack Aircraft Program." Ph.D. diss., Syracuse University, 1970.

Hillman, James, Lt. Col., USA. "Task Force 1-41 Infantry: Fratricide Experience in Southwest Asia." Study project, USAWC, 1993.

Hinds, Robert, Col., USAF. "Replacing the A-10." Defense analytical study, USAFAWC, 1989.

Kasbeer, Charles, Maj., USAF, and Maj. Thomas Lyon, USAF. "The A-10 Pilot SAR Handbook." Research report, ACSC, 1981.

Lewis, Michael. "Lt. Gen. Ned Almond, USA." Master's thesis, School of Advanced Airpower Studies, Air University, 1996.

Littlejohn, Edward, Lt. Col., USA. "Close Air Support: Battle in the Fourth Dimension." Military studies project, USAWC, 1990.

Martin, Jerome. "Reforging the Sword." Ph.D. diss., Ohio State University, 1988.

Mekkelsen, Norman, Maj., USA. "Close Air Support in Europe." Research report, Air Command and Staff College, 1980.

Mullendore, Lauren, Lt. Col., USAF. "The Future of the Joint Air Attack Team in the Air-Land Battle." Defense analytical study, USAFAWC, 1989.

Nelson, Randy, Maj., USA. "The Combat Use of Apache Helicopters in the Kuwaiti Theater of Operations—Effective or Not?" Master's thesis, C&GSC, 1992.

Nophsker, H. G., Maj., USAF. "Perceptions of Fighter Strikes." Master's thesis, C&GSC, 1976.

Odgers, Peter, Col., USAF. "Design-to-Cost: The A-X/A-10 Experience." Research report, USAFAWC, 1974.

Proffitt, Glenn A., Maj., USAF. "The FAC." Research program paper, Armed Forces Staff College, 1978.

Ratley, Lonnie O., Capt., USAF. "A Comparison of the USAF Projected A-10 Employment in Europe and the Luftwaffe Schlachtgeschwader Experience on the Eastern Front in World War Two." Master's thesis, Naval Postgraduate School, March 1977.

Reeves, Raymond, Maj., USAF. "Close Air Support, AH-56 vs. A-X." Research study, ACSC, 1972.

Riley, Michael, Maj., USA. "The A-10 Thunderbolt as an Organic Army Asset." Master's thesis, C&GSC, 1991.

Ritchie, L. Michael, Maj., USAF. "Technological Solutions to the Problem of Fixed-Wing Air-to-Ground Fratricide." Master's thesis, C&GSC, 1993.

Roper, Robert, Maj., USAF, and Gail Tatum, Maj., USAF. "The A-10 Close Air Support Mission in the NATO Environment." Research report, ACSC, 1980.

Smith, Douglas, Lt. Col., USA. "Army Aviation in Operation Just Cause." Research Paper, USAWC, 1992.

Smith, Ross, Maj., USAF. "Close Air Support—Can It Survive the 80s?" Master's thesis, C&GSC, 1979.

Spencer, James, Maj., USAF. "Army Field Manual 100-5 vs. Sun Tzu." Research paper, Naval War College, 1988.

Terrien, Max, Maj., USA. "Close Air Support: Retrospective and Perspective?" Master's thesis, C&GSC, 1982.

Welsh, Mark, Lt. Col. (service unknown). "Just Give Me the FACs!" National security policy process research paper, National War College, 1993.

White, Emmitt, Lt. Col., USA. "Close Air Support—A Case for Divestiture Planning in the Department of Defense." Military studies program paper, USAWC, 1989.

Ziemke, Caroline. "In the Shadow of the Giant." Ph.D. diss., Ohio State University, 1989.

GENERAL BOOKS CONSULTED

Adkin, Mark. *Urgent Fury.* Lexington, Mass.: Lexington Books, 1989.

Air War—Vietnam. New York: Arno Press, 1978.

Ambrose, Stephen. *Citizen Soldiers.* New York: Touchstone, 1997.

Andradé, Dale. *Trial by Fire.* New York: Hippocrene Books, 1995.

Armitage, M. J., and R. A. Mason. *Air Power in the Nuclear Age.* Urbana: University of Illinois Press, 1983.

Atkinson, Rick. *Crusade.* New York: Houghton Mifflin, 1993.

Barlow, Jeffrey. *Revolt of the Admirals.* Washington, D.C.: Brassey's, 1998.

Barlow, Jeffrey, ed. *Reforming the Military.* Washington, D.C.: Heritage Foundation, 1981.

Beaumont, Roger. *Joint Military Operations: A Short History.* Westport, Conn.: Greenwood Press, 1993.

Bergerson, Frederic. *The Army Gets an Air Force.* Baltimore: Johns Hopkins University Press, 1980.

Bergerud, Eric. *Fire in the Sky.* Boulder, Colo.: Westview Press, 2000.

Bonds, Ray, and others, eds. *The Great Book of Modern Warplanes.* New York: Salamander Books, 1987.

Bowden, Mark. *Black Hawk Down.* New York: Atlantic Monthly Press, 1999.

Boyne, Walter. *Beyond the Wild Blue.* New York: St. Martin's Press, 1997.

———. *Clash of Wings.* New York: Simon & Schuster, 1994.

Bradin, James. *From Hot Air to Hellfire.* Novato, Calif.: Presidio Press, 1994.

Bradley, Omar, Gen., USA. *A Soldier's Story.* New York: Henry Holt, 1951.

Bright, Charles D. *The Jet Makers.* Lawrence: Regents Press of Kansas, 1978.

Broughton, Jack, Col., USAF (ret.). *Going Downtown.* New York: Orion Books, 1988.

Brown, Harold. *Thinking about National Security.* Boulder, Colo.: Westview Press, 1983.

Brown, Michael. *Flying Blind.* Ithaca, N.Y.: Cornell University Press, 1992.

Brulle, Robert. *Angels Zero.* Washington, D.C.: Smithsonian Institution Press, 2000.

Buckley, John. *Air Power in the Age of Total War.* Bloomington: Indiana University Press, 1999.

Bugos, Glenn. *Engineering the F-4 Phantom II.* Annapolis, Md.: Naval Institute Press, 1996.

Builder, Carl. *The Masks of War.* Baltimore: Johns Hopkins University Press, 1989.

Burton, James, Col., USAF. *The Pentagon Wars.* Annapolis, Md.: Naval Institute Press, 1993.

Caidin, Martin. *Black Thursday.* New York: Ballantine Books, 1960.

Cassidy, John. *Dot.Con.* New York: HarperCollins, 2002.

Clancy, Tom, with Gen. Chuck Horner, USAF (ret.). *Every Man a Tiger.* New York: G. P. Putnam's Sons, 1999.

Clark, Asa, and others, eds. *The Defense Reform Debate.* Baltimore: Johns Hopkins University Press, 1984.

Clodfelter, Mark. *The Limits of Air Power.* New York: Free Press, 1989.

Coates, James, and Michael Killian. *Heavy Losses.* New York: Viking Press, 1985.

Cockburn, Andrew. *The Threat.* New York: Random House, 1980.

Colgan, Bill. *World War II Fighter-Bomber Pilot.* Manhattan, Kans.: Sunflower University Press, 1985.

Copp, DeWitt. *A Few Great Captains.* McLean, Va.: EPM Publications, 1980.

———. *Forged in Fire.* McLean, Va.: EPM Productions, 1980.

Cordesman, Anthony. *The Lessions and Non-Lessons of the Air and Missile Campaign in Kosovo.* Westport, Conn.: Praeger, 2001.

Cordesman, Anthony, and Abraham Wagner. *The Lessons of Modern War.* Vols. 1 and 3. London: Mansell Publishing, 1990.

Corum, James. *The Luftwaffe.* Lawrence: University Press of Kansas, 1997.

Corum, James, and Richard Muller. *The Luftwaffe's Way of War.* Baltimore: Nautical and Aviation, 1998.

Coulam, Robert. *Illusions of Choice.* Princeton: Princeton University Press, 1977.

Cowan, Ruth Schwartz. *More Work for Mother.* New York: Basic Books, 1983.

Crabtree, James. *On Air Defense.* Westport, Conn.: Praeger, 1994.

Crane, Conrad. *American Airpower Strategy in Korea.* Lawrence: University Press of Kansas, 2000.

Cross, Gary, and Rick Szostak. *Technology and American Society.* Englewood Cliffs, N.J.: Prentice Hall, 1995.

Davis, Larry. *MiG Alley.* Carrollton, Tex.: Squadron/Signal Publications, 1978.

Dörfer, Ingemar. *Arms Deal.* New York: Praeger Publishers, 1983.

Dorr, Robert. *Skyraider.* New York: Bantam Books, 1988.

Drendel, Lou. *A-10 Warthog in Action.* Carrollton, Tex.: Squadron/Signal Publications, 1981.

Dupuy, R. Ernest, and Trevor N. Dupuy, Col., USA (ret.). *The Encyclopedia of Military History, from 3500 B.C. to the Present.* New York: Harper & Row, 1986.

Durden, Rodney, and others. *Falklands, the Air War.* Poole, Dorset, U.K.: Arms and Armour Press, 1986.

Ethell, Jeffrey, and Alfred Price. *Air War, South Atlantic.* London: Sidgwick & Jackson, 1983.

Etzold, Thomas. *Defense or Delusion?* New York: Harper & Row, 1982.

Everett-Heath, John. *Helicopters in Combat.* Poole, Dorset, U.K.: Arms and Armour Press, 1992.

Fallows, James. *National Defense.* New York: Random House, 1981.

Fredericksen, Donald, and others. *Selected Papers.* McLean, Va.: Hicks & Associates, June 1996.

Gabriel, Richard. *Military Incompetence.* New York: Hill and Wang, 1985.

Galvin, John R. *Air Assault.* New York: Hawthorne Books, 1969.

Gooderson, Ian. *Air Power at the Battlefront.* London: Frank Cass, 1998.

Gordon, Michael, and Bernard Trainor, Lt. Gen. USMC (ret.). *The Generals' War.* Boston: Little Brown, 1995.

Green, William. *Famous Bombers of the Second World War, Volume One.* New York: Doubleday, 1959.

Gunston, Bill. *AH-64 Apache.* London: Osprey Publishing, 1986.

———. *Attack Aircraft of the West.* New York: Charles Scribner's Sons, 1974.

———. *Bombers of the West.* New York: Charles Scribner's Sons, 1973.

———. *Early Supersonic Fighters of the West.* New York: Charles Scribner's Sons, 1976.

Hackett, Sir John, Gen., British Army. *The Third World War: The Untold Story.* Toronto: Bantam Books, 1982.

Hackett, Sir John, Gen., British Army, and others. *The Third World War: A Future History.* New York: Macmillan, 1978.

Hackworth, David, Col., USA (ret.), and Tom Matthews. *Hazardous Duty.* New York: William Morrow, 1996.

Hadley, Arthur T. *The Straw Giant.* New York: Random House, 1986.

Halberstam, David. *War in a Time of Peace.* New York: Charles Scribner's Sons, 2001.

Hallion, Richard. *The Naval Air War in Korea.* Baltimore: Nautical and Aviation, 1986.

———. *Storm over Iraq.* Washington, D.C.: Smithsonian Institution Press, 1992.

———. *Strike from the Sky.* Washington, D.C.: Smithsonian Institution Press, 1989.

Hallion, Richard, ed. *Air Power Confronts an Unstable World.* London: Brassey's, 1997.

Harding, Stephen. *Air War Grenada.* Missoula, Mont.: Pictorial Histories Publishing, 1984.

Heitman, Helmoed-Römer. *War in Angola.* Ashanti Publishing, 1990.

Herzog, Chaim. *The War of Atonement.* London: Greenhill Books, 1998.

Higham, Robin. *Air Power.* New York: St. Martin's Press, 1972.

Hoeppner, Ernest Wilhelm von. *Germany's War in the Air.* Translated by J. Hawley Larned. Nashville, Tenn.: Battery Press, 1994.

Hogg, Ian. *Tank Killing.* New York: Sarpedon, 1996.

Huenecke, Klaus. *Modern Combat Aircraft Design.* Annapolis, Md.: Naval Institute Press, 1994.

Hughes, Thomas Alexander. *Over Lord.* New York: Free Press, 1995.

Isenberg, Michael T. *Shield of the Republic.* New York: St. Martin's Press, 1993.

Johnson, Dana, and James Winnefield. *Joint Air Operations.* Annapolis, Md.: Naval Institute Press, 1993.

Kelly, Orr. *King of the Killing Zone.* New York: W. W. Norton, 1989.

Kotz, Nick. *Wild Blue Yonder.* New York: Pantheon Books, 1988.

Krepinevich, Andrew. *The Army and Vietnam.* Baltimore: Johns Hopkins University Press, 1986.

Kruzel, Joseph, ed. *American Defense Annual, 1987–1988.* Lexington, Mass.: D. C. Heath, 1987.

———. *American Defense Annual, 1989–1990.* Lexington, Mass.: Lexington Books, 1989.

Lambeth, Benjamin. *The Transformation of American Air Power.* Ithaca, N.Y.: Cornell University Press, 2000.

Lavell, Kit. *Flying Black Ponies.* Annapolis, Md.: Naval Institute Press, 2000.

Lee, Arthur. *No Parachute.* London: Arrow Books, 1969.

Macksey, Kenneth. *Technology in War.* New York: Prentice-Hall, 1986.

Marcus, Alan, and Howard Segal. *Technology in America.* Fort Worth, Tex.: Harcourt Brace Jovanovich, 1989.

Mason, R. A. ("Tony"), AVM, RAF. *Air Power: A Centennial Appraisal.* London: Brassey's, 1994.

Mason, R. A. ("Tony"), AVM, RAF, ed. *War in the Third Dimension.* London: Brassey's, 1986.

Mason, R. A. ("Tony"), AVM, RAF, and John Taylor. *Aircraft, Strategy, and Operations of the Soviet Air Force.* London: Jane's, 1986.

McDaid, Hugh, and David Oliver. *Smart Weapons.* New York: Welcome Rain, 1997.

McMaster, H. R. *Dereliction of Duty.* New York: Harper Collins, 1997.

McNamara, Robert. *In Retrospect.* New York: Random House, 1995.

Meilinger, Phillip. *Hoyt Vandenberg.* Bloomington: Indiana University Press, 1989.

Millett, Allan, and Peter Maslowski. *For the Common Defense.* 2d ed. New York: Free Press, 1994.

Moore, Harold, Lt. Gen., USA (ret.), and Joseph Galloway. *We Were Soldiers Once . . . and Young.* New York: Harper Perennial, 1992.

Murphy, Paul, ed. *The Soviet Air Forces.* Jefferson, N.C.: McFarland, 1984.

Murray, Williamson, and Allan Millett, eds. *Military Innovation in the Interwar Period.* Cambridge, U.K.: Cambridge University Press, 1996.

Neubeck, Ken. *A-10 Warthog.* Mini Number 4. Carrollton, Tex.: Squadron/Signal Publications, 1995.

Nolan, Keith. *Into Laos.* Novato, Calif.: Presidio Press, 1986.

Nordeen, Lon O. *Air Warfare in the Missile Age.* 2d ed. Washington, D.C.: Smithsonian Institution Press, 2002.

Prados, John. *The Soviet Estimate.* New York: Dial Press, 1982.

Price, Alfred. *Air Battle Central Europe.* New York: Free Press, 1986.

Pursell, Carroll. *The Machine in America.* Baltimore: Johns Hopkins University Press, 1995.

Puryear, Edgar F., Jr. *Destined for Stars.* San Francisco: Presidio Press, 1983.

Rasor, Dina, ed. *More Bucks, Less Bang*. Washington, D.C.: Fund for Constitutional Government, April 1983.

Rhodes, Richard. *Dark Sun*. New York: Simon & Schuster, 1995.

Rich, Ben, and Leo Janos. *Skunk Works*. Boston: Little, Brown, 1994.

Richardson, Doug. *Modern Fighting Aircraft: F-16 Fighting Falcon*. New York: Arco, 1983.

Ridgway, Matthew, Gen., USA (ret.). *Soldier*. New York: Harper Brothers, 1956.

Rodman, Peter. *More Precious than Peace*. New York: Charles Scribner's Sons, 1994.

Rosen, Stephen Peter. *Winning the Next War*. Ithaca, N.Y.: Cornell University Press, 1991.

Rudel, Hans Ulrich. *Stuka Pilot*. New York: Ballantine Books, 1958.

Schwarzkopf, H. Norman, Gen., USA (ret.), and Peter Petre. *It Doesn't Take a Hero*. New York: Bantam Books, 1992.

Shapley, Deborah. *Promise and Power*. Boston: Little, Brown, 1993.

Sheehan, Neil. *A Bright Shining Lie*. New York: Vintage Books, 1988.

Sherrod, Robert. *History of Marine Corps Aviation in World War II*. Washington, D.C.: Combat Forces Press, 1952.

Simpkin, Richard. *Antitank*. Oxford: Brassey's, 1982.

Slessor, John. *Air Power and Armies*. Oxford: Oxford University Press, 1936.

Smallwood, William. *Warthog*. Washington, D.C.: Brassey's (U.S.), 1993.

Smith, Perry, Maj. Gen., USAF (ret.). *The Air Force Plans for Peace*. Baltimore: Johns Hopkins Press, 1970.

Smith, Peter. *Close Air Support*. New York: Orion Books, 1990.

Stephens, Alan, ed. *The War in the Air, 1914–1994*. Fairbairn, Australia: Air Power Studies Center, 1994.

Stevenson, James. *The Pentagon Paradox*. Annapolis, Md.: Naval Institute Press, 1993.

Stevenson, Jonathan. *Losing Mogadishu*. Annapolis, Md.: Naval Institute Press, 1995.

Straubel, James. *Crusade for Airpower*. Washington, D.C.: Aerospace Education Foundation, 1982.

Sumida, John Tetsuro. *In Defense of Naval Supremacy*. London: Unwin Hyman, 1989.

Sweetman, Bill. *Aviation Fact File, Modern Fighting Aircraft: A-10*. London: Salamander Books, 1984.

Sweetman, Bill, and others. *The Great Book of Modern Warplanes*. New York: Portland House, 1987.

Tenner, Edward. *Why Things Bite Back*. New York: Alfred A. Knopf, 1996.

Tillman, Barrett. *The Dauntless Dive Bomber of World War Two*. Annapolis, Md.: Naval Institute Press, 1976.

Travers, Tim, and Christon Archer, eds. *Men at War*. Chicago: Precedent, 1982.

Warden, John, Col., USAF. *The Air Campaign*. London: Pergamon-Brassey's, 1989.

Weigley, Russell. *The American Way of War*. Bloomington: Indiana University Press, 1973.

———. *Eisenhower's Lieutenants.* Bloomington: University of Indiana Press, 1981.
Whiting, Kenneth. *Soviet Air Power.* Boulder, Colo.: Westview Press, 1986.
Woodward, Bob. *The Commanders.* New York: Simon & Schuster, 1991.

ARTICLES IN PERIODICALS AND JOURNALS

"AF, Army Agree on Role of Copters." *AFT,* 17 September 1975, 4.
"AF Concerned over AX Opposition." *AFJ* (11 October 1969): 14–15.
"AFTI F-16 Integrates Airborne Targeting Data during CAS Mission." *AW&ST* (7 November 1988): 51.
"AFTI F-16 Testbed to Evaluate Sensors for Close Air Support." *AW&ST* (2 November 1987): 76.
"Air Force Pilot Tests A-10's Toughness in Battle." *AW&ST* (5 August 1991): 43.
"Air Force Studies Fighter Pilot Workload." *Electronic Warfare/Defense Electronics* (April 1978): 55–58.
Alberts, Donald, Maj., USAF. "Tactical Air Power in NATO." *AUR* (March–April 1980): 67.
Alberts, Donald, Maj., USAF, Maj. Price Bingham, USAF, and Lt. Col. Owen Wormser, USAF. "Fire-Counterfire." *AUR* (March–April 1980): 59–73.
Alder, Konrad. "Airborne Anti-Tank Warfare, Part I." *Armada International* (July–August 1980): 42–44.
Amouyal, Barbara. "AF to Cease F-16 Buys as Wings Are Pared." *DN,* 10 December 1990, 1, 51.
———. "DAB to Settle Close Air Support Debate before Competitive Flyoff." *DN,* 9 October 1989, 26.
———. "Gen. Galvin Favors F-16 for CAS Role." *DN,* 10 April 1989, 21.
———. "Pentagon Expected to Assign Role of CAS to F-16." *DN,* 18 June 1990, 45.
———. "Pentagon Panel Rejects F/A-16 for Air Force CAS Mission." *DN,* 3 December 1990, 3, 44.
———. "Powell Taps Air Force for Close Air Support." *DN,* 13 November 1989, 1, 43.
———. "Prospects of A-7F in Air Support Role Appear Dim." *DN,* 8 January 1990, 29.
———. "Tactical Wings Fight for Parity in Face of Historic Reductions." *DN,* 12 November 1990, 6, 22.
Anderson, Casey. "Close Air Support: A-10 or F-16?" *AFT,* 15 October 1990, 28.
Anderson, Jack. "Secrets Leak Out of Foggy Bottom." *Washington Post,* 23 August 1980, E3.
Anderson, Jack, and Dale Van Atta. "Air Force Plan Encounters Army Flak." *Washington Post,* 28 December 1990, E5.
———. "The Hero That Almost Missed the War." *Washington Post,* 5 March 1991, C9.
Andrews, Walter. "The 'Known Unknowns' and the 'Unknown Unknowns.'" *AFJ*

(14 December 1968): 17.

———. "Rugged Tests and a Rugged Machine." *AFJ* (14 December 1968): 17.

———. "There Are No Red Flags." *AFJ* (14 December 1968): 20–23.

———. "A Weapons Ship for All Environments." *AFJ* (14 December 1968): 12–13.

Armour, Richard, Maj., USA. "Soviet Aviation Increases Offensive Threat." *Air Defense Magazine* (October–December 1982): 15–18.

"Army's Analysis of LHX Helicopters Will Omit Close Air Support Issue." *AW&ST* 126 (16 February 1987): 18.

"A-7D, A-10 'Fly-Off' Suggested." *AFT*, 15 August 1973, 29.

Asprey, Robert, Capt., USA. "Close Air Support." *Army* (November 1961): 36–37.

Astor, Gerald. "The World's Toughest Air Force: It Keeps Israel Alive." *Look*, 30 June 1970, 20–21.

"A-10: A NATO Preview." *NATO's Fifteen Nations* (April–May 1977): 90–91.

"A-10 Assembly Line Geared for 1982 Peak." *AW&ST* (6 March 1977): 15.

"A-10 at Exercise Reforger." *Flight International* (22 October 1977): 1193–94.

"A-10 Deployment." *AW&ST* (19 September 1977): 53.

"A-10 Program Unaffected by Crash at Paris Airshow." *AW&ST* (13 June 1977): 36.

"A-10 Reliability Expands Crew Training." *AW&ST* (6 February 1978): 229.

"A-10 Testing to Continue in Europe." *AFT*, 20 June 1977, 43.

"A-10A Capabilities Set Basis for Flyoff." *AW&ST* (7 January 1974): 53.

"A-10/AAH Together Pack More Punch Than Separately." *AFJI* (June 1977): 18.

"A-10/A-7 Flyoff Rules Drafted, Await Defense Chief's Approval." *AW&ST* (15 October 1973): 23.

"A-10/Helicopter Tactics Prove Effective." *AW&ST* (6 February 1978): 217–18, 229.

"A-10s Set for European Deployment in January." *AFJI* (December 1978): 15.

"A-10s Take on the Night." *AFT*, 5 February 1996, 25.

Aubin, Stephen. "*Newsweek* and the 14 Tanks." *AFM* (July 2000): 59–61.

Austen, Bruce, and others. "Lessons from Desert Shield." *US News & World Report*, 10 September 1990, 36.

"AX Fighter Paved Way for Prototyping." *AW&ST* (26 June 1972): 103–4.

"A-X Takes Flak Head On." *AFJI* (September 1973): 22–23.

Bacon, Lance. "Saving the Day." *AFT*, 1 April 2002, 8–9.

Bahnsen, John, Brig. Gen., USA (ret.). "A New Army Air Corps or a Full Team Member?" *AFJI* (October 1986): 62–80.

Baker, Caleb. "CAS Test Plan Dispute Settled on Surface." *DN*, 12 June 1989, 10.

Baker, Peter. "Afghans Strengthen U.S. Force." *Washington Post*, 8 March 2002, A01.

Barger, Millard. "What USAF Has to Do to Put the 'Air' Back in AirLand Battle." *AFJ* (June 1986): 58–64.

Barrett, Mark, Capt., USAF. "Close Air Support." *Infantry* (May–June 1988): 32–34.

"Battle Brews over Follow-On Close Air Support Aircraft." *AW&ST* (2 February 1987): 19.

Bavaro, Edward. "Tank Busters." *USAAD* (June 1985): 2–6.

Beach, Joe, Maj., USA. "JAWS." *USAAD* (September 1978): 44–47.

Beavers, Jim, Lt. Col., USAF (ret.). "The A-10—Monstrosity with a Mission." *AFM* (February 1980): 66–70.

Berry, F. Clifton, Jr. "The 81st Tactical Fighter Wing." *AFJ* (June 1979): 17–18.

———. "JAWS Chew Armor in Different Ways." *AFJI* (January 1978): 32–34.

———. "TAC and TRADOC Talk a Lot, But Do They Communicate?" *AFJI* (September 1979): 27–36.

"Better USAF/Army Coordination Sought." *AW&ST* (6 February 1978): 197.

Beyers, Dan. "Russ' Support of F-16 Revamp over New Ground Support Plane Shocks AFA." *DN*, 2 February 1987, 4.

Bingham, Price. "Air Interdiction and the Need for Doctrinal Change." *Strategic Review* (Fall 1992): 24–33.

———. "Dedicated, Fixed-Wing Close Air Support—A Bad Idea." *AFJI* (September 1987): 58–62.

———. "Let the Air Force Fight Future Land Battles." *AFJI* (August 1993): 42.

———. "Revolutionizing Warfare through Air Interdiction." *AJ* (Spring 1996): 29–35.

Blackwell, James. "American Close Air Support—The Next Model." *NATO'S 16 Nations* (December 1988–January 1989): 59.

Blohm, Michael, Lt. Col., USAF. "AGM-65D Algorithms for Aircrews: An Aircrew Oriented Look." *FWR* (Spring 1990): 10–14.

Bloom, Allan. "Attack Aircraft." *Journal of Defense and Diplomacy* (April 1987): 12–16.

Boatman, John. "ACC: Stand-Off Fights Will Mean Less CAS." *IDR* (November 1992): 1081.

———. "USA Redefines CAS Doctrine for a New Era of Warfare." *Jane's Defense Weekly*, 12 December 1992, 17.

———. "USAF and the DAB Debate the Price of Close Air Support." *Jane's Defense Weekly*, 6 February 1993, 5.

Bond, David. "Budget Cuts, Cost Growth Cloud F-16 Plans for 1990s." *AW&ST* (4 December 1989): 18–19.

———. "Congress Eases F-16 Production Curbs." *AW&ST* (20 November 1989): 32.

———. "Defense Dept. Plans CAS Review." *AW&ST* (11 September 1989): 31–33.

Bond, David, and John Morrocco. "Close Air Support Move in Congress Might Disrupt Production of F-16." *AW&ST* (13 November 1989): 22–24.

Booda, Larry. "USAF, Army Air Roles Evolving Slowly." *AW&ST* (27 May 1963): 30–31.

Boyne, Walter. "The Evolution of Jet Fighters." *AUR* (January–February 1984): 35–45.

Braddon, J. R., Lt. Col., USMC. "What Every Marine Should Know about Harrier." *Marine Corps Gazette* (May 1971): 20–24.

Brindley, John. "Fairchild's A-10 Is Entering Service." *Interavia* (November 1976): 1089–90.

Brown, David. "A-10 Crews Sharpen Air Support Skills." *AW&ST* (2 April 1979): 37–39.

———. "A-10 Pilots Stress Navaid Reqirement." *AW&ST* (19 September 1977): 52–53.

———. "LTV Begins Flight Tests of Reengined A-7F for Air Guard Mission." *AW&ST* (4 December 1989): 19–21.

Brown, Fredric, Maj. Gen., USA. "Attack Helicopter Operations on the Heavy Battlefield." *USAAD* (July 1985): 2–11.

Brown, Merrill. "Bill Expands U.S. Purchase of Fairchild's A10 Aircraft." *Washington Post*, 8 December 1980, WB3A.

Brown, Robert, Capt., USAF. "A Real Hog War." *AJ* (Winter 1990): 54–66.

Brownlow, Cecil. "Burgeoning U.S. Use of Air Power Aimed at Forestalling Ground War with Chinese." *AW&ST* (26 April 1965): 26–31.

———. "Senate Unit Presses A-7/A-10 Flyoff." *AW&ST* (20 August 1973): 18–19.

Bunyard, Jerry Max, Brig. Gen., USA. "Tomorrow's War Today." *AFM* (October 1980): 64–67.

Canan, James. "More Flak in the AirLand Battle." *AFM* (February 1988): 76–80.

———. "Sorting Out the AirLand Partnership." *AFM* (April 1988): 50–59.

Canby, Steven. "Tactical Air Power in NATO Armed Warfare." *AUR* (May–June 1979): 2–20.

Carlson, Bruce, Lt. Col., USAF. "Close Air Support." *MR* (June 1989): 51–59.

"CBO Views Europe Defenses." *AFT*, 31 January 1977, 12.

"Center Seeks New Combat Techniques." *AW&ST* (6 February 1978): 81–82.

Chapman, Robert, Maj., USAF. "Technology, Air Power, and the Modern Theater Battlefield." *AJ* (Summer 1988): 42–51.

Charlton, John, Capt., USAF. "Heating the Hog." *FWR* (Summer 1984): 16–18.

Clay, Lucius, Jr., Lt. Gen., USAF. "Shaping the Air Force Contribution to National Strategy." *AUR* 21 (March–April 1970): 3–9.

"Close Air Charade." *AW&ST* 128 (28 March 1988): 9.

"Close Air Support: A-10 or A-16?" *AFT*, 15 October 1990, 28.

"Close Air Support: Cheyenne vs. Harrier vs. A-X?" *AFJ* (19 October 1970): 19.

Cockerham, Samuel, Brig. Gen., USA (ret.). "The Fighting Hind." *Defense & Foreign Affairs* (July/August 1987): 27–30.

"Competitive Prototype Program Planned for USAF AX Aircraft." *AW&ST* (4 May 1972): 25–26.

Coniglio, Sergio. "Close Air Support." *Military Technology* (October 1993): 30–38.

Coram, Robert. "The Case against the Air Force." *Washington Monthly*, July–August 1987, 17–24.

Corley, Stephen, Capt., USMC. "Letters." *AW&ST* (20 April 1987): 84.

Correll, John. "Battle Damage from the Budget Wars." *AFM* (April 1988): 38–44.

"Cost Cutter." *Time*, 7 March 1983, 29.

Covault, Craig. "Alaskan A-10s Assume Growing Role in European, South Korean Defense." *AW&ST* (18 July 1988): 47–51.

Coyne, James. "Frontal Aviation's One-Two Punch." *AFM* (March 1985): 74–75.

"Darts and Laurels: General Robert D. Russ." *AFJI* (August 1988): 95.

Daskal, Steven. "Adapt Tactical Air to AirLand Battle." *Journal of Defense and Diplomacy* (April 1987): 17–20.

Dean, Stewart. "Correspondence." *AW&ST* (12 July 1999): 6.

"Decision on A-10 Production May Sway Action in Congress." *AW&ST* (1 July 1974): 23.

"Defense Dept. Asks USAF to Broaden Design Options for New CAS Aircraft." *AW&ST* (23 November 1987): 28–29.

Deitchman, Seymour J. "The Implications of Modern Technological Developments for Tactical Air Tactics and Doctrine." *AUR* (November–December 1977): 36–43.

"Desert Storm Performance Sparks Program to Extend Active Service of A-10 'Warthog.'" *AW&ST* (22 July 1991): 44.

Dikkers, Gary, Lt. Col., USAF. "Overcoming Poor Attitude Is Key to Effective CAS." *AFT*, 3 December 1990, 23, 61.

Dilger, Robert, Col., USAF. "Tank Killer Team." *National Defense* (November–December 1976): 190–91.

Dougherty, Tim, Tech. Sgt., USAF. "Bombers: A Formidable Weapon in Fight against Terrorism." *Air Force Link* (7 March 2002). Database on-line, cited 9 March 2002. http://www.af.mil/news/n20020307_0370.shtml.

Drew, Dennis, Col., USAF. "Two Decades in the Air Power Wilderness." *AUR* (September–October 1986): 2–13.

Dryden, Joe Bill. "Design Philosophy Comparisons." *Code One* (January 1989): 22–23.

Dudney, Robert. "Battle of the F-22." *AFM* (September 1999): 16–17.

———. "Washington Watch: The Services Take Their Cuts." *AFM* 71 (April 1988): 16–20.

Dupuy, Trevor, Col., USA (ret.). "Why Deep Strike Won't Work." *AFJI* (January 1983): 56–57.

"Each Service Has Its Own Views on Close Air Support." *Interavia* (September 1971): 1034–35.

"Eastern Bloc Augments Attack Force." *AW&ST* (6 February 1978): 57–61.

Eggers, Jeff, Capt., USAF. "Freefall Weapons Delivery with LASTE." *FWR* (Winter 1991): 8–11.

Elko, Edward, and T. R. Stuelpnagel. "The 30-mm GAU-8A Ammunition Story." *National Defense* (September 1981): 51.

Emme, Eugene, Ph.D. "Some Fallacies Concerning Air Power." *Air Power Historian* (July 1957): 137.

Engle, Kenneth, Maj., USAF, and Orin Patton, Col., USAF. "Basing the A-10 in Europe." *AUR* (March–April 1980): 82–85.

Ericson, Edward. "Recycling the Army Way." *E/The Environmental Magazine* (March/April 1997). http://www.emagazine.com/march-april_1997/0397curr_army.html.

Ethell, Jeff. "Close Air Support." *AFJI* (September 1988): 97–100.

Evans, Michael. "History of Arms Is the Difference." *Naval Institute Proceedings* (May 1998): 75.

"The Fairchild Can-Opener." *Air International* (June 1979): 267–82.

"Fairchild Tightens Procedures Following Air Force Review." *AW&ST* (9 September 1985): 18–20.

Faletti, Matthew, Lt. Cdr., USN. "Close Air Support Must Be Joint." *Naval Institute Proceedings* (September 1994): 56–57.

Famiglietti, Len. "A-10, Thunderbolt II, Harks Back to P-47." *AFT*, 17 April 1978, 10.

Fedorchak, Scott, Capt., USA. "Close Air Support" *AJ* (Spring 1994): 22–33.

Ferrell, Mark, Capt., USAF, and Capt. Scott Reynolds, USAF. "Apache Thunder." *USAAD* (May 1989): 2–10.

Fialka, John. "A-10 'Warthog,' A Gulf War Hero, Would Fly to Scrap Heap If Air Force Brass Has Its Way." *Wall Street Journal*, 29 March 1991, A12.

Fink, Donald. "A-10 Survivability in Attack Role Shown during Simulated Combat." *AW&ST* (20 June 1977): 88–93.

———. "Joint Combat Exercise Tests Tactics." *AW&ST* (23 May 1977): 24–25.

"First A-10 Squadron Operational Ahead of Schedule despite Transition Difficulties." *AW&ST* (6 February 1978): 208–12.

"Five Lessons of the War." *Newsweek*, 5 November 1973, 54.

"Fly Two before Deciding." *Government Executive* (November 1970): 30.

"Flyoff Decision Favoring A-10 Expected." *AW&ST* (17 June 1974): 16.

Foglesong, Robert H., Col., USAF. "Defense Forum." *AFJI* (November 1988): 7.

Foglesong, Robert, Col., USAF, and Lt. Col. W. Edward Shirron, USA. "Close Air Support." *Defense* 89 (July–August 1989): 28–35.

Ford, Barry, Lt. Col., USMC. "The Future Is Attack Helicopters." *Naval Institute Proceedings* (September 1994): 54–55.

Foster, Harold Edward. "Defense Forum." *AFJI* (October 1988): 10.

Fukuyama, Francis. "The End of History." *National Interest* (Summer 1989): 3.

Fulghum, David. "Air Force Calls CAS Test Plan Redundant, Expensive." *AW&ST* (18 June 1990): 21–22.

———. "Air Force Is Equipping F-16 Squadron with a 30mm Antitank Weapon." *DN*, 19 June 1989, 24.

———. "A-7F Makes Comeback as Temporary Solution." Defense Trends Section. *AFT*, 25 September 1989, 4.

———. "A-10s Find Targets Harder to Locate after Initial Easy Successes in Iraq." *AW&ST* (11 February 1991): 23.

———. "Congress Holds AF's Favorite Fighter Hostage." *AFT*, 27 November 1989, 25.

———. "Defense Trends, Air Force Times Special Section: Close Air Support." *AFT,* 25 September 1989, 1–15.

———. "Doubts Cast on AF Role." *AFT,* 25 September 1989, 2.

———. "Isolated, Serb Army Faces Aerial Barrage." *AW&ST* (19 April 1999): 26–28.

———. "Night-Fighting CAS Force Gains Preliminary Approval." *AW&ST* (8 February 1993): 52.

———. "Pentagon Dissecting Combat Data." *AW&ST* (26 July 1999): 68–69.

———. "Senate Panel Backs Marine Corps' Postwar Upgrade Plan for AV-8B Harrier." *AW&ST* (5 August 1991): 44–45.

———. "Signature Reduction Key to A-10 Survival." *AW&ST* (7 June 1993): 135–36.

———. "Study Touts A-10 Wing for Norway's Defense." *Air Force Times Defense Trends,* 13 June 1988, 31.

Fulghum, David, and John Morrocco. "Air War Pays Off." *AW&ST* (14 June 1999): 63–64, 73.

Fulghum, David, and Robert Wall. "Joint Strike Fighter Price Creep Triggers Pentagon Review." *AW&ST* (16 March 1998): 27.

Gamboa, Suzanne. "Lockheed Martin to Build Fighter Jet." In Yahoo! News. Database on-line. Cited 26 October 2001. http://www.dailynews.yahoo.com/h/ap/20011026/pl/future_fighter_jet.html

Garrett, Thomas, Lt. Col., USA. "Close Air Support: Which Way Do We Go?" *Parameters* (December 1990): 42.

Gavin, James, Gen., USA. "Cavalry, and I Don't Mean Horses." *Harper's* (April 1954): 54–60.

Geddes, J. Phillip. "AH-64." *IDR* (August 1978): 1249.

———. "The A-10—USAF Choice for the Close Air Support Role." *IDR* (January 1974): 76.

Gilson, Charles. "Can the A-10 Thunderbolt II Survive in Europe?" *IDR* (February 1979): 186.

Ginovsky, John. "GAO Report Urges Air Force to Delay A-10 Modification, Test Information System." *DN,* 4 June 1990, 54.

Gonseth, Jules, Col., USA. "Tactical Air Support for Army Forces." *MR* (July 1955): 3–16.

Goodman, Glenn. "Airborne Goliath." *AFJI* (March 2002). Database on-line. Cited 13 March 2002. http://www.afji.com/AFJI/Mags/2002/March/specops.html.

———. "Army Aviation Community Ponders Its Future Fleet Structure." *AFJI* (October 1991): 16.

———. "Close Air Support." *AFJI* (January 2002). Database on-line. Cited 13 March 2002. http://www.afji.com/AFJI/Mags/2002/Jan/CloseAir.html.

———. "Juggling Air Dominance." *AFJI* (September 1999): 50–60.

Gorton, William, Maj., USAF. "Of Mudfighters and Elephants." *AFM* (October 1988): 102–7.

Greeley, Brendan, Jr. "Let the Army Fly Its Own Close Air Support." *AW&ST* (9 February 1987): 11.

———. "Speed, Stealth, Maneuverability Sought for Survivable Close Air Support." *AW&ST* (27 April 1987): 143–45.

———. "TAC to Modify A-7s, F-16s for Close Air Support." *AW&ST* (29 August 1988): 40–41.

———. "USAF, Army Grapple with Key Issues of Close Air Support." *AW&ST* (23 March 1987): 50–55.

———. "USAF Reviewing Contractor Proposals for Attack Aircraft." *AW&ST* (22 July 1985): 16–18.

Greenhous, Brereton. "Close Air Support Aircraft in World War I: The Counter Anti-Tank Role." *Aerospace Historian* (Summer 1974): 90–91.

Grossman, Elaine. "USAF May Expand A-10 Mod Effort at Same Cost—and Get It Sooner." *Inside the Pentagon*, 9 November 2000. Inside Washington Publishers database on-line. Cited 9 November 2000. http://insidedefense.com.

"Grumman: Moving beyond the Wild Blue Yonder." *Business Week* (1 February 1988): 54–55.

"Grumman, Rockwell Team to Bid on A-10 Upgrade." *DN*, 8 May 1989, 19.

"Grumman Team Evaluates Buy of Fairchild Republic Co." *AW&ST* (25 November 1985): 17.

Gulick, John, Maj., USAF. "The A-10 Does It Better." *AFM* (July 1976): 75–79.

Hackworth, David. "Air Power Just Won't Work." *Newsweek*, 17 May 1993, 32.

Hain, Raymond, III, Lt. Col., USAFR. "Letters, CAS/BAI Aircraft Selection." *APJ* (Fall 1990): 7.

Halbert, John, Hugh Lucas, and Ron Sherman. "The Warthog." *NATO's Fifteen Nations* 25 (June–July 1980): 72.

Hall, Mike. "Viewpoint." *AW&ST* (17 August 1998): 94.

Hallion, Richard. "Battlefield Air Support." *AJ* (Spring 1990): 8–28.

———. "A Troubling Past." *AJ* (Winter 1990): 4–23.

Hansen, Woods. "A-10 Prototype Designed for Production." *AW&ST* (26 June 1972): 14–20.

Harris, Lorenzo, MSgt, USAF. "Those Non-Stop Knuckle-Busters." *Airman* (April 1985): 40–41.

———. "Wart Hogs!" *Airman* (April 1985): 36–39.

Harrison, Benjamin, Maj. Gen., USA (ret.). "The A-10: A Gift the Army Can't Afford." *Army* 41 (July 1991): 36–39.

Harrison, Richard, Lt. Cdr., USN. "JSF and UCAV Aren't the Answer for the Navy." *Naval Institute Proceedings* (December 2002): 59–61.

Hartman, Richard. "Airborne Anti-Tank Capabilities Scrutinized, Part I." *Defense Electronics* (November 1979): 53–54.

———. "Airborne Anti-Tank Capabilities Scrutinized, Part II." *Defense Electronics* (December 1979): 54–64.

Harvey, David. "Aircraft for Export." *Defense & Foreign Affairs* (April 1980): 45–46.

———. "Dangerous Midnight." *Defense and Foreign Affairs* (November 1980): 12–15, 43.

"HASC Restores Most Senate Weapons Cuts." *AFJ* 107 (4 October 1969): 2.

Haynes, William, Lt. Col., USAF (ret.), Maj. Charles Brown, USAF, and Ed Flynn. "Airmail" *AFM* (September 1988): 17–18.

Head, Richard, Lt. Col., USAF. "The Air Force A-7 Decision: The Politics of Close Air Support." *Aerospace Historian* (Winter 1974): 224.

Hedberg, Scott. "Role of A-10 in Army." *Army* (November 1991): 6.

Helbling, Tony, Maj., USAF. "Do It in the Spinich." *Aerospace Safety* (April 1978): 1–3.

Henderson, Breck. "A-10 'Warthogs' Damaged Heavily in Gulf War But Survived to Fly Again." *AW&ST* (5 August 1991): 42–43.

Henry, Ron, Lt. Col., Maryland ANG. "LASTE Employment in Olympic Bombing." *FWR* (Spring 1992): 20–22.

Herbert, Adam. "Smaller Bombs for Stealthy Aircraft." *AFM* 84 (July 2001). Database on-line. Cited 13 March 2002. http://www.afa.org/magazine/July 2001/0701small.html.

Hessman, James, and others. "The Army Gets a New Weapons System." *AFJ* (14 December 1968): 7.

Higgins, Alexander. "Tank-Killing 'Warthog' Proves Itself." *Huntsville Times*, January 1991, 1A.

Holmquist, C. O., Capt., USN. "Developments and Problems in Carrier-Based Attack Aircraft." In *1969 Naval Review*, ed. Frank Uhlig, 195–215. Annapolis, Md.: U.S. Naval Institute, 1969.

Hotz, Robert. "The Mideast Surprise," *AW&ST* (15 October 1973): 7.

"House Unit to Urge Enforcer Flight Test." *AW&ST* (23 September 1974): 27.

Hubbert, Bob. "JAAT." *USAAD* (October 1979): 61–64.

Hudson, Neff. "AF Said to Slight Close Air Support: Could Lose Mission." *AFT*, 9 November 1992, 11.

———. "Fewer Than 20 AF Fighter Wings Seen Soon." *AFT*, 9 November 1992, 11.

Hughes, David. "Syracuse Wing Finds F-16A Effective in Close Air Support." *AW&ST* (18 June 1990): 36–43.

———. "USAF Firing 100 Mavericks per Day in Current Air-to-Ground Missions." *AW&ST* (11 February 1991): 24–25.

Icarus [pseud.]. "Choice of New CAS Aircraft Not a Simple Matter." *AFJI* (September 1988): 114–17.

Isaacson, Walter. "The Winds of Reform." *Time* (7 March 1983): 12–30.

Isherwood, Mike, Capt., USAF. "Combat Hammer." *FWR* (Winter 1988): 5–8.

Jones, Brian, Lt. Col., USAF. "Close Air Support: A Doctrinal Disconnect." *AJ* (Winter 1992): 60–71.

Jordan, Bryant, and Andrew Compart. "'Global Mission.'" *AFT*, 1 April 1996, 22.

"The Journal Interviews: BG Sidney L. Davis." *Field Artillery Journal* (September–October 1975): 11–13.

Kaplan, Fred. "Beast of Battle: The Tale of the Warthog." *Boston Globe*, 21 July 1991, 12.

Katzman, Jim, T Sgt., USAF. "Groundwork for the Future." *Airman* (January 1986): 44–47.

Kocivar, Ben, and John Rhea. "A-10 Ready for Attack." *Flight International* (22 October 1977): 1191.

Kolcum, Edward. "USAF to Retire A-7s and F-4s, Boost F/A-16 Role in Tactical Fighter Fleet." *AW&ST* (25 February 1991): 54–55.

"Korea Eyeing A-10." *AW&ST* (17 September 1979): 48.

Kropf, Roger, Maj., USAF. "MR Letters." *MR* (October 1989): 90.

———. "The U.S. Air Force in Korea." *AJ* (Spring 1990): 39–40.

"Lack of Clear Flyoff Decision Could Postpone Funds for A-10." *AW&ST* (3 June 1974): 22.

Ladd, Steve, Lt. Col., USAF. "The Pulse, the Word, and the Difference." *FWR* (Fall 1984): 20.

Lambert, Mark. "A-10s in the German Front Line." *Interavia* (November 1981): 1146.

LaRue, Lowell, Lt. Col., USA. "TASVAL." *Air Defense* (January–March 1980): 26.

"Laughter, the Best Medicine." *Reader's Digest* (July 1967): 73–74.

Leibstone, Marvin. "Gone Are the Halcyon Days." *Military Technology* (March 1988): 43–45.

"Let the Buyer Beware!" *IDR* 18 (December 1985): 1945.

Levens, LuAnne, and Benjamin Schemmer. "Loyalty Down Is as Important as Loyalty Up." *AFJI* (February 1979): 28.

Levine, H. D. "Some Things to All Men." *Public Policy* (Winter 1977): 120–67.

Lieberherr, John, Maj., USAF. "JAAT." *USAAD* (January 1979): 42–43.

Licven, Anatol. "The World Turned Upside Down." *AFJI* (August 1998): 40–43.

Loeb, Vernon. "Technology Changes Air War Tactics." *Washington Post*, 28 November 2001, A16.

Loeb, Vernon, and Bradley Graham. "Seven U.S. Soldiers Die in Battle." *Washington Post*, 5 March 2002, A01.

Loida, Mike, Capt., USAF. "The Viper and High Threat Close Air Support." *FWR* (Fall 1990): 14.

"Look But Don't Shoot." *Command: Military History, Strategy and Analysis* (December 1997): 10–14.

Lopez, Ramon. "Night/Adverse Weather A-10 at the Crossroads." *Interavia* (August 1980): 707–9.

Ludvigsen, Eric. "Up from the Ground or Down from the Sky?" *Army* (July 1971): 49–50.

Luttwak, Edward. "The American Style of Warfare and the Military Balance." *AFM* (August 1979): 86–88.

Lynch, E. M., Brig. Gen., USA. "Close Air Support: Its Failed Form and Its Failing Function." *AFJI* (August 1986): 72–78.

Machos, James, Maj., USAF. "Air-Land Battles or AirLand Battle?" *MR* (July 1983): 33–40.

Madelin, Ian, Grp. Capt., RAF. "The Emperor's Close Air Support." *AUR* (November–December 1979): 82–86.

Maduras, Lawrence. "Correspondence: CAS Is Best Use for A-10." *AW&ST* (26 October 1998): 6.

Mann, Edward, Lt. Col., USAF. "Beyond AirLand Battle." *AFT,* 16 December 1991, 31.

———. "Operation Desert Storm? It Wasn't AirLand Battle." *AFT,* 30 September 1991, 27, 61.

Mann, Paul. "Senate Girds for Fight with USAF over Close Air Support Aircraft." *AW&ST* (22 August 1988): 21–22.

Maze, Rick. "Aspin: Budget Tactics Rooted in Need to Teach DoD a Lesson." *AFT,* 23 April 1990, 10.

McAdoo, Patrick, Capt., USAF. "Letters to the Editor: A-7D/A-10A Debate." *AW&ST* (1 October 1973): 126.

McCarthy, Geoffrey, Lt. Col., USAF. "Human Factors in F-16 Mishaps." *Flying Safety* (May 1988): 17–21.

McConnell, John, Gen., USAF. "The Quest for New Orders of Capability." *Air Force/Space Digest* (June 1968): 121–23.

McDermont, Robert, Lt. Col., USAF. "The Avenger Cannon." *FWR* (Fall 1978): 15.

———. "Engineering for Survivability." *FWR* (Fall 1978): 13–14.

McMullen, Thomas, Maj. Gen., USAF. "TACAIR Outlook." *Astronautics and Aeronautics* 16 (September 1978): 21–22.

McPeak, Merrill A. "The Kosovo Result." *AFJI* (September 1999): 62, 64.

———. "The Roles and Missions Opportunity." *AFJI* (March 1995): 32–34.

———. "TACAIR Missions and the Fire Support Coordination Line." *AUR* (September–October 1985): 65–72.

Menetrey, Louis, Gen., USA. "The U.S. Military Posture on the Korean Peninsula." *Asian Defense Journal* (August 1988): 14.

Meyer, Carlton, Maj., USMCR. "For the Guys on the Ground." *AFJI* (June 1995): 58–59.

Miller, Annetta, and others. "Peace Brings Tough Times." *Newsweek,* 6 August 1990, 44–45.

Momyer, William, Gen., USAF. "Close Air Support in the USAF." *IDR* (February 1974): 79.

Moon, Gordon, II, Col., USA. "Needed: Joint Doctrine on Close Air Support." *MR* (July 1956): 8–13.

Morrocco, John. "Air Force Drops Plan for New F-16s, Will Modify F-16C/Ds for Close Air Support." *AW&ST* (3 December 1990): 28.

———. "Air Force, Pentagon Unable to Agree on New CAS Aircraft." *AW&ST* (28 November 1988): 84–86.

———. "Pentagon Proposes Testing Plan to Resolve CAS Aircraft Dispute." *AW&ST* (20 February 1989): 97–99.

———. "Pentagon to Review Air Force Study Supporting Modified F-16 for CAS Role." *AW&ST* (31 October 1988): 30.

———. "Study Supports Call for Design of New Close Air Support Aircraft." *AW&ST* (28 September 1987): 29–30.

———. "USAF Plans to Introduce A-10s into Forward Air Control Fleet." *AW&ST* (9 February 1987): 23.

Multhopp, Hans. "The Challenge of the Performance Spectrum for Military Aircraft." *AUR* (May–June 1966): 30–41.

Myers, Charles E. ("Chuck"), Jr. "Air Support for Army Maneuver Forces." *AFJI* (March 1987): 46–47.

———. "Airmail." *AFM* (June 1988): 9. New, Noah, Col., USMC. "Perspectives of Close Air Support." Marine Corps Gazette (May 1973): 14.

"A New Pilot Tries to Pull Fairchild Out of a Nosedive." *Business Week* (16 September 1985): 100–102.

"New York Legislators Ask for Fairchild Reprieve." *AW&ST* 123 (9 September 1985): 18.

Nihart, Brooke. "Close Air Support: Sixty Years of Unresolved Problems." *AFJ* (25 April 1970): 19–24.

———. "Packard Review Group." *AFJ* (19 April 1971): 20, 39–40.

Nordeen, Lon. "The Close Air Support Debate." *Military Technology* (February 1990): 36–42.

North, David. "Congressional Impasse on T-46A Prompts NGT Recompetition." *AW&ST* (27 October 1985): 16–17.

"Northrop Streamlines A-9A Management." *AW&ST* (26 June 1972): 107–13.

Oliveri, Frank. "The Warthog Round at Gunsmoke." *AFM* (March 1992): 32–37.

O'Rourke, Gerald, Capt., USN (ret.). "Is TacAir Dead?" *Naval Institute Proceedings* (October 1976): 35–41.

Otis, Glenn, Gen., USA (ret.). "Shuffling Forces Isn't the Answer.'" *Armed Forces International* (May 1995): 36–37.

Patton, Orin, Col., USAF, and Kenneth Engle, Maj., USAF. "Basing the A-10 in Europe." *AUR* (March 1980): 80–85.

Paulsen, James. "The Air Force Wasn't Even Close." *Naval Institute Proceedings* (July 2000): 72–76.

"Pentagon Gave Short Shrift to Improving Older Aircraft to Save Pilot Lives." *AW&ST* (5 August 1991): 45.

"Pentagon Taps Odd Mix of Weapons in Afghan Campaign." *AW&ST* (5 November 2001). Database on-line. Cited 13 March 2002. http://www.aviationnow.com/avnow/news/channel_military.jsp.

Pentland, Patrick ("Doc"). "A-10 Completes FOL Validation Test." *TAC ATTACK* 18 (December 1978): 18–20.

———. "Close Air Support—A Warfighting View." *AFJI* (September 1988): 92–96.

Pivarsky, Carl, Jr., Capt., USAF. "30 mm Gun Employment." *FWR* (Fall 1984): 26–32.

"A Plan for Action." *Jane's Defense Weekly* (23 October 1993): 32.

Porth, Jacquelyn. "Enforcer Out of Mustang Came." *Defense & Foreign Affairs* (October 1981): 6–14.

Powell, Craig, Col., USAF (ret.). "The A-10." *Army* (March 1976): 43–47.

———. "A-10 Source Selection." *Government Executive* (November 1973): 17.

Quinn, Jimmie, Maj., USA. "Deadly as a Praying Mantis." *Armor* (July–August 1977): 13–16.

Ralph, John, Maj. Gen., USAF, Patrick Finneran, Maj., USMC, and Lawrence Kelley, Maj., USMC. "Commentary on 'The Emperor's Close Air Support,' by GC Ian Madelin, RAF." *AUR* (May–June 1980): 94–105.

Rasmussen, Robert, Col., USAF. "The A-10 in Central Europe." *AUR* (November–December 1978): 26–44.

———. "The Central Europe Battlefield." *AUR* (July–August 1978): 2–20.

Ray, Charles, Maj., USA. "JAAT." *USAAD* (January 1979): 42–47.

Reinburg, J. Hunter ("J.H."), Col., USMCR. "Close Air Support." *Army* (April 1962): 60–61.

———. "The Flying Tank." *Infantry* (May–June 1965): 66–67.

———. "For Maximum Effort: LACAS." *Infantry* (September–October 1963): 24–26.

———. "Trials of Close Air Support Aircraft." *Army* (March 1966): 67–68.

Rennspies, Norman, Capt., USAF. "LASTE and the GAU-8 at 15,000 Feet." *FWR* (Winter 1990): 2–7.

"Report: Armed UAV's Fire First Missiles in Combat." *AW&ST* (18 October 2001). Database on-line. Cited 13 March 2002. http://www.aviationnow.com/avnow/news/channel_military.jsp.

Reynolds, Scott, Capt., USAF. "Letters." *AW&ST* (21 September 1987): 116.

Rhodes, Jeffrey. "Close Support Test Bed." *AFM* (April 1990): 56–59.

———. "Improving the Odds in Ground Attack." *AFM* (November 1986): 48–52.

Ricks, Thomas. "Bull's-Eye War: Pinpoint Bombing Shifts Role of GI Joe." *Washington Post*, 2 December 2001, A01.

Ricks, Thomas, and Bradley Graham. "U.S. Special Forces Stepping Up Pursuit of Taliban Leaders." *Washington Post*, 14 November 2001, A01.

RisCassi, Robert, Lt. Gen., USA. "Army Aviation in the 1980s: The Successes of the First 5 Years, the Challenges of the Second." *USAAD* (January 1986): 2–8.

Robinson, Clarence. "A-10A Future Keyed to Gun Development." *AW&ST* (29 January 1973): 16–17.

———. "Flight Test Program Sought for Enforcer." *AW&ST* (12 August 1974): 50–55.

———. "USAF Plans A-10 Deployment to Europe." *AW&ST* (1 March 1976): 44–47.

———. "USAF Unveils War Game Study to Blunt A-10 Contract Attacks." *AW&ST* (5 March 1973): 14.

Romjue, John. "AirLand Battle." *MR* (March 1986): 52–55.

———. "The Evolution of the AirLand Battle Concept." *AUR* (May–June 1984): 4–15.

Ropelewski, Robert. "Affordability to Take Priority in USAF Advanced Fighter Design." *AW&ST* (23 September 1985): 14–16.

———. "A-16 Stretches Lead in Race for Close Air Support Role," *AFJI* (April 1989): 17.

———. "Congress Stirs Pot, Air Force Simmers as Close Air Support Decision Nears." *AFJI* (February 1990): 22.

———. "Night/Adverse Weather Version of A-10 Provides Confident Low-Level Operations." *AW&ST* (17 September 1979): 46–56.

———. "USAF Seeks Fighter Derivative as A-10, A-7 Replacement." *AW&ST* (4 March 1985): 14–15.

Rubinstein, Murray. "Ugliest Plane in the Air Force." *Mechanix Illustrated* (February 1977): 64–65, 132.

Russ, Robert, Gen., USAF. "The Air Force, the Army, and the Battlefield of the 1990s." *Defense* 88 (July–August 1988): 12–17.

———. "No Sitting Ducks." *AFM* (July 1988): 92–97.

———. "An Open Letter on Tacair Support." *Armor* (May–June 1988): 45.

"Russ' Support of F-16 Revamp over New Ground Support Plane Shocks AFA." *DN*, 2 February 1987, 4.

Saint, Crosbie, Lt. Gen., USA, and Walter Yates, Col., USA. "Attack Helicopter Operations in the AirLand Battle: Deep Operations." *MR* (July 1988): 2–9.

Saye, Jeremy, Wing Cdr., RAF. "Close Air Support in Modern Warfare." *AUR* (January–February 1980): 2–21.

Schaeffer, Will. "A-7 Strikefighter—Rebirth of the Corsair II." *Asian Defense Journal* (December 1986): 39.

———. "The LTV YA-7F: Updated CAS/BAI Aircraft Nearing Reality." *Asian Defense Journal* (May 1988): 82–86.

Schemmer, Benjamin. "Army's YAH-64 Attack Helicopter Survives DoD Attempt to Kill It." *AFJI* (24 January 1980): 24.

———. "Congress Mandates Air Force Transfer Its A-10s to Army and Marines." *AFJI* (December 1990): 20.

———. "Packard Personally Heads New Group to Resolve Inter-Service Close Support." *AFJ* (15 March 1971): 15.

———. "Tactical Air Command: Getting Ready for Blitzkrieg War." *AFJ* (March 1977): 18–19.

Schlitz, William. "Getting A-10 Firepower Forward." *AFM* (July 1983): 44–51.

———. "Lethal against Armor." *AFM* (September 1980): 45–53.

———. "Rutan's Scaled Composites to Develop Close Air Support Aircraft, Business Jet." *AW&ST* (21 November 1988): 27–28.

———. "Thunderhog II: Flying from the FOLTA and the DOLS." *AFM* (January 1981): 84–88.

Schraeder, Rod, Maj., USAF. "Augustine's Law XXIII: Hardware Works Best When It Matters Least." *FWR* (Fall 1987): 29–32.

Scott, William, Col., USAF. "The Rise and Fall of the Stuka Dive Bomber." *AUR* (May–June 1966): 46–63.

Scott, William. "AFTI F-16 Completes CAS Test, Demonstration Phase." *AW&ST* (2 April 1990): 38–40.

———. "Joint Strike Fighter Balances Combat Prowess, Affordability." *AW&ST* (3 August 1998): 51–54.

———. "Rutan's Scaled Composites to Develop Close Air Support Aircraft, Business Jet." *AW&ST* (21 November 1988): 27.

"Senate Unit Cuts A-10A Funds; Recommends Flyoff with A-7D." *AW&ST* (23 July 1973): 22.

Shifrin, Carole. "LTV Receives Air Force Contract to Modernize, Reengine A-7Ds." *AW&ST* (1 June 1987): 85–88.

———. "Sale Will Enable Fairchild Aircraft to End Uncertainty, Pursue Growth." *AW&ST* (14 December 1987): 105–11.

———. "TAC Demonstration Bolsters Support for F-16 in CAS Role." *AW&ST* (17 April 1989): 49–50.

Shrader, Charles. "Friendly Fire." *Parameters* (Autumn 1992): 29–44.

Sights, Albert, Col., USAF. "Lessons of Lebanon." *AUR* (July–August 1965): 42.

"6 A-10 'Tank Killers' Win NATO Acclaim in Europe." *AFT*, 3 October 1977, 38.

Smallwood, William. "Old 'Hog,' New Tricks." *Air and Space* (February–March 1999): 24.

"Special Avionics, Weapons Needed for F-16s in CAS Role." *AW&ST* (18 June 1990): 43.

"Stopping the Incredible Rise in Weapons Costs." *Business Week*, 19 February 1972, 60–61.

Stout, Joe. "CAS: Close Air Support Role." *Code One* (January 1989): 24–27.

Strode, Rebecca. "Soviet Design Policy and Its Implications for U.S. Combat Aircraft Procurement." *AUR* (January–February 1984): 46–61.

"Sukhoi Incorporates Changes to Su-25 Based upon Afghan Combat Experience." *AW&ST* (19 June 1989): 31.

Suydam, Edmund. "GAU the Giant Killer." *National Defense* (September–October 1977): 132–33.

Sweetman, Bill. "Close Air Support: Fighters High, Helicopters Low." *IDR* (November 1992): 1077–81.

———. "Frogfoot: A New Shturmovik on Trial." *Interavia* (August 1983): 830–32.

———. "Sukhoi Su-25 Frogfoot: STOL Striker for the Pact." *IDR* (November 1985): 1759–62.

Tabor, Dale, Lt. Col., USAF. "An Open Letter to A-10 Pilots." *TAC ATTACK* (August 1976): 5–7.

"Targets Lacking." *AW&ST* (2 April 1979): 38.

Taylor, John. "Attack and Observation Aircraft." *AFM* (May 1978): 117.

———. "Attack and Observation Aircraft." *AFM* (May 1984): 160–61.

———. "Gallery of Soviet Aerospace Weapons: Surface-to-Air Missiles." *AFM* (March 1975): 74–75.

———. "Gallery of Soviet Aerospace Weapons: Surface-to-Air Missiles." *AFM* (March 1978): 109.

———. "Gallery of USAF Aircraft." *AFM* (May 1983): 150–51.

———. "Gallery of USAF Aircraft." *AFM* (May 1992): 142.

———. "Gallery of USAF Weapons." *AFM* (May 1975): 115–16.

———. "Gallery of USAF Weapons," *AFM* (May 1979): 119–20.

Taylor, John, and Susan Young. "Gallery of USAF Weapons." *AFM* (May 1982): 155–57.

Taylor, John, Capt., USAF. "The Mean Machine." *Airman* (September 1977): 20–28.

"T-46 Termination Will Force Closure of Fairchild Facility on Long Island." *AW&ST* (23 March 1987): 27.

"T-46A—Dead or Alive?" *AW&ST* (18 August 1985): 13.

Tirpak, John. "Scoping Out the New Strike Fighter." *AFM* (October 1998): 36–41.

———. "Short's View of the Campaign." *AFM* (September 1999): 43–47.

Trimble, Stephen. "Air Campaign Spurs Upgrade Market for Versatile C-130." *AW&ST* (8 November 2001). Database on-line. Cited 13 March 2002. http://www.aviationnow.com/avnow/news/channel_military.jsp.

Tuttle, Henry, Maj., USA. "Is AirLand Battle Compatible with NATO Doctrine?" *MR* (December 1985): 4–11.

Ulsamer, Edgar. "The A-10 Approach to Close Air Support." *AFM* (May 1973): 41–45.

———. "AX: Lethal, Accurate, Agile , and Cheap." *Air Force/Space Digest* (January 1970): 33–39.

———. "New Muscle for the Tactical Air Arm." *AFM* (October 1974): 22–23.

———. "New Roadmap for AirLand Battle." *AFM* (March 1987): 108–13.

"USAF to Accept Plan on Modifying F-16, A-10 for Close Air Support." *AW&ST* (13 March 1989): 28.

"USAF Agrees to Flyoff for A-10, A-7D." *AW&ST* (1 October 1973): 22–23.

"USAF, Defense Dept. Reach Compromise on New CAS Aircraft." *AW&ST* (7 December 1987): 33.

"USAF Examines T-46 Alternatives in Face of More Budget Restraints." *AW&ST* (21 October 1985): 23.

"USAF Panel Urges Separate CAS, Interdiction Aircraft." *AW&ST* (27 July 1987): 22–23.

"USAF Plans to Introduce A-10s into Forward Air Control Fleet." *AW&ST* (9 February 1987): 23.

"USAF Seeks Designs for Close Air Support Successor to A-10." *AW&ST* (22 June 1987): 37.

"USAF to Begin Evaluating FLIR Candidates for A-10." *AW&ST* (30 April 1990): 47.

"USAF Urges Cancelling Fairchild T-46, Extending Cessna T-37 Service Life." *AW&ST* (17 March 1986): 31–32.

Volz, Joseph. "AX vs. AH-56: Competitive or Complementary?" *AFJ* (25 April 1970): 25–27.

Von Spangenberg, Marconi [pseud. of George Spangenberg]. "USAF's 'Fighter Mafia' Needs a New Design Philosophy." *AFJI* (March 1986): 64–68.

Wacker, Rudolph, Brig. Gen., USAF. "A-10 Close Support Operations in Europe." *AFM* (August 1979): 73–77.

Walker, Sam, Capt., USAF. "GAU-8A." *FWR* (Spring 1982): 2–5.

Wall, Robert. "Airspace Control Challenges Allies." *AW&ST* (26 April 1999): 30–31.

———. "Joint STARS Changes Operational Scheme." *AW&ST* (3 May 1999): 25–26, 29.

———. "NATO Endgame, Helos Stir Debate." *AW&ST* (24 May 1999): 37–38.

———. "NATO Shifts Tactics to Attack Ground Forces." *AW&ST* (12 April 1999): 22–24.

———. "USAF Eyes New Tools for Korean Fight." *AW&ST* (22 May 2000): 58–60.

———. "USAF Steps Up Operational Training." *AW&ST* (22 May 2000): 64.

Wanstall, Brian. "Squabbles over Warthog Successor." *Interavia Aerospace Review* (February 1990): 151–55.

Warrick, Joby. "Air Force Steps Up Attacks of Iraqi Troops." *AFT,* 11 February 1991, 8.

Warwick, Graham. "A-10: Europe's Anti-Tank Vanguard." *Flight International* (24 February 1979): 523–24.

"Washington Roundup: Endless Saga." *AW&ST* (10 December 1990): 19.

"Washington Roundup: Eternal Trainers." *AW&ST* (17 August 1987): 17.

"Washington Roundup: Seeing Red." *AW&ST* (13 June 1988): 15.

"Washington Roundup: Strike Delay." *AW&ST* (19 February 1973): 71.

"Washington Roundup: Thunderbolt Retirement." *AW&ST* (1 August 1988): 13.

Weiss, George. "CO 2/17 AirCav: 'Gunships Took Tanks, Survived Flak' in Laos." *AFJ* (19 April 1971): 22–23.

Welch, William, Lt. Col., USA (ret.). "Is Fixed-Wing CAS Worth It?" *Naval Institute Proceedings* (September 1994): 51–55.

Wetmore, Warren. "A-10 Program Approach Reshaped." *AW&ST* (10 February 1975): 44–46.

———. "Israelis' Air Punch Major Factor in War." *AW&ST* (3 July 1967): 21–23.

"Where USAF's A-10's Fit in NATO's Defense." *AFJI* (June 1979): 18.

Whitcomb, Darrel, Maj., USAFR. "Hogs in the Rear." *AJ* (Winter 1987–1988): 16–21.

Whitley, Al, Capt., USAF. "A-10 OT&E TD&E." *FWR* (Fall 1977): 34.

Wilson, George. "Opponents Find Vought Corporation's A7 Impossible to Shoot Down." *Washington Post,* 13 March 1981, A3.

Wilson, Michael. "Fairchild A-10." *Flight International* (20 March 1976): 708–17.

Winton, Harold, Lt. Col., USA (ret.). "Partnership and Tension." *Parameters* (Spring 1996): 100–119.

———. "Persian Gulf War: Campaign without Doctrine." *AFT,* 18 November 1991, 23.

Wright, Brooks ("Kato"), Maj., USAF. "Urban Close Air Support: The Dilemma." *Weapons Review* (Summer 1998): 15–19.

Zeybel, Henry, Lt. Col., USAF (ret.). "Truck Count." *AUR* (January–February 1983): 36–45.

Video Productions

AH-64 Apache. *Wings* series. Produced and directed by Daphna Rubin. 60 min.
   Discovery Channel Productions, 1995.
Air Power Showdown, Part III: The Best Attack Aircraft. Produced by *AW&ST*. 60
   min. Aviation Week Group, 1997. Videocassette.
MiGs vs. America. Produced and directed by Adam Friedman and Monte
   Markham. 45 min. U.S. News Productions in association with Perpetual
   Motion Films, 1993. Videocassette.

Miscellaneous

BDM Corporation. "A Study of Statistical Lessons Learned in Vietnam." Vol. 6.
   BDM/W-78-128-TR-VOL-6-1. *Conduct of the War*. 9 May 1980. AMHI
   Library.
Mahon, George. Interview by Kent Hance. c. 1980. Interview 00048. Southwest
   Collection, Texas Tech University.
Reis, V. H. "Close Air Support Systems: A First Order Analysis." Technical Note
   1980-5. Lexington, Mass.: MIT Lincoln Laboratory, 19 February 1980.
Supporting documents for George Watson Jr., *The A-10 Close Air Support Aircraft:
   From Development to Production, 1970–1976*; and Edward Mischler, *The A-X Spe-
   cialized Close Air Support Aircraft: Origins and Concept Phase, 1961–1970*.
   WPAFB: AMC History Office.

ABOUT THE AUTHOR

Douglas N. Campbell has had a lifelong interest in history, military affairs, and combat aviation. He majored in political science at North Georgia College, where he participated in Army ROTC, and received a bachelor of arts degree in 1976. During his career in combat aviation, he received his Navy commission in 1977 via Aviation Officer Candidate School, and a year later he was awarded his Navy pilot's wings. As a Navy pilot he flew A-7E attack jets off the USS *Kitty Hawk* on Western Pacific and Indian Ocean cruises, and off the USS *Enterprise* on precruise workup deployments.

After a follow-on tour as a TA-4J advanced jet-flight instructor and landing signal officer, he transferred to the U.S. Air Force in 1984. During his tenure in the Air Force, Campbell flew A-10s in Europe as part of an operational fighter wing, and in Nevada as part of a weapon and tactics test operation. He concluded his military career flying F-117s both operationally and as commander of a training detachment.

Campbell received his master's degree in history from the University of Nevada, Las Vegas, while serving in the U.S. Air Force. After he retired from the military in 1995, he entered the doctoral program in history at Texas Tech and received his degree in 1999. Along with jobs involving flying, he has taught online college courses at both the graduate and undergraduate levels. He has also written numerous encyclopedia articles, but this is his first book. He lives in Wyoming with his wife, Teri, and their daughter, Patty.